The Best We Share

The Best We Share

Nation, Culture and World-Making in the UNESCO World Heritage Arena

Christoph Brumann

berghahn
NEW YORK • OXFORD
www.berghahnbooks.com

First published in 2021 by
Berghahn Books
www.berghahnbooks.com

Library of Congress Cataloging-in-Publication Data

Names: Brumann, Christoph, author.
Title: The best we share : nation, culture and world-making in the UNESCO World Heritage arena / Christoph Brumann.
Description: New York : Berghahn, 2021. | Includes bibliographical references and index.
Identifiers: LCCN 2020042385 (print) | LCCN 2020042386 (ebook) | ISBN 9781800730441 (hardback) | ISBN 9781800730458 (ebook)
Subjects: LCSH: World Heritage Committee. | World Heritage areas—Management. | Cultural property—Protection—International cooperation. | Cultural property—Protection—Decision making.
Classification: LCC G140.5 .B76 2021 (print) | LCC G140.5 (ebook) | DDC 353.7/211—dc23
LC record available at https://lccn.loc.gov/2020042385
LC ebook record available at https://lccn.loc.gov/2020042386

British Library Cataloguing in Publication Data
A catalogue record for this book is available from the British Library

ISBN 978-1-80073-044-1 hardback
ISBN 978-1-80073-045-8 ebook

To Lisa, Sven and Niklas and the worlds they will build

Contents

List of Illustrations viii
Acknowledgements x

Introduction 1

Chapter 1. A Day in the Life of the UNESCO World Heritage Committee 23

Chapter 2. The Promise of World Heritage 38

Chapter 3. Fulfilling the Promise 47

Chapter 4. Rebellion and Peace 83

Chapter 5. The Nation State 98

Chapter 6. Procedures 155

Chapter 7. Concepts 179

Chapter 8. Global North and South 224

Conclusion. Utopian Remnants and the Logic of Growth 257

References 271
Index 287

Illustrations

Figures

0.1 Google Ngram values for 'World Heritage' phrase and equivalents in printed books. 5
4.1 World Heritage Committee in session, Brasilia, 2010. 84
4.2 World Heritage Committee in session, Bonn, 2015. 94
5.1 Information booth of South Korean NGO outside meeting venue, Bonn, 2015. 130
5.2 State name plates in the World Heritage General Assembly, Paris, 2009. 142
6.1 Micro-editing decision texts live on screen, Saint Petersburg, 2012. 173

Tables

0.1 Internet search hits for major international competitions and awards. 6
3.1 Cultural World Heritage inscriptions of the 2008 Committee session. 62
4.1 World Heritage upgrade inscriptions in the Committee sessions of 2010–15. 91
4.2 ICOMOS recommended decisions for cultural and mixed World Heritage candidate properties, 2004–19. 96
5.1 States Parties with the longest World Heritage Committee presence, 1977–2019. 117
5.2 Number of contenders in recent World Heritage Committee elections. 119
5.3 World Heritage inscriptions of Committee members states and observer states compared. 120

5.4 Recommended and upgrade inscriptions for Committee member states and observer states compared. 122
6.1 State of Conservation Reports up for Committee decision in 2015–19. 166
8.1 States Parties with most World Heritage List inscriptions in 2009. 229
8.2 Properties with State of Conservation Reports and Danger-Listed properties discussed and/or decided by the Committee in 2009. 231
8.3 Populations, surface areas and World Heritage property numbers of 2010 Committee members. 235
8.4 World Heritage inscriptions per UNESCO region before and since 2010. 239
8.5 States Parties with the most World Heritage inscriptions before and since 2010. 240

ACKNOWLEDGEMENTS

UNESCO World Heritage has occupied me for more than a decade. Initial funding came from a Heisenberg Fellowship of the German Research Association (DFG) that began in 2008, while I was still based at the University of Cologne. Since 2010, I have continued the project as a Head of Research Group at the Max Planck Institute for Social Anthropology (MPI) in Halle, Germany, within the 'Resilience and Transformation in Eurasia' department headed by Chris Hann, whom I thank for his unfailing support. Colleagues and staff at the MPI have been a constant source of inspiration and help, contributing to what for me has been a uniquely stimulating environment. Also at the MPI, I headed the research group 'The Global Political Economy of Cultural Heritage' running from 2014 to 2018, where Leah Cheung Ah Li, Pierpaolo De Giosa and Vivienne Marquart were a pleasure to work and discuss UNESCO intricacies with.

Outside the MPI, the ideas in this book have particularly benefited from conversations and cooperation with David Berliner, Chiara Bortolotto and Aurélie Élisa Gfeller, to whom I am grateful. I was able to refine my arguments through dozens of presentations in academic settings around the world. For providing these opportunities or co-organizing them with me, I wish to thank (in alphabetical order) Saskia Abrahms-Kavunenko, Nicolas Adell, Björn Ahl, Cristina Amescua, Lourdes Arizpe, Fanny Badache, Bettina Beer, Yazid Ben Hounet, Regina Bendix, David Berliner, Lotta Björklund Larsen, Michael Bollig, Chiara Bortolotto, Hannah Brown, John Collins, Emilie Dairon, Thomas Hylland Eriksen, Corinna Forberg, Christina Garsten, Pauline Garvey, Aurélie Élisa Gfeller, Laurence Gillot, Florence Graezer-Bideau, Maria Gravari-Barbas, Valdimar Hafstein, Brigitta Hauser-Schäublin, Ellen Hertz, Carola Hommerich, Cyril Isnart, Leah Kimber Arye, Susanne Klien, Kristin Kuutma, Kathryn Lafrenz Samuels, Peter Bille Larsen, Chang Liu, Lucile Maertens, Irène Maffi, Wolfram Manzenreiter, Ron Niezen, Johan Nilsson, Julia Obinger, Diamantis Panagiotopoulos, Joanna Pfaff-Czarnecka, Andreas Ranft, Adam Reed, Aike Rots, Richard Rottenburg, Maria Sapignoli, Wolfgang Schenkluhn, Reinold Schmücker,

Burkhard Schnepel, Matthias Schumann-Brandau, Tansen Sen, Gabor Sóos, Adrienne Sörbom, Guido Sprenger, Philipp Stockhammer, Cornelia Storz, Hana Synkova, Markus Tauschek, Mark Teeuwen, Anne-Christine Trémon and Tom Yarrow. I also wish to thank two anonymous readers for their exceptionally thorough and encouraging reviews. At Berghahn Books, I am grateful to Caryn Berg, Jon Lloyd, Caroline Kuhtz and Sulaiman Ahmad for speedy and meticulous processing. All errors and shortcomings, of course, remain my own.

Most of all, heartfelt thanks are due to the people occupied with or committed to UNESCO World Heritage who took the time to share their views and insights with me, both during my ethnographic forays into the World Heritage arena and in interviews and conversations on separate occasions. Following standard anthropological practice, they remain anonymous; where it appeared to be called for, I also changed personal details to obstruct identification. Without their candour and support, this book would not have seen the light of day.

Parts of the text incorporate revised sections of my publications 'Shifting Tides of World-Making in the UNESCO World Heritage Convention: Cosmopolitanisms Colliding', *Ethnic and Racial Studies* 37(12) (2014): 2176–92; 'Conclusion. Imagining the Ground from Afar: Why the Sites Are So Remote in World Heritage Committee Sessions', in *World Heritage on the Ground: Ethnographic Perspectives,* edited by myself and David Berliner (Oxford: Berghahn Books, 2016), pp. 294–317; 'Anthropological Utopia, Closet Eurocentrism and Culture Chaos in the UNESCO World Heritage Arena', *Anthropological Quarterly* 91(4) (2018): 1203–33; and 'Slag Heaps and Time Lags: Undermining Southern Solidarity in the UNESCO World Heritage Committee', *Ethnos* 84(4) (2019): 719–38. Permission is gratefully acknowledged. Names of Japanese individuals follow the Japanese name order, with the family name preceding the given name.

Introduction

On 25 June 2009, the UNESCO World Heritage Committee assembled in Seville, Spain, for its annual session takes a step that in its thirty-two years of operation, it has never taken before: contrary to the German delegation's wishes, it deletes 'Dresden Elbe Valley' from the World Heritage List. There has been a first delisting of a World Heritage property two years earlier but at that time, the concerned state, Oman, had itself asked for 'Arabian Oryx Sanctuary' to be removed, as that nature reserve had been given to oil prospecting and the oryx antelopes were largely gone. Now in 2009, however, Germany opposes the step, arguing that the controversial new bridge under construction over the Elbe, although much criticized, will not suffice to destroy the World Heritage value of a river valley and cultural landscape more than 20 km in length. Why not wait just one more year and then assess the completed structure without prejudice, the Mayor of Dresden and other German delegates plead. But after more than two hours of tortured and often confused debate, the Committee goes for a vote, using the secret ballot that two state members have demanded. The deletion is decided with exactly the two-thirds majority required.

Several times during the following days, Committee member delegates I talk to feel urged to justify the decision when learning of my German nationality. It has not been an easy step, they say, but since the building project has simply been pushed forward even though the World Heritage Committee had been threatening deletion for already three years, they had owed it to themselves to take a tough decision. The debate shows the Committee divided, as

German lobbying in the preceding weeks has brought some member states over to their side. Of the other states too, none appears to have a particular score to settle with Germany and were it only for their national self-interest, a softer course against one of the world's economic leaders might suggest itself. Yet still, fourteen out of twenty-one states vote for what they say they hate to do, namely delisting a World Heritage property.

Arguably, the delegates are thereby going by one specific interpretation of the Committee's mandate. The World Heritage Convention postulates that the most important cultural and natural heritage sites on earth are everyone's sites, not just those of the nation state in which they are located, and that such universal ownership includes both rights and responsibilities. It follows that when a state neglects its conservation duties, the others must step in to defend the common interest. The debate about the Arabian Oryx Sanctuary in 2007 has thrown up the question how this is best done: several states claimed that Oman had no right to ask for deletion, given that the site had become humanity's heirloom, not just the country's. Now in Seville, delegates could also argue for keeping Dresden on the List – severing the formal link with a site does not increase the World Heritage Committee's influence on it. But as that influence has not prevented the construction project – and has not even significantly slowed it down – inaction would amount to rewarding a treaty state that is trampling the Committee's authority, as the critics see it, further encouraging such behaviour from others. Also, the Committee has kept asserting that with the bridge in place, the site will irredeemably lose its 'outstanding universal value', the precondition for World Heritage status. Therefore, with much rhetorical hand-wringing and the inclusion of vague gestures towards a possible future re-nomination of Dresden heritage in the decision text, the deletion of the Dresden Elbe Valley is eventually sealed. In the online version of the List, the entry still features, just like that of the Omani site, but as '~~Dresden Elbe Valley~~ Delisted in 2009'.[1]

Six years later in 2015, when a reconciled Germany hosts the Committee's thirty-ninth session in Bonn, the mode of interaction has changed. There is certainly no lack of confrontation behind the substantive issues discussed, such as the destructions of World Heritage properties in war-torn Syria, Iraq, Libya and Yemen. After a major diplomatic drama, Committee member South Korea agrees to adding a series of early industrial sites in Japan to the World Heritage List, on the condition that the wartime forced labour of Koreans in some of these sites be properly acknowledged. And in their usual closed-door negotiations, the Israeli, Palestinian and Jordanian delegates fail to come to terms on the Old City of Jerusalem. This leads to the adoption by secret ballot of a decision leaning towards the Arab viewpoint and angry statements that the three delegations subsequently read out. Yet on all other issues, the Committee is in utter harmony. The Seville meeting went to formal votes over three decisions aside from Dresden and the trend ballooned in subsequent sessions, making these the most contentious in the history of the Convention. Yet in Bonn, only the single vote on Jerusalem is called and all other decisions are by consensus. Delegates also tone down their attacks on the expert bodies advising the Committee, with the belligerence of previous years all but gone.

By returning peace to the meeting room, the Bonn session goes for a different interpretation of global ownership, namely that of unencumbered access to, and use of, the benefits of the World Heritage title for all treaty states. In most cases, the concerned countries receive the lenient decisions they favour, including World Heritage titles for candidate sites not approved by the expert bodies and softened conservation demands for already-listed properties. States thus have no reason to put up a fight – strong wishes are heard now, whereas ignoring a Convention state's pleas to the extent of the Dresden case is out of the question. To this day, no other property has been deleted from the World Heritage List and even the preceding steps – declaring a site as being 'in danger' or the mere threat of this possibility for the future – have rarely been taken against the concerned state's will.

The Puzzle

So what is behind this paradigm change? Why has the World Heritage Committee 'lost its teeth', as some would have it, or why – as others prefer to see it – has it finally begun to operate in a reasonable and inclusive manner? Where has the ambition to be more than the assembled national interests gone and what are the reasons for quite a few participants to believe that such an ambition is misguided to begin with? Who was responsible for crafting the tacit nonaggression pact now in place or was this beyond anyone's control? Building on an ethnography of the World Heritage arena and the way its discourses and practices have developed, this book tries to answer these questions and explain the momentous and, in my view, lasting and irreversible transformation that occurred in the six turbulent years between Seville and Bonn.

This transformation, the book argues, goes beneath the surface level. I identify its root cause in yet another interpretation of universal ownership, namely that of the World Heritage List representing humankind and its achievements in a comprehensive way that reflects their full diversity. More than heretofore realized, lingering imbalances between the Global North and South have contributed to the current sense that the proper way to share the best at a global level is by letting nation states have their way. This is not because the World Heritage arena has shied away from addressing global inequality and Northern bias; rather, and crucially, reform was promised, but then went only halfway, making the sense of deficiencies and contradiction all the more acute. More than anything else, I argue, a feeling of having been withheld their due has encouraged Southern countries to push for immediate rewards rather than for time-consuming fundamental revisions. The parallels with other global bodies, I will show, are striking, suggesting a larger trend in international governance that presents a serious challenge to present-day multilateralism. World Heritage is not becoming a world affair easily.

Our world has never been more densely interconnected and people were never more at the mercy of long-distance processes, forces and factors reaching into the remotest corners of the globe – the spread of COVID-19 is yet another demonstration. This has certainly increased the common person's awareness of being part of larger entities. However, whether it has boosted planetary solidarity is a different matter. The fortification of national borders, the rise of xenophobia, populism and religious extremism, the unilateralist turn of powerful countries, and the slow progress of global climate policy leave little room for illusions here. Still, universalism, cosmopolitanism and the dream of a brotherhood of men/siblinghood of humans have a long and venerable history, sometimes restricted to communities of religious or political faith, such as Catholicism or communism, but also in a more encompassing fashion. This book is about one such universalist project. This one has a particularly idealistic bent, as it symbolically converts parts of national territories into the property of humankind. It is also administered by the United Nations (UN) agency with the most idealistic mission, the Educational, Scientific and Cultural Organisation (UNESCO). With such utopian overtones, to what degree has World Heritage managed to create a connected world of heritage? What does the world own in common here, who is made to represent its interests and how is the world being represented when the World Heritage Committee meets? The title of the book can be read with three different emphases and all three – 'the best', 'we' and 'share' – are put under scrutiny.

The Significance of World Heritage

Engaging with World Heritage is encouraged by the importance that it has acquired in the contemporary world, 'grown beyond anyone's wildest dreams', as one veteran of the arena phrased it to me. I myself recall that when I first saw the bronze plaque with the World Heritage logo on the walls of Cologne Cathedral, this appeared as an obscure honour to me – what could this designation I had not heard of possibly add to one of Germany's most-visited sites? Twenty years on, few people would voice similar doubts. World Heritage has

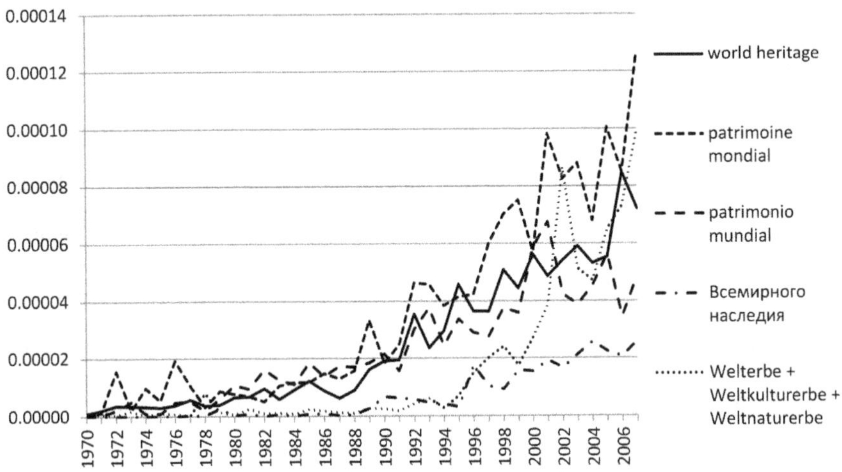

Figure 0.1. Google Ngram values for 'World Heritage' phrase and equivalents in printed books. Note: Created by the author based on own Google n-gram search on https://books.google.com/ngrams. The numerical values are the percentages of the respective bigram (i.e. two-word compound) or unigram (i.e. word) in relation to the total numbers of bigrams or unigrams that occur in the printed books of the respective language, retrieved with the option 'case insensitive' ticked. For better visibility, the table excludes the Chinese 世界遗产 that slowly rose to just under 0.00006 until 2000 but then soared to 0.00039 in 2007 (more than three times as high as any other value), the cutoff year for the Google Ngram data. In German, 'Weltkulturerbe' (cultural World Heritage) is used more often than the offical term 'Welterbe', so I added up the totals for these two terms and 'Weltnaturerbe' (natural World Heritage).

become a prominent global distinction and a World Heritage title is rarely ever left unmentioned, even when the likes of the Great Barrier Reef, Angkor or Mount Fuji are being discussed. In Cologne itself, the World Heritage title was key in bringing down a high-rise development plan just across the river that in the eyes of critics would have affected the dominance of the cathedral over the city skyline. The advertising industry marvels how a brand with such traction could be created with so little investment, as a former World Heritage Centre official told me. World Heritage documentaries, tours, websites, apps and publications – from coffee-table books and atlases to serious scholarly treatises – abound. 'World Heritage' and its equivalents in other languages have greatly surged in print presence, particularly since the late 1990s (see Figure 0.1), and this is also when the phrase became more prevalent than other 'world' compounds such as 'world map', 'world record' or 'world news' – 'World Heritage' even managed to beat 'world peace' in 2006, perhaps a sign

Table 0.1. Internet search hits for major international competitions and awards.

Olympics/Olympic Games	324,000,000
Academy Award(s)/Oscars	152,000,000
FIFA World Cup	71,500,000
World Heritage	71,200,000
Guinness World Record/s	45,800,000
Nobel Prize/s	43,800,000
Grammy Award/s	42,700,000
Eurovision Song Contest	14,400,000
Paralympics/Paralympic Games	13,400,000
ISO certified/certification	12,500,000
Michelin star/s	11,300,000
Booker Prize	8,690,000
Prix Goncourt	6,290,000
(New) Seven Wonders of the World	6,060,000
Pritzker Prize	1,350,000
Right Livelihood Award/s	332,000

Note: Numbers retrieved from www.google.com on 12 November 2019, using phrase search and inserting 'OR' between the alternative options that are indicated by slashes or parentheses above (such as in the search entry <'Olympics' OR 'Olympic Games'>).

of the times.[2] The World Heritage Committee sessions have grown from the congregations of a few dozen conservationists to global events attended by thousands of participants, including ambassadors, government ministers and the occasional head of state. On the internet too, World Heritage holds its own when compared to other globally known competitions and awards (see Table 0.1). It is clearly the one activity most on people's mind when UNESCO is being mentioned, despite the agency's primary commitment to education, and is crucial for the agency's image (cf. also Schmitt 2011: 142). Often, 'UNESCO' becomes a shorthand for World Heritage in popular reference, such as in talk of 'UNESCO sites', showing how much the two things have merged in the public consciousness. And so have the several other UNESCO-administered initiatives concerning heritage, including the sister Convention for the Safeguarding of the Intangible Cultural Heritage adopted in 2003 that developed in a sort of love-hate dynamic with the older treaty, which it both imitates and rejects (Brumann and Berliner 2016a: 11–12; Hafstein 2009, 2018: 70–80, 134–35, 162–63).

World Heritage has become important enough to go to war: when the ancient Khmer temple Preah Vihear was inscribed on the World Heritage List in 2008, this provoked a series of bloody clashes between the armies of Cambodia, which nominated the site, and Thailand, which claims the temple grounds as its own territory, but had failed to thwart the inscription (see Chapter 5). World Heritage has also become important enough that its deliberate targeting brings publicity. This was demonstrated and clearly intended when Islamic fundamentalists mutilated World Heritage sites in Timbuktu, Mali, in 2012 (see Chapter 1 and the Conclusion) and when 'Islamic State' forces followed in their footsteps by blasting the archaeological remnants of Hatra, Iraq, and Palmyra, Syria, in 2015. The very fact that heritage veneration has become almost universal by now, due in no small part to the World Heritage Convention, provokes such spectacular iconoclasms meant to demonstrate the superiority of other authorities and commitments. Nothing has done more to promote the global significance of heritage than the World Heritage endeavour, complete with the risks that go along with celebrity status.

World Heritage can be transformative at and around the sites graced with the title. Often, the designation changes little in terms of actual conservation, particularly when nation states have decided to accept the honour, but not the related obligations, or when they lack the capacity to live up to the latter. What is generated instead may be massive increases in tourism and the corresponding development, lining some or even many people's pockets, but threatening rather than augmenting protection. It is extremely difficult to generalize about the effects of World Heritage inscription since the more than one thousand properties and their specific national and local conditions are so diverse. World Heritage status has been crucial in stopping high-rise developments in Cologne, Vienna, Riga or Saint Petersburg; giving pause to infrastructural and mining projects in the Serengeti National Park or Kakadu National Park, Australia; or motivating international aid for conserving the ruins of Angkor. If it weren't for their World Heritage status and how it helps to mobilize a large number of international actors, a World Heritage Centre official told me, the nature reserves in the Democratic Republic of the Congo, such as Virunga National Park, would be long destroyed. There are thus a considerable number of cases where the Convention has served its official purpose. But there is at least an equally long list of sites where the

World Heritage status has made hardly any difference in terms of conservation or has even added challenges in its own right.

As for the people living within, near or from World Heritage properties, a general assessment is even more difficult: in terms of how the prestigious designation generates or reroutes revenues and rights, almost every site has both winners and losers. Insights from ethnographic field studies such as those assembled in a volume I coedited (Brumann and Berliner 2016b) call for caution in expecting what others have hailed as 'benefits beyond borders' (Galla 2012). The distribution of the material and immaterial perks created through World Heritage honours is often skewed, and control over sites tends to shift away from local horizons, not necessarily to the Convention's global bodies but to national, often newly established institutions and their understandings of proper World Heritage management and presentation (Brumann and Berliner 2016a: 24). At the same time, 'UNESCO myths' proliferate: people expect wonders from the World Heritage institutions, which these, given their limited power and resources, are almost certain to disappoint (e.g. Istasse 2016: 43, Marquart 2015: 85–95). A detailed assessment of the local consequences of the World Heritage title is beyond the scope of this book. However, without any doubt, the designation is a weighty factor in the transformation of many locations in the contemporary world, including some of the most prominent ones.

The rise of World Heritage also feeds the academic interest in heritage and the formation of a new interdisciplinary field. Overall, the bulk of heritage research is still done on heritage items proper and on the practical problems of their conservation and protection, in such fields as art history, architectural history and archaeology for cultural heritage, and geology and biology for natural heritage; geography features in both camps. Yet, more so than applied concerns, it is the reflection on the social, political and economic contexts of cultural heritage and its conservation that fuels the recent expansion. For the purposes of distinction from the more technical and/or affirmative studies, this approach is often labelled 'critical heritage studies' (e.g. Winter and Waterton 2013), for example, in the Association of Critical Heritage Studies,[3] which grew to a membership in the thousands within just a couple of years. Not all that is discussed in this context and in such venues as the *International Journal of Heritage Studies*, the *International Journal of Cultural Property* and the *Journal of Social Archaeology* is new; older arguments about the invention of tradition (Handler and Linnekin 1984; Hanson 1989; Hobsbawm and Ranger 1983; Lenclud 1987; Ranger 1993), for example, are sometimes reiterated with surprisingly little awareness of these predecessors. And not all of it is quite so critical either, with quite a few protagonists still committed to the intrinsic value of heritage and its conservation, at least if properly cleansed of nationalist, capitalist and classist abuses.

One may question how much of this is due to World Heritage proper, as a general 'heritage boom' was already diagnosed when World Heritage was not yet the big buzzword (e.g. Merriman 1989: 14; Walsh 1992: 94) and modernity's obsession with culture loss has been described as a more encompassing phenomenon (Berliner 2018, 2020). However, if only for the space that general introductions to and overviews of cultural heritage dedicate to World Heritage (e.g. Graham and Howard 2008; Harrison 2012; Logan, Nic Craith and Kockel 2016; Meskell 2015a; Smith 2006; Tauschek 2013; Waterton and Watson 2015) and the frequency with which educational programmes and book series prefix 'world' to 'heritage studies' – such as at the Universities of Minnesota, Dublin, Birmingham, Cottbus, Turin and Tsukuba – I think its impact can hardly be overestimated. World Heritage is perhaps less momentous in Europe, North America and Japan, where institutionalized conservation goes back to the nineteenth century. There, World Heritage tends to add prestige rather than concrete protection measures to what are already fairly elaborate legal, technical and administrative apparatuses. However, World Heritage plays a larger role in the considerable number of often postcolonial countries where conservation regimes were substantially enhanced or even freshly established in the last few decades. The missionary effects of World Heritage have spread everywhere, and likewise around the world, the wish to be part of a global movement and to meet global standards has become a strong motivational force. Yet in 'old Europe' too, World Heritage continues to stir high-flying ambitions and emotions: for example, the candidacy of the Francke Foundation in my workplace, the city of Halle in East Germany, was accompanied by a highly visible public relations campaign and event calendar, down to announcements of the candidacy in every tram that passed by the buildings.[4] Similar phenomena of mobilization abound, particularly at sites whose fame stands to gain significantly from a World Heritage title. Without doubt, the venture is more than a purely academic matter, and large numbers of people expect something from it, although not necessarily all the same thing.

The Research Context

In dealing with World Heritage and addressing the puzzle outlined above, I build on prior studies of UNESCO and other organizations within the UN system. Primarily investigated by specialists in political science and international relations, these bodies have seen increasing anthropological scrutiny in recent years, building on earlier work about the European Union (EU) (Abélès 1992; Shore 2000) and on a more general trend in anthropology to explore modern organizations. In addition to articles and monographs, this has brought forth specialized meetings and conference panels, edited overview volumes (Müller 2013; Niezen and Sapignoli 2017) and ethnographic team studies of the World Trade Organization (Abélès 2011) or the 'Rio+20' global summit of 2012 (Foyer 2015). Aside from human rights processes (e.g. Billaud 2014; Cowan 2014; Kelly 2011; Merry 2006), anthropologists have been most drawn to UN processes that touch on what is commonly seen as their area of competency, particularly the very interesting and legally innovative activities on indigenous peoples and their rights and traditional knowledge (e.g. Bellier 2013; Groth 2012; Koester 2005; Muehlebach 2001; Oldham and Frank 2008; Rößler 2008; Sapignoli 2017, 2020: 159–207; Siebert 1997) and the aforementioned intangible cultural heritage convention and the way it deals with more ephemeral 'cultural expressions', so the term goes, such as performative arts, rituals, folk crafts and cuisines (Arizpe 2011; Arizpe and Amescua 2013; Bortolotto 2007, 2011b; Bortolotto et al. 2020; Hafstein 2004, 2007, 2009, 2018; Kuutma 2007; Nas 2002; Rudolff 2010; Savova 2009). The '2003 Convention' and its list entries are frequently referred to as World Heritage too, but despite such slippages and other mutual influences, it is nonetheless a legally and administratively independent venture.

The central organs of World Heritage proper – the '1972 Convention' – have attracted prior ethnographic curiosity as well (see below). However, few of these studies have been conducted by social and cultural anthropologists. The fact that World Heritage sites are often seen as falling within the province of other disciplines may play a role here. Yet many anthropologists are also unaware how World Heritage has shifted perspective over the years by turning to the remnants of everyday life such as vernacular architecture, industrial plants, trade routes, railway lines or 'cultural landscapes' – rice terraces, sacred groves, former maroon hideouts – instead of just focusing on palaces and pyramids. Conservation architects certainly have no edge over anthropologists when it comes to, say, understanding Chief Roi Mata's domain, a collection of sites connected with the eponymous semi-mythical figure that became Vanuatu's first World Heritage property in 2008. And some World Heritage bids build on anthropologists' active involvement, such as that of Marilyn Strathern, who privately contributed funds for nominating the Kuk Early Agricultural Site, which has been Papua New Guinea's only World Heritage site since 2008.[5] World Heritage is thus closer to disciplinary concerns than many anthropologists might expect.

Such closeness is also fostered by World Heritage being dominated by cultural rather than natural heritage. The culture concept has been key to the anthropological endeavour: the very formation of the discipline built on extending a term previously reserved for elite achievements to all the everyday ideas, habits, customs, rules and material products shared by the members of a given group or society. Subsequently in their crusade against scientific racism, Franz Boas and his famous students such as Ruth Benedict and Margaret Mead convinced the educated world that these shared features are in fact socially transmitted rather than biologically inherited. Yet precisely because such a broad, non-elitist notion of 'culture' has spread in everyday usage so that people no longer think only of going to the opera when hearing the word, many anthropologists are uneasy with the term. They argue that it lends itself to drawing boundaries and exaggerating differences between groups of people in quite the same way as 'race' once did, and there are calls for dropping the concept altogether from the disciplinary toolkit. The lively debate about this issue (Abu-Lughod 1991; Boggs 2004; Borofsky et al. 2001; Brightman 1995; Fox and King 2002; Goody 1994; Keesing 1994; Kuper 1999; Lentz 2016; Rodseth 1998; Sahlins 1999; Trouillot 2002; Wikan 1999; Wimmer 2005), to which I myself have contributed a cautious position (Brumann 1999), may never reach a consensus. Yet it attests to the fact that popular appropriations of 'culture' and its uses and abuses through 'cultural fundamentalism' (Stolcke 1995) and 'culturespeak' (Hannerz 1999) are a relevant concern for most anthropologists, as of course they also are to many related disciplines. World Heritage is a key institution shaping popular understandings of culture today and an intellectual project that, by having opened up to the everyday life of ordinary people and its vestiges, has unwittingly followed anthropology's own foundational trajectory.

World Heritage, just like other UN bodies and initiatives, also invites reflections on the nature of contemporary globalization. Much discussed in the 1990s (e.g. Appadurai 1996; Brumann 1998a; Friedman 1994; Hannerz 1996; Kearney 1995), globalization seems to have become more of an expected presence in the new millennium, and anthropological debate has shifted towards specific constituent phenomena such as migration, transnationalism, export production, commodity chains and neoliberal regimes, and global mass media. But the principal questions such as whether the cultures of the world are converging or not remain with us, and here the study of the World Heritage arena has insights to contribute. It has been argued that globalization has raised people's imaginary potential to a new level (Appadurai 1996: 3–11). In this vein, World Heritage properties and their media and virtual representations are key anchors for 'world-making' (Brumann 2014b; Meskell 2016: 72; Meskell, Liuzza and Brown 2015: 438), that is, creating and structuring the world in people's minds and directing their actual and imagined movements through it. World Heritage is a 'global dream' in the way outlined by Anna Tsing (2000: 342) and as a virtual global collection and mapping exercise, it has few equals. In the words of Ulf Hannerz, it is one of 'those interfaces where the confrontations, the interpenetrations and the flowthrough are occurring, between clusters of meaning and ways of managing meaning; in short, the places where diversity gets, in some way and to some degree, organized' (1989: 211). For an anthropological understanding of globalization, World Heritage therefore has much to offer.

Multilateral Ethnography

For my own approach to the World Heritage arena, I have employed anthropology's special methods, most of all ethnographic field research. In a play on 'multisited ethnography' (Marcus 1995), I have called this 'multilateral ethnography' (Brumann 2012a), given that UNESCO and the governing organs of its conventions are multilateral bodies that transcend the bilateral framework of state-to-state interaction. Of the official meetings of the World Heritage institutions, I attended the World Heritage Committee sessions of Seville (2009), Brasilia (2010), Paris (2011), Saint Petersburg (2012) and most of the Committee session of Bonn (2015), all of which lasted between nine and thirteen days; the three-day World Heritage General Assemblies in the Parisian UNESCO headquarters of 2009 and 2011; one day of the Workshop on the Future of the Convention, also in Paris (2009); and the week-long Closing Event of the Celebration of the Fortieth Anniversary of the World Heritage Convention in Kyoto (2012). I also participated in one of the week-long annual sessions of the Intergovernmental Committee for the Safeguarding of Intangible Cultural Heritage in Istanbul (2008) and one day of the UNESCO General Conference in Paris (2011). For all these events except for the General Conference, where my World Heritage General Assembly badge sufficed to get past the doormen, I applied for participation as an observer, a recognized status that many others – including academics interested in heritage and the colleagues mentioned below – also use.[6] Universal media access to World Heritage Committee sessions was only granted in 2012, but even before this, I did not have the feeling that observer participation was tightly controlled. If anything, lack of space slowed down acceptance for what is the one group of participants on which the proceedings least depend, and I had to appeal to a sympathetic permanent delegation (i.e. embassy of a member state to UNESCO) to speed up admission in one year. The identification badges of all participants are colour-coded, making it obvious to everyone that observers such as me are not state representatives, let alone Committee members with decision-making powers. But once granted access, there are no particular strings attached to participation in the public functions such as the plenary session – where theoretically I could even have asked for the floor – and those receptions and special events that are open to all participants. To what degree this also included food, drinks, shuttle services and the customary excursion differed from year to year, with

Turkey, Spain, Brazil and Russia being particularly generous hosts, whereas other countries appeared to be more closely watched by the taxpayer. For uncovered needs and for travel and hotel accommodation, I used my own research funding.

None of what I did in my field research was particularly unique for an anthropologist. I followed the plenary session from the back seats or the gallery, usually typing a summary of what was happening on my notebook computer, and attended the sessions of working groups and many of the special 'side events' outside plenary time. I also made my observations and engaged in informal interaction and countless conversations with other participants over coffee in the lobby outside, in the shuttle buses to and from the events, over lunches, dinners and the conference hotel breakfast, during receptions, on excursions, in bars after hours, in hotel corridors or at the front desks and at the airport. The exchange of emails and text messages came on top of this.

Contrary to other researchers of UN organizations, I made no attempt to do fieldwork in the UNESCO Secretariat or any other contributing organization. I am sure that doing so might have yielded additional insights, but it would also have placed me with the body in question, constraining my interaction with the members of others. I was sometimes included in more special circles, such as social events of the German Committee delegation or the 'Green Machine' social meeting of the World Heritage nature experts, but cannot claim to have become an insider to any of them. On average, I reached less closeness to my informants than in my earlier, more continuous fieldwork in Kyoto, even when some friendships developed. However, there is hardly anyone at the World Heritage meetings for whom this is the most intimate social environment.

The meetings are nonetheless very intense events. Precisely because they are finite, people have full schedules and socialize almost around the clock. For many participants, these are much-anticipated high points of their years, where they go to interesting cities, reconvene with colleagues, friends and acquaintances of long standing, and enjoy what is often rather good food, drink, music and artistic performances in inspiring locations such as Oscar Niemeyer's Palácio Itamaraty, the Alcazar of Seville or the top-floor restaurants of the UNESCO headquarter building with their spectacular view of Eiffel Tower. Most participants will also rarely be in a more multinational environment. I find this comparable to international academic conferences, although World Heritage Committee sessions (sometimes) offer more luxury, a greater professional diversity of participants and perhaps less sense of hierarchy: outside the meeting hall, there is no spatial segregation and – if it is not a private reception to begin with – no gradation of access. Also, since people inhabit different national and professional spheres, the details of one another's rank are not obvious to everyone. Parts of the meeting have a festive, sometimes even effervescent atmosphere that the slowness of the official session procedures does not entirely kill off, and striking up conversations with strangers is expected behaviour in a setting where most participants spend much more of their time talking than in their usual workdays. Many of them are also deeply invested in specific issues arising and are eager to share their triumphs and frustrations. Time spent in the 'World Heritage village' is thus significantly fuller and more communicative than in less special and high-strung settings. This also meant that in almost all sessions, I ended up dictating rather than typing my fieldnotes in order to catch at least some sleep.

My participant observation was complemented by more than fifty formal interviews with individuals from all contributing organizations, sometimes at the sessions proper, but most often on separate occasions. For these encounters, I often made use of other research and conference travel to locations as diverse as those of the meetings, from Mexico City to Tokyo. I used English, German, Japanese and (for one interview each) Spanish and French, and I taped most of these conversations, assuring my informants that I would not identify them in my presentations and publications. In most cases, I had a list of topics to cover, but was also very willing to follow my interlocutors to whatever they considered meaningful.

When around 20,000 pages of documents are prepared for a single Committee session, much of it in two languages (English and French), the study of texts is indispensable. Alongside the official meeting documents, reports, draft decisions and records, these include a vast production of anything from manuals over newsletters and magazine articles to protest letters by the involved individuals and organizations, self-reflexive documents such as internal audit reports and retired officials' reminiscences and the unending trail of media coverage and comment with which the outside world honours World Heritage activities. Quite a lot of this is available online. Skim reading is essential for everyone intent on not drowning in this flood of words and numbers, and there is no hope of exhausting it, much as there is no hope of capturing everything that goes on in parallel at a Committee session. Just like the conversations and interviews, texts are important sources of factual information and explanatory hypotheses, but they are also part of a World Heritage discourse, which I examine for its recurring tropes and rhetorical figures, tacit assumptions and blind spots. The statutory documents also embody the institutional memory of the World Heritage arena, given that its personnel continues to be replaced over time and that people tend to forget the details. However, texts are not all-powerful here and certainly less so than in judicial institutions. Diplomats heading Committee delegations can be active and effective players while having read only a minimum of the session documents, relying instead on networking skills, the input of the delegation's experts or simply the political weight of their country. Also, and as in any other institutional environment, texts reflect only part of what goes on: much is too delicate or too self-evident to be put into writing, and the actual meaning and import of documents can only be gauged by observing the social practice they both reflect and engender. There are substantial synergies between these three groups of sources, such as when reading the draft decision text helps in understanding the session debate or when individuals encountered at the meetings are open to interview requests. However, some may be less obvious, such as when hints about the meaning of texts and where to find them in the first place stem from informal session conversations. As an anthropologist, I see my participant observation and interviews as crucial, but it is evidently the triangulation of all sources that has led me to the presented insights.

In general, I felt attracted to the cosmopolitan environment of the World Heritage meetings and not too intimidated by it. I do not mind dressing up and using my foreign languages, including the ones in which I am more limited. I did have the occasional moment of self-consciousness in the beginning when it seemed I was the only person not to know anyone in the crowd, but this was mild when compared with the initial stages of other ethnographic fieldwork. Only once did I have a moment of acute panic of feeling out of place. This was when I joined a couple of delegates who, one evening of the World Heritage General Assembly of 2011 in Paris, suggested attending the reception at 'the Orsay'. I thought they meant the Musée d'Orsay, but when our taxi arrived, it dawned on me that it was the French Foreign Ministry we were about to crash. When the initially sceptical security staff had let us in – a phone call inquiry seemed to produce no clear reason to the contrary – an Australian delegate taunted me that as an anthropologist worth my salt, I had to stick this through, adding that we were all impostors anyway. And so I did, reassured by the observation that other participants no more eminent than me had also made their way into the majestic rooms where Foreign Minister Alain Juppé would welcome us.

This was one of several occasions where my interlocutors engaged with my professional identity. I was open about my anthropological research interest and my home institutions throughout, and most people seemed to find this mildly interesting. To what degree they actually understood my agenda no doubt varied, but the presence of researchers – most of them more strongly invested in what goes on in the arena (see below) – is expected and does not surprise anyone. Occasionally, people would play with my identity, such as the World Heritage Centre official who pointed to other Max Planck Institutes' research on primates when introducing me to the participants of a nature experts' social gathering. Also, a sympathetic International Council on

Monuments and Sites (ICOMOS) official joked about people such as herself being the mice in my lab; an ICOMOUSE that would be, I replied, and pointed out that unlike in a laboratory setup, I was a fellow mouse.

However, commonality went beyond the shared humanity of observer and observed in ethnographic fieldwork, in that most participants have a university education and are familiar with modern organizational settings too. Many of them write and speak publicly as well, travel widely and are impressively multilingual – in Laura Nader's (1972) time-honoured terms, I was 'studying up' (cf. also Hølleland and Niklasson 2020). And then also, there is a large grey zone between the social actors in the World Heritage arena and academia, with many moving back and forth; sometimes, I thus wondered whether the scholarly conferences on cultural heritage I attended were occasions to present my research results or rather extensions of my ethnographic field. When a professor inviting me to an international workshop turned out to be a past UNESCO Assistant Director-General for Culture, when a high-ranking ICOMOS official came to listen to and comment on an academic conference presentation of mine or when a state representative to the Committee, after moving to a university, invited me to give a guest lecture in his heritage studies programme, the feedback circuits were significantly shortened. And so they were when a diplomat of a Committee state told me that after having read one of my publications, his delegation would be a bit more cautious with me from now on. One participant cited me with regard to the contents of a previous incarnation of Table 0.1 (Terrill 2012: 69) and through a natural heritage expert whom I had sent them, my statistics about how current Committee member states receive a disproportionate share of new World Heritage listings (see Table 5.3) ended up in a critical external audit of the World Heritage venture,[7] to then be taken up repeatedly in subsequent Committee debates I witnessed.

Similar to earlier fieldwork in Kyoto (Brumann 2012b: 9), I thus contributed to and influenced the social and discursive field I was observing. However, there are myriad other influences on the UNESCO World Heritage arena, both from within academia and the wider world, and nothing indicates that mine has been particularly momentous. I doubt that what I write will please everyone in the arena, but several informants encouraged me to be candid, particularly the more critical minds who hoped that what I will disclose will provoke reform.

Important as participant observation was, it is a relative term in meetings such as these. As I did not belong to any of the involved organizations, I took part in informal interaction, but not in 'backstage' processes such as the specialized working meetings of these groups or their internal electronic communication. In particular, I was not privy to the diplomats' and other state representatives' informal interaction and lobbying, on which a lot of the decision-making rests, learning about it only through hearsay and candid participants. However, this does not differ from the experience of most others at the sessions, and the scope and span of my fieldwork still led to a wealth of information. Institutional and financial independence also spared me from the dilemmas discussed at the 'Collaborative Dilemmas' workshop in 2016, convened by Chiara Bortolotto at the École des Hautes Études en Sciences Sociales in Paris.[8] Fellow participants at that event included researchers with all kinds of engagement with the global organizational field – alternating between detached observation of and paid work for their UN body, coming with a specific political agenda, acting as their country's official representative, even doing an ethnographic study commissioned by the target organization. Obviously, these approaches offer valuable opportunities. What is gained in terms of insider knowledge of specific contributing organizations may then be lost in terms of capturing the bigger picture and of the license to write freely, however, so I do see merits for my own kind of positioning as well.

In the burgeoning literature on UNESCO World Heritage, there are works that share intellectual concerns or methodological approaches with this book and my prior publications (Brumann 2011, 2012a, 2013, 2014b, 2015, 2016a, 2017a, 2017b, 2019a, 2019b; Brumann and Berliner 2016a, 2016b; Brumann and Meskell 2015), so I should outline how I relate to these. As already stated, this is not a study of one or several World Heritage sites; rather, I approach the World Heritage arena through its central institutions. In doing so, I place less emphasis on the exegesis of official texts than some others (e.g. Harrison 2012: 63–67, 114–39; Smith 2006: 88–114). I work from the assumption that what these texts mean in practice and what binding power they have can only be uncovered by scrutinizing institutional practice in depth, not by treating World Heritage or 'UNESCO' as a kind of black box to which a unified agency and voice can be unproblematically ascribed (as in much of Di Giovine (2009)). Others have employed fieldwork too, but at earlier stages that do not extend to the period at issue here (Schmitt 2009, 2011; Turtinen 2000, 2006), with different disciplinary backgrounds that place less emphasis on ethnography (Rudolff 2010; Schmitt 2009, 2011) or over a shorter timespan (Hølleland and Wood 2019; James 2016; James and Winter 2015; Schäfer 2016) and with a more limited thematic interest (Larsen and Buckley 2018). There are sophisticated statistical studies of trends in World Heritage Committee decision-making (Bertacchini, Liuzza and Meskell 2017; Bertacchini and Saccone 2012; Bertacchini et al. 2016; Meskell et al. 2014; Meskell, Liuzza and Brown 2015; Reyes 2014) and the underlying expert evaluations (Schmutz and Elliott 2017); others combine statistics with session documents (Claudi 2011) or rely exclusively on the latter (Braun 2007). Still other works try to read World Heritage policies indirectly, from the selection and treatment of World Heritage sites (van der Aa 2005) or from the nomination dossiers that treaty states submit for their candidate properties (Labadi 2010, 2012). There is a book-length legal commentary on the World Heritage Convention (Francioni 2008). The early period of the Convention is the subject of historical studies and personal reminiscences (Batisse and Bolla 2005; Cameron 2008; Cameron and Rössler 2011, 2013; Gfeller 2013, 2015; Logan 2013, Meskell 2013b; Stott 2011, 2012, 2013; Titchen 1995). And finally, arena insiders and officials have at various points reflected about or defended its contemporary policies in academic contexts (e.g. Cameron 2008; Cameron and Rössler 2013; Rao 2010; Rössler 2005, 2006a, 2006b; Terrill 2014a, 2014b), with this particular genre itself becoming an object of study (Hølleland and Johansson 2019).

The single-authored part of archaeologist Lynn Meskell's work (2012b, 2013a, 2014, 2015b, 2015c, 2016, 2018) is closest to my own approach, as it is also based on ethnographic fieldwork at the sessions and with the contributing organizations, and makes many similar observations, so much so that we have coauthored one overview (Brumann and Meskell 2015). However, Meskell points to a very large number of explanatory factors for recent transformations, and nation-state 'pacting', corporate interests, global power differentials, regional geopolitical alignments as well as transregional ones (such as the Brazil, Russia, India, China and South Africa (BRICS) alliance) take turns in being emphasized in individual publications, with no clear theory as to how precisely they interact. I believe that more precision is possible. In her book (2018) in particular, Meskell also treats World Heritage and UNESCO as a continuous social field, suggesting at least implicitly that long-term trends within UNESCO can explain what happens inside the World Heritage Committee – here, I also disagree. Yet, what most distinguishes Meskell's work from my own is that she does not accord the question I raise and the period I highlight special significance; rather, the developments she describes are undated or attributed to longer time periods. If the latter are subdivided, then with different separating dates than the all-important 2010 Committee session that is the key turning point in my account. Needless to say, I draw on all this scholarship and will engage with it in more detail while I proceed, often in the notes; however, I am convinced that none of it explains the puzzle initially outlined.

Given my prior interest in globalization, the anthropological concept of culture and its public usage, and cultural heritage policies in Japan (Brumann 1998a, 1999, 2002, 2009, 2012b; Brumann and Cox 2010), World Heritage has been an obvious topic to get involved with. The commons aspect of what is stylized as the 'common heritage of mankind' was also important in my studies of the survival of common-property arrangements in utopian communes and of the problems of protecting Kyoto's historic townscape as a collective resource (Brumann 1998b, 2000, 2001, 2003, 2005, 2012b). Moreover, introducing the cultural heritage of Cologne to its visitors as a tour guide back in the 1980s and 1990s has given me some understanding for the position of 'heritage believers', my term for people sympathetic to conservation and convinced of the intrinsic merits of cultural heritage (Brumann 2014a). But much as all this has drawn me to the World Heritage arena, I align myself with the venerable Boasian anthropological tradition of cultural relativism, that is, providing ethnographic description and explanation without passing judgement. I try to take other people's investments in and/or criticisms of it seriously – practising the 'heritage agnosticism' I preach (Brumann 2014a), not 'heritage atheism' – but I am not committed to the World Heritage endeavour myself. As is expected from an anthropological study, this book refrains from accepting the self-presentation of the World Heritage organizations and their personnel at face value, and looks behind the formal side of things. Inevitably, this leads to the deconstruction of received understandings and the identification of gaps and discrepancies between official representations and actual workings. Readers invested in the official representations may therefore feel under attack. At times in my account, I also wonder why specific obvious alternatives to the established measures and procedures were never tried out, which readers might understand as promoting these alternatives. However, this is not my intention, and neither do I wish to formulate a 'critique' of the World Heritage system, write 'for' or 'against' World Heritage or its components, subject them to a systematic assessment or evaluation, or launch a reform agenda. If 'critical heritage studies' implies any of these commitments, this book is not a contribution to that field.

I think that such a detached position is unremarkable for a social anthropologist, but I am all the more surprised that it is far from common when World Heritage comes into play. Take, for example, the comments to a recent contribution to one of anthropology's leading journals (Meskell 2016): David Byrne warns that 'the field of heritage conservation practice … needs to take a big step back from World Heritage' (ibid.: 85); Martin Hall claims that 'an insistence that the World Heritage list is "above" and "outside" everyday conflicts … is dangerous' (ibid.: 87); Charlotte Joy wonders if 'the World Heritage List [is] merely an empty and cynical exercise' (ibid.: 88); and Chiara de Cesari asks: 'Do we really need such a list?' (ibid.: 86). Meskell herself, in other work, states that 'World Heritage … may be deeply imperfect and in need of serious revision' (2013a: 492) and sees it as in 'gridlock' (2015b) and a 'current crisis' (Meskell et al. 2014: 13). If the topic were, say, cross-cousin marriage, pastoralism or shop-floor strategies of resistance, I would be surprised about a similar urge to ring alarm bells and I wonder where it comes from. Note that neither of these comments bothers to define the standard against which World Heritage is dangerous, empty, not needed, imperfect, cynical or in crisis, as if this were self-evident. There seems to be an assumption that heritage conservation is principally desirable or could be so if only conducted properly, and that a less politicized World Heritage system allowed to fulfil its official function would be preferable to the current one. Also, I find it striking that the onus is mostly on perceived deficiencies regarding the uses and abuses of heritage sites and site communities – the World Heritage arena might be accused of letting down global equity just as well, but this is raised much more rarely.

Many a reader, particularly from the heritage field, may be disinclined to question the intrinsic value of heritage and of a multilateral body dedicated to its protection. However, let me be clear that I do not share this view. Whether conservation in general and World Heritage in particular benefit the world is a complex question I do not intend to answer; anyone venturing into doing so would need a clear sense of what is good to begin with. I admit to have been rather pleased when an ICOMOS official, after listening to one of my presentations, expressed her surprise at how fond my description of the arena appeared to her. A sympathetic relationship to the people populating the research field is a precondition for thorough ethnographic research, as even colleagues working with the most challenging subjects (Hedlund 2019; Smith 2011) personally confirmed to me. Yet I neither root for World Heritage nor do I long to bring it down, and while global fairness strikes a stronger chord in me, this penchant should also be no criterion for judging this book. My simple goal is to explain how the World Heritage arena works and why it has changed in the way it has.

The Structure of the Book

For some initial familiarity with the setting, Chapter 1 takes the reader through the twists and turns of a World Heritage Committee day, including both the formal proceedings and the informal interaction unfolding around it. Chapter 2 spells out the utopian premise of World Heritage, namely the idea that the world's most important sites do not just belong to the country in which they are located, but also to humanity as a whole, a new idea in international law. It then describes the path towards the adoption of treaty based on this idea in 1972. Chapter 3 tells the story of how the utopian premise and promise was put into practice, through increasing growth and elaboration, reform of what was perceived as Eurocentric bias and increasing Committee self-assertion against recalcitrant nation states. This culminated in the deletion of Dresden in 2009, which probably saw World Heritage at its multilateral apex. Chapter 4 turns to the immediate aftermath, the Brasilia session of 2010, which redefined the World Heritage game, and how a new mode of Committee operation more in line with treaty state wishes was normalized over the next few years.

The search for the causes of this transformation begins in Chapter 5, which describes the strong identification of actors in the World Heritage arena with their nation state, the ascendancy of career diplomats, which has boosted national self-assertion, the clear evidence of national self-serving and the lengths the Committee goes to in terms of trying to stay out of other nations' business, particularly when the latter is contested between two or more of them. Chapter 6 shows that established procedures for submitting World Heritage nominations, watching the condition of listed sites and taking decisions in the Committee session are vulnerable, offering multiple entry points for vested interests. Chapter 7 demonstrates that vaguely defined and applied core concepts and a surprising degree of case-based improvisation likewise do little to give nation states pause. Chapter 8 penetrates to what I see as the root cause of dissatisfaction and rebellion, the vastly uneven representation of Global North and South on the World Heritage List and the reasons why even in the aftermath of the rebellion, fundamental revision is pending and Southern solidarity is weak. The Conclusion then looks at the root cause of the root cause, the absolute definition of World Heritage value, which, in the absence of numerical limitations, was destined to provoke 'prize inflation' at some point, as it also did in the allocation of other famous distinctions. After all, the World Heritage arena and its actors depend on growth, and despite all complaints, nobody strongly presses for the closure of the World Heritage List, making a fundamental turnaround unlikely.

But let us first take a closer look at Committee business, travelling to the Russian summer for this purpose.

Notes

1. http://whc.unesco.org/en/list (retrieved 25 August 2020).
2. Cf. https://books.google.com/ngrams with input 'world heritage,world map,world record,world news,world peace'; option 'case-insensitive' ticked; time range from 1970 to 2007, corpus 'English' and smoothing of '0'.
3. www.criticalheritagestudies.org (retrieved 25 August 2020).
4. Disappointment was all the greater when the expert evaluation was so negative that the bid had to be withdrawn before the 2016 Committee session.
5. Cf. http://whc.unesco.org/uploads/nominations/887.pdf, p. iii (retrieved 25 August 2020).
6. Representatives of 'non-profit-making institutions having activities in the fields covered by the Convention, according to criteria defined by the World Heritage Committee, may be authorized by the Committee to participate in the sessions of the Committee as observers' (Rules of Procedure, 2015 version, § 8.3). Different versions of the Rules of Procedure from the first to the current 2015 version can be found at http://whc.unesco.org/en/committee (retrieved 25 August 2020). Curiously, the admission of observers is

only fully approved in the respective Committee session proper, as item 2 on the standard agenda, when observers are usually already in the room.

7. WHC-11/35.COM/INF.9A, p. 83. Wherever possible here and in the following, I cite the official document code, not the URL. These documents are online, but their URLs tend to change over time, so that a web search for the code is the surest route to the document.
8. See http://frictions.hypotheses.org/47 (retrieved 25 August 2020).

Chapter 1

A Day in the Life of the UNESCO World Heritage Committee

Saint Petersburg on Sunday, 1 July 2012. After little sleep and a hurried breakfast with a Kenyan delegate, I board a small shuttle bus arranged by the organizers, which takes us to the conference venue across the Neva River. Our large hotel has conserved its socialist heritage rather well in terms of Spartan rooms and intransigent staff. Although by no means cheap, it is still one of the least costly options, so on the bus ride I am surrounded by participants from the not quite so affluent countries such as Cuba, Chile and Slovakia; the Saudi Arabian delegate may just have booked too late for the posher places. On the short ride, I joke with the Kenyan delegate about his own possible inscription on the World Heritage List, given that so much is listed these days, and chat with a South Korean university professor about the lunchtime event her delegation organized yesterday. Participants have dressed formally, with my own suit and tie not standing out.

The driver drops us off at the Tauride Palace, which usually houses the meetings of the members of the Commonwealth of Independent States (CIS). Passing by the registration tent, we enter the grandiose classicist building through an airport-style security check with metal detectors. The tables lining the walkway to the large foyer are filled with promotional materials about World Heritage sites and candidates, and there are also the TV screens where the Japanese private station TBS Tokyo shows its acclaimed World Heritage documentaries. Special desks have been set up for such purposes as booking excursions and arranging return flights. There is no coffee yet in the foyer, so I walk up the stairs to the meeting hall where the pews are slowly filling up. I chat with Japanese participants about the upcoming fortieth anniversary celebration of the Convention, which they will organize in the autumn in Kyoto. Delegates of the International Council of Monuments and Sites (ICOMOS) complain to me about the new style of decision-making that ignores their expert advice. I install myself in the back rows reserved to ordinary, nonstate observers. On my notebook computer and with the Wifi access offered by the organizers, I see that the deliberate destructions of World Heritage properties in Timbuktu, Mali, which Islamist insurgents have been committing for several days now, are dominating global news headlines. World Heritage is clearly in the limelight.

The Morning Session

Shortly past 10AM and despite the troubling news, chairperson Eleonora Mitrofanova – the Russian Federation's 'Permanent Delegate' (i.e. ambassador) to UNESCO – opens the morning session of the seventh day of the thirty-sixth World Heritage Committee session to routine business. This is the examination of this year's nominations to the World Heritage List and, continuing with the cultural properties from the previous day, she calls up agenda item 36 COM 8B.37, 'Schwetzingen: A Prince Elector's Summer Residence', a candidate submitted by Germany. As is usual when cultural heritage is concerned, she first hands over to ICOMOS whose representative is also seated on the podium facing the semi-circular pews rising in front of them. Supported by a PowerPoint presentation, this retired French professor of technical history takes a couple of minutes to summarize the evaluation, which has been online for six weeks. He reiterates that the nominated property lacks 'outstanding universal value' or 'OUV', the precondition for World Heritage listing. Instead, it does not distinguish itself from many other Baroque palaces and parks, so that ICOMOS advises rejection. As everyone is aware, adopting this verdict would seal the fate of the candidate by precluding submission of a revised nomination file in the future. Most states therefore quietly withdraw such bids ahead of the session, thereby avoiding a binding decision. But Germany has already done so in 2009 when Schwetzingen was up the first time and was also deemed unworthy by ICOMOS. Therefore, since extensive revisions of the nomination file have not improved the judgment, the delegation is determined to put up a fight.

Once the presentation is finished, the chair opens the debate, inviting the twenty-one Committee member delegations in the first rows to make their comments. To her right, the Director of the World Heritage Centre (the Convention secretariat) – an Indian nature conservation expert – assists her in giving the floor to the delegations in the order in which they raise their state name plates, turning them from horizontal to vertical in the groove of a small wooden block on their desks. The chair always calls up the state name, not that of the individual. The delegates then speak into their microphones for up to three minutes, using one of the two official working languages of English and French or the other languages for which treaty states have volunteered to sponsor interpretation (this time Russian, Spanish and Arabic). At the entrance, the several hundred participants in the hall have been equipped with small broadcasting devices with headphones for this purpose.

Colombia speaks first, followed by France, and soon the delegates find themselves embroiled in contention: Germany – itself on the Committee – complains that the ICOMOS evaluation, in addition to missing the full significance of the palace theatre, passes over the eighteenth-century mosque in the palace gardens, the oldest in Western Europe. The ICOMOS expert objects that this is just a small and unremarkable structure reflective of the Orientalist leanings of the time. Yet Qatar, Algeria and the United Arab Emirates declare themselves impressed by this symbol of religious tolerance and even propose immediate World Heritage inscription, a rather extreme turnaround; it looks as if the Germans have asked them for their support. In contrast, Switzerland, France and Senegal are sceptical, insisting on the difference to a real, functioning mosque, and this clearly upsets the German ambassador. Nobody brings up Timbuktu, where a World Heritage-listed mosque in continued usage is being mutilated while the Committee speaks. The Indian ambassador suggests 'deferring' consideration of the Schwetzingen property, the one of four customary decision options that allows for the resubmission of a substantially revised nomination file no sooner than two years from now. With other delegations voicing support, a compromise seems in sight. But then, the Swiss ambassador loses his patience and calls for a vote. After some confused back and forth, the legal advisor – a UNESCO official on the podium that the chair consults over procedural matters – clarifies that the Swiss demand is for a vote on the original draft decision, which is the provisional decision text drafted by ICOMOS and the World Heritage Centre that was put online ahead of the session.

The German ambassador hastens to declare that her delegation would be quite happy with the suggested deferral, but this does not prevent the Committee from sinking into rare depths of confusion for the better part of an hour, with participants forgetting their most basic procedural routines. For one thing, the Swiss proposal diverges from the usual practice to vote not on the decision text proper, but on proposed *amendments* to it. As a two-thirds majority is required for decisions regarding World Heritage inscription or noninscription, this ends up turning the numerical advantage against the strict line Switzerland has been demanding, a fact of which the delegation appears mysteriously unaware. When others point out the breach of usual practice, the legal advisor insists that no amendments to the draft decision have been received from the Committee members, and the usual recipient of such submissions, the rapporteur – a diplomat from member state Mexico, also on the podium, who is tasked with recording the decisions – does not interfere. However, the decision text on the big screens in the hall has an edit marked in blue that foresees inscription for Schwetzingen and gives the aforementioned Arab states as supporters. It is common practice for the typists working in the back of the podium to add such edits in track-changes mode while the Committee is speaking, grasping the delegates' intent even without special prodding. But nobody points out the obvious, namely that these edits are usually treated as amendments by the Committee. And if considered an amendment, the inscription proposal by the Arab states would have to be voted on first, as this is the decision option most removed from the original draft text, which foresaw noninscription. However, unlike almost everyone in the room, the legal advisor cannot see the screen from her seat, as she later tells me.

It is strange to see that many speakers sense that something is unusual, with incredulous laughter rising at times, while nobody manages to put their finger on what exactly is wrong. The chair as the person best placed to do so – as she can speak any time, not just when her turn in the queue comes up – is confused too. She reminds herself belatedly that substantive debate must end after the Swiss call for a vote, forgetting that the motion must first be supported by a second delegation, only to then let the debate continue the very next moment; she leaves some delegates perplexed about exactly on what they are voting; and she claims that after the noninscription of Schwetzingen fails to receive the required two-thirds majority (unusually, no count of the show of hands is announced), the other decision options must be voted too, just to again drop that (incorrect) idea the very next moment. Several times, confused delegates, often signalling a point of order by forming a T with the state name plate and their arm, weigh in, usually only for interventions that reveal their own puzzlement. In the end and after having regained her signature self-assurance, the chair convinces the Committee that the deferral option is now their consensus, dropping her gavel to mark adoption when no objections are raised. The decision thus returns to India's much earlier suggestion and had Germany withdrawn the bid entirely, the practical consequences (major revisions before resubmission) would have been the same. Small wonder then that the chair declares her intention to avoid further such 'difficult and unpleasant situations'.

It is only now that the chair suspends the discussion of nominations and hands over to the Malian Minister of Culture who reads out an emotional – and, in the end, tearful – statement, in which she reports fresh destructions in Timbuktu, denounces them as running counter to the spirit of

Islam, laments her country's plight, pleads for everyone's support, given that such tragedies could happen everywhere and closes with: 'God help Mali!' Long applause follows and the Senegalese ambassador, chair of the Group of Islamic Cooperation within UNESCO, confirms their solidarity. The chair then suggests that she express the grief felt by everyone in the hall rather than having more interventions on the matter, given that 'we have a lot of work ahead'. Somewhat piqued by the chair's suspicion that she might wish to return to Schwetzingen, the German ambassador calls for a moment of silence. 'We have to interrupt our work for this one minute', she insists and the chair obliges, with the hall rising for the brief gesture. It is difficult to say what is more strange – a minute of silence honouring lost heritage rather than lost lives (casualties have not been reported) or the fact that the Committee does not have more than a moment for a World Heritage property under attack while it speaks, with the world watching the destructions on YouTube and the perpetrators giving UNESCO interference as one of their reasons for their actions.

'Street fighting!' is a Dutch delegate's comment on the Schwetzingen scuffle when I leave the room for a quick coffee in the foyer. I return to more orderly proceedings. For the 'Vineyard Landscape of Piedmont: Langhe-Roero and Monferrato' presented by Italy, which is up next, ICOMOS misses clear selection criteria and overall coherence among the nine spatially discrete components of this so-called 'serial' property, proposing a deferral decision. No sustained attempts to amend it are made, indicating Italy's commitment not to lobby for 'upgrades' of Advisory Body judgments, and the deferral is adopted in just 20 minutes' time, not the 90 minutes it took for Schwetzingen. Next, the chair announces several reshuffles of the agenda to accommodate key participants' flight bookings – a new trend of recent sessions – and moves to the 'Mining Sites of Wallonia', Belgium, another serial candidate site. For the first time today, ICOMOS supports inscription of what is also a revised nomination from deferral two years earlier. Citing time pressure, the South African vice-chair – briefly pitching in for Mitrofanova – suggests moving to the decision right away and since he sees no objections, he declares the property inscribed on the World Heritage List and congratulates Belgium. From the rows of the observer States Parties not currently on the Committee, a Belgian delegate offers kudos to the Committee, ICOMOS and the nomination team, emphasizes the importance of coal for Belgian history, and hopes to share the sites and Belgian multiculturalism with visitors from all over the world. All this is expected content in the acceptance speech, for which the concerned states have two minutes. Proceedings move on while delegates walk over to the Belgian delegation to congratulate them in person.

For the 'Decorated Farmhouses of Hälsingland', a Swedish nomination of a series of seven such structures and likewise a revised earlier bid, ICOMOS misses a joint management body, a more extensive 'buffer zone' around one of the houses and better fire protection. It recommends 'referral', meaning minor revisions with possible resubmission already by next year, but the Swedes assert that the issues have been resolved in the meantime, with Switzerland and Estonia backing them up. South Africa proposes immediate inscription and in the absence of objections, this is what the chair adopts, to more applause. In his acceptance speech, the Swedish delegate mentions that the residents of six of the houses are sitting next to him. One cannot help wondering whether these would have travelled to Saint Petersburg only to see just another postponement – the Swedish delegation must have counted on the Committee overruling ICOMOS. Yet much as this has become a common occurrence in recent sessions, the Swedes are usually strongly opposed to this practice, as are Switzerland and Estonia, whom they must have asked for their support. This means that even the paragons of virtue within the Committee remain aware of their national advantage.

Obliging a further request for accelerated treatment, the chair turns to 'Rio de Janeiro, Carioca Landscapes between the Mountain and the Sea' (Brazil). This is yet another revised nomination, for a cultural landscape embedded in a megacity. Coming just months after the UNESCO General Conference adopted the 'Recommendation on the Historic Urban Landscape', it is a kind of test case for a new approach to urban heritage. However, the Rio property only includes hills, green areas, parks and beaches such as the Corvocado peak with the famous Christ the Redeemer statue, the Copacabana beach and the Sugar Loaf. There is hardly a built structure included – the actual urban fabric of Rio is declared the buffer zone of the nominated components, even though these only become connected through this buffer zone and, of course, take their name from the city. The wish to have the World Heritage title with no conservation strings attached is obvious, practical as this will be in an urban environment preparing for the upcoming FIFA World Cup and Summer Olympics. Accordingly, ICOMOS bases its referral recommendation on the missing specifications as to how the buffer zone will be monitored and protected. Yet when the debate opens, it is the other Committee members from the same UNESCO 'electoral group' – Latin America and the Caribbean – which rush to Brazil's support, with Mexico waxing lyrical about how the property embodies the future course of the World Heritage Convention and the symbiotic dialogue between a city and its surrounding landscape. A Colombian delegate makes an attempt to hasten the decision, but the chair, having been assured that the Brazilian Minister of Culture can stay a bit longer, interrupts proceedings for lunch.

A World Heritage Centre official takes over to make the usual announcement of the meetings and events during the break – the working group on the budget of the Convention, a meeting of African ministers of culture (meaning that a sufficient number have travelled to Saint Petersburg) and an information event about one of the Centre's 'Thematic Initiatives', the one for prehistoric sites. The chair closes almost exactly on time at 1 PM, and after sorting their belongings and the session documents piling up on their desks, the delegates slowly file out of the hall, most of them engrossed in conversation.

The Lunch Break

After brief exchanges with the legal advisor and a Swiss representative – both of them confident of having made no mistakes – I run into the Dutch delegate again, who shares my impression that the Swiss 'shot themselves in the foot', as he puts it. Crossing the foyer and walking out into the courtyard, I enter a large white tent, one of several places to partake of the lunch buffet. Russia spares no expenses to host us in style and all participants – not just the state representatives – are offered free lunches and invited to splendid receptions. The last one was just the night before, in the garden of the Peter and Paul Fortress on the banks of the Neva, complete with a sumptuous dinner buffet, freely flowing drinks, top-class ballet and somewhat more debatable pop performances, social dancing and shuttle bus services to the location and then back to our hotels. All this comes with endless sunshine on top, now that the famous 'White Nights' have just ended. It is rumoured that the whole session costs Russia to the tune of €10 million or even more, exceeding the annual budget of the World Heritage Fund.

A delegate in the line tells another how cumbersome the confidential negotiations about the old city of Jerusalem are, with Israel and the United States sitting in one room and Jordan and Palestine in the other. As one of the agreed mediators, he is on his knees all the time, he jokes, but a settlement is in sight. With my plate filled, I join several ICOMOS representatives at one of the large round tables. The mosque in the Schwetzingen park was just a last straw, one of them claims, and played hardly any role in the submitted nomination file. They update me about the current state of affairs of an ambitious candidacy of Le Corbusier buildings – twice referred already, but still up for resubmission in a future session – and how the French claim unquestioned leadership in the multinational bid, down to the use of their language rather than English. I also cannot resist walking up to the Swiss ambassador I know from an interview and ask him if his call for a vote was wisely put. He claims that demanding and then losing the vote was strategic, with the deferral decision his real objective. But wasn't the Committee heading there anyway, I wonder; he throws up his hands with a flourish and turns away. A blunder it was, I cannot help thinking, but who could possibly stay alert for days on end?

The Afternoon Session

I am back at my seat when the session resumes shortly after 3 PM with further discussion of the Rio property. More Committee member support for World Heritage inscription follows. While nominating states may only respond to specific questions asked by Committee members and may not engage in advocacy, a young Brazilian diplomat ventures into a long eulogy, which the chair interrupts only in the seventh minute, reminding everyone that two is the maximum for non-Committee states. The rapporteur says she has already received an amendment for immediate inscription from Colombia this morning, a detail that lays bare how minds were made up already before the debate opened. So in the absence of objections, Rio is declared World Heritage, with the 'Statement of OUV' adopted only provisionally, since ICOMOS has had no time yet to talk this mandatory text through with the Brazilian delegation. But they have drafted one, as they do whenever they propose a referral, an option that usually presupposes the presence of OUV. The Brazilian Minister of Culture in her acceptance speech and Colombia and India, which unusually request the floor another time, all stress the pioneer character of this urban site.

The pattern of the Committee 'upgrading' recommended decisions continues through the next agenda items. For the 'Russian Kremlins', ICOMOS misses a satisfactory comparative analysis: it is neither clear how the three kremlins included in that serial property have been chosen nor how they relate to those four kremlins that, as part of other World Heritage properties encompassing larger areas, such as in Moscow and Kazan, are already on the List. Problems of authenticity for reconstructed sections come on top so that the recommendation is for outright rejection, just as with Schwetzingen. Estonia, Colombia and Switzerland are on ICOMOS's side, but the other Committee members put forth counter-arguments, some of them rather tangential: ICOMOS only requested additional information once, not twice as with other candidates; ICOMOS assesses authenticity in too orthodox a way; ICOMOS reduces the kremlins to a purely military function; with 8,000 km between the nominated components, 'a continent in one country' is covered; inscription of the nominated kremlins does not rule out adding further deserving kremlins in the future if these have been missed out on this time, and so on. The time limit is ignored again for the Russian statement and as so often in recent sessions, personal testimony is brought to bear, with the German ambassador asserting the greatness of what she has visited herself in its entirety, as she claims. The decision is for a referral, together with a 'consultative mission'. In this further recent adaptation to nation-state impatience, ICOMOS experts will visit to help improve the nomination file rather than critically assess the property as during the evaluation visit that, because of the referral decision, cannot take place in this case. But as Russia still wishes for advice, it will invite the experts at its own cost.

For 'Çatalhöyük' in Turkey, the OUV of the famous Neolithic archaeological site and oldest urban settlement is not in doubt, but ICOMOS still proposes a referral, as it misses a solid conservation framework – no dedicated budget, no management plan, and too much dependency on the foreign archaeological teams and their project funding.[1] But Committee members' praise for the site is enthusiastic; the World Heritage List would be incomplete without it, the Indian ambassador contends. Turkish delegates claim that the conservation issues have all been addressed in the meantime, an assurance that must be taken at face value, coming only in the session, not in writing – if the Committee were to follow its own rules, it must therefore be ignored. Still, Çatalhöyük is awarded the World Heritage title, to more applause and congratulations.

The next item is the first today where ICOMOS has its way against the concerned country: the proposal to extend the already listed property 'Kiev: Saint Sophia Cathedral and Related Monastic Buildings, Kiev-Pechersk Lavra' (Ukraine), is deferred, given unclear boundaries and buffer zones, and the absence of construction rules for the surrounding city. While no one points it out, the state of conservation of the very same property has been up earlier in the session, and a decision clause that threatened Danger Listing for the following year because of rampant construction all around the protected zone was only narrowly averted. This might explain why Russia's and Serbia's referral proposal finds no support.[2] But this is also a mere extension of an already listed site: in such cases, lobbying is not quite as intense as for the new bids with their much higher stakes.

This decision ends the list of cultural candidate properties and the chair moves on to the natural sites. Just like when the discussion of the cultural candidates started two days earlier, the Advisory Body in charge, the International Union for Conservation of Nature (IUCN), summarizes its evaluation procedure, going over what is familiar terrain for session regulars. The head of IUCN's World Heritage Unit, a British geologist, stresses that 'dialogue' with the States Parties – a key demand of Committee members during the last few days – cannot extend to sharing evaluation results ahead of time and that only the deferral decision option allows for extensive support for the revision of insufficient nomination files while the referral option rules this out (as mentioned above). He also sees a need for further reflection concerning indigenous communities and their rights, something IUCN will also take up at its World Conservation Congress later in the year.

The statement of a civil society representative coming next continues this topic. Infrequent though such interventions are, the chair may give the floor to anyone whom the Committee does not object to. I recognize a German anthropologist who had been anxious if and when his turn would come. Standing in front of a microphone in the back rows, he introduces himself as representing the seventy nongovernmental organizations (NGOs) united in the 'International Workgroup for Indigenous Affairs' and deplores the lack of consideration for indigenous rights in World Heritage processes. Last year, 'Kenyan Lake System in the Great Rift Valley' was listed even though indigenous groups had demanded a deferral. This year, he says, affected indigenous groups have been insufficiently consulted about two other candidates, 'Western Ghats' and 'Sangha Trinational', so he urges the Committee to defer these. Reports of monitoring missions to World Heritage properties and evaluations of candidate sites should be made accessible to indigenous stakeholders. Also, the Committee should make their 'free, prior and informed consent' a mandatory prerequisite for World Heritage listing, thus abiding by the requirements of the 'Declaration of the Rights of Indigenous Peoples' (UNDRIP) adopted by the UN General Assembly in 2007. Two days earlier already, the chair herself had raised the topic, reporting receipt of a letter from the same NGO coalition and saying that she had invited representatives of the UN Permanent Forum on Indigenous Issues (PFII) to the session. This means that the indigenous challenge to the nation-state system has reached the World Heritage Committee (cf. also Disko and Tugendhat 2014).

However, nobody requests the floor in response, so the chair moves on to 'Lakes of Ounianga' nominated by Chad. This is a unique hydrogeological system where a saline lake in the centre prevents a corona of freshwater lakes from salinization, and this in the driest part of the Sahara where annual rainfall is in the millimetres. IUCN has only minor concerns and sees OUV as proven; also, threats are largely absent and the national commitment to conservation is described as exemplary. The recommended inscription is adopted and the Chadian delegation cheers their first World Heritage property. Among the celebrants, I recognize a former university colleague of mine, a German geologist who is pleased with the success of what, he tells me, he himself had kicked off.

Yet, the activist's concerns return with the next property, 'Sangha Trinational', three adjoining nature reserves straddling the border triangle of Cameroon, the Central African Republic and Congo. IUCN recognizes that many of the issues that made it advise deferral the previous year – upgraded to referral by the Committee – have been addressed in the meantime, so that inscription is now in order. However, the advice to also include the cultural heritage of the BaAka hunter-gatherers has not been taken up and their exclusion from previously used lands within the property is noted, with the hope – but apparently no guarantees – that they can sustain themselves in the buffer zone. Obviously, adding the World Heritage protective layer to a large and relatively undisturbed reserve counts more for IUCN than respect for indigenous rights, and with no objections raised by the Committee, Sangha Trinational goes on the List.

Nevertheless, the activist's statement does make a difference for the Kenyan ambassador, who is next given the floor, interrupting the nominations agenda item. The allegations about the Rift Valley lakes are unfounded, she says, and she would welcome the critics' visit to show them how intense the cooperation with local communities is. No comments follow and the chair then resumes nominations. The 'Chengjiang Fossil Site in China' is quickly dealt with, as IUCN finds this early Cambrian community of fossils to clearly stand out on a global scale. With inscription decided, there are even two TV cameras in the room, documenting the commotion of well-wishers around the Chinese delegation.

Then finally, 'Western Ghats' in India is up, referred from last year's session. IUCN notes that the selection and boundaries of the no less than thirty-nine components of this serial property are unchanged, even though some of them include settlements, dams and plantations, and about 40 per cent lie outside of nationally protected areas. Sites in this mountain range might well have OUV, their expert says, but this particular collection is questionable and should be deferred. IUCN receives some support, mostly because of the unchanged components. However, Russia – clearly reciprocating India's favour over the kremlins – reminds the experts that they themselves have qualified the Western Ghats as missing on the World Heritage List (which, of course, is no contradiction to what IUCN said). The Indian ambassador insists on the scientific rationale of their national selection process and rules out changes to a bid on which six federal states had to agree. With no clear picture emerging, the chair urges the Committee members to position themselves individually, and only Switzerland and Estonia keep up their opposition, while twelve other members voice support for the inscription proposed by Russia. This is then adopted and at the very last moment, it is again Russia which has the recommended consultation with local indigenous groups expunged from the decision text.

After inscription, the chair unusually gives the floor to an Indian observer, who introduces himself as the representative of a local policy centre. He is full of praise for the dialogue between experts and village communities and for how nature protection, poverty reduction and sustainable development come together here, in the overall quest against climate change and for fulfilling the UN Millennium Development Goals – the statement does indeed tick all the boxes. There is no way for this statement to precede the Indian delegation's acceptance speech without the latter's approval, and clearly it is meant to dispel allegations of government disregard for local rights. Yet including this intervention appears to be the only consequence of the indigenous rights activist's challenge. Instead of the deferrals he demanded, Sangha Trinational and the Western Ghats end up where their nation states wanted them, on the World Heritage List.

The chair now suggests postponing the last natural site to tomorrow; she must have her reasons, as this is a controversial Russian nomination. Instead, she returns to unfinished agenda item 7B, the state of conservation of the properties on the List of World Heritage in Danger and to the 'Old City of Jerusalem and its Walls'. Since its simultaneous inscription on the World Heritage List and the 'List of World Heritage in Danger' on the initiative of Jordan – in 1981, eighteen years before Israel joined the Convention – this property, due to its contested territorial status, has been a most protracted headache for the Committee. The chair announces that two separate decisions have been negotiated between the concerned states, with draft texts distributed in the room. One decision is for the property in general and one is specific to the Mughrabi ascent, the ramp-like structure with which Israel wants to replace the collapsed pathway to one of the Temple Mount gates, against Jordan's and Palestine's vehement opposition. Stating for the record that Israel disagrees with and further (unmentioned) State Parties wish to dissociate themselves from the first of the two decisions, the chair then quickly drops the gavel for both items, knowing that opening a debate would result in hours of mutual accusations. The decision texts reiterate prior unanswered demands, such as Israel hosting a monitoring mission by the World Heritage Centre and ICOMOS. This means no progress in the subject matter of conservation, but an open confrontation has been avoided.

With the scheduled business finished, the Russian chair finally returns to Timbuktu. She says that she is saddened to report that, as the Malian Minister of Culture informs her, three more of the World Heritage-listed mausoleums have been destroyed, and suggests a fundamental reflection about how to address such challenges in the future. The Minister adds that the insurgents intend to destroy all sixteen components of the World Heritage property (in actual fact, only the mausoleums, holy places and a mosque entrance, not the mosques proper) and is sure they will follow through on their announcement. 'They know that we're in session, they know we're discussing this topic. They want to push us to the brink. But what can we do?', she asks.[3] The other African Committee members – Senegal, Algeria and Ethiopia – go first in sharing her exasperation and supporting the emergency fund the Minister proposed, and South Africa – a key source of financial aid for the Timbuktu sites under President Thabo Mbeki's 'African Renaissance' policy (Meskell 2012a: 47) – says it has drafted a resolution. An international appeal is being proposed and the chair points out that the international press, to which it might be read, is assembled. Then, the UNESCO Assistant Director-General for Culture and former Director of the World Heritage Centre speaks extensively. While he sometimes appears absent-minded when sitting on the podium, he now delivers a reasoned summary of the situation in the mixture of empathy and realism that appears to be called for. He recalls the precedent of the Taliban blasting the two great stone buddhas of Bamiyan in 2001,[4] the long years of UNESCO engagement in Timbuktu and the UNESCO Director-General's emergency mission to the capital Bamako earlier this year; however, he also says that the divided political situation on the ground, with several armed groups competing for hegemony, should not lead anyone to expect miracles.[5]

The chair then informs us that after five hours, the interpreters' shift is over. Senegal is last to speak and appeals to the African Union and the UN to do all they can to liberate the north of Mali. The chair announces further discussion of the Timbuktu declaration as soon as a draft statement is ready and then closes the session shortly after 8 PM.

The Reception

I move to the courtyard, where the reception of Côte d'Ivoire celebrating the World Heritage inscription of the 'Historic Town of Grand-Bassam' two days earlier is taking place. A State Party representative greets us, but then the Malian Minister of Culture speaks again, claiming that the destroyers in Timbuktu are foreigners and her compatriots would never contemplate such acts. I chat with a Hungarian delegate, the delegate of the Holy See and a German heritage studies professor. The latter notes that we are the only light-skinned delegates among several dozen people present – most non-African participants of the session do not find the event worth their attention. The Ivorian ambassador tells us about the symbolic importance of inscribing Grand-Bassam, just one year out of their civil war and with different ethnic groups and both blacks and whites collaborating for the bid. Everyone back home was eagerly following the news from Saint Petersburg, she assures me, and certainly the jubilation and flag-waving in the hall the other day were lively.

The Assistant Director-General for Culture joins us and finds this the best-organized session ever. With the circular pews, we can see each other when speaking; 'we're finally a parliament', someone else throws in. The 'ADG Culture' continues to praise the details; food is also better and more plentiful than in previous sessions. Cambodia offers to host the Committee next year, but with the Preah Vihear issue unresolved, this is still uncertain. He himself is all for using the UNESCO headquarters in Paris, just like last year; there at least, everything is in place. Schwetzingen comes up again and how the legal advisor was in the dark.

When boarding the shuttle bus to my hotel, a couple of African delegates wish to be dropped off at a different place. This provokes a spontaneous round of Committee joking, applying the special lingo to unlikely subjects: a Kenyan delegate suggests to 'defer' that other hotel: 'I'm against the inscription of that site!' Others respond in kind, with the merriment based on the shared awareness that this banter would be completely lost on ordinary mortals. 'A memorable day', a veteran UK delegate confirms, and he too has rarely seen the chaos of the Schwetzingen debate. The Kenyan underlines what the Malian Minister claimed: the Timbuktu insurgents are foreigners and no neighbouring government supports them, afraid as they all are of similar turmoil at home. Back in the hotel, the Kenyans who wanted to go and watch football with me decide to first go for dinner. I therefore find a nearby pub on my own, seeing a brilliant Spanish side take apart Italy and clinch the European Championship. When I finish taping my observations of the day, based on the jottings in my notebook, it is past midnight.

Around the World in a Day

More so even than on other session days, I have the impression that this Sunday has it all. The timeframe of heritage ranges from the Cambrian to twentieth-century industrial facilities and the spatial focus from the Rio megacity to the remotest corners of the Sahara. Palaces are up, but so are peasants' and miners' abodes, and topics include high art, but also agriculture, geology and wildlife. Hotspots of political conflict and international concern such as Jerusalem and Timbuktu force themselves on to the agenda. They are approached with all the reticence of an intergovernmental forum of limited weight where everyone is aware that political solutions – if any there are – must be found elsewhere. This reaches a degree that even a conscious and globally reported challenge to Committee authority does not fully register: shielded by its procedural routines and tied down by the inertia these produce, the Committee ends up debating what is properly Islamic over a princely folly in a German Baroque park, not over Timbuktu where World Heritage is being mutilated for *not* being properly Islamic. Time for giving the latter provocation due attention is found only during the very last session hour. All this unfolds in an environment where what is spent on the meeting, not just by the Russian hosts but also by the governments and other organizations that pick up participants' airfares and hotel bills, is at least double of what the World Heritage system has for conservation.

Expressions of sorrow and exasperation by what happens at some sites appear truly felt, but they are sweetened by free food and reception champagne in securitized spaces. There is high-minded universalism and there are appeals to the international community, invoking the largest possible 'we', the pan-human one. But there is also the ease with which nation states brush aside transnational advisory organizations, subnational challenges such as those by indigenous peoples, and demands from their peer states, modest though the latter have become over recent years. And while processes for expert fact-finding and appraisals are increasingly sophisticated, their results can easily be passed over once diplomats agree that they do not matter for the issue at hand, however improvised their reasons. With the up-and-coming countries such as India being most assertive about fulfilling their wishes and defending national sovereignty, the day also illustrates ongoing geopolitical power shifts. In short, it shows that the World Heritage Committee is very much a mirror of the world we live in, rather than a world apart.

Is this the kind of world the creators of the World Heritage Convention were imagining? In order to answer this, let us next move to the history and rationale of its conception.

Notes

1. For a close-up analysis of the national and local intricacies, cf. Human (2015).
2. Russia's de facto annexation of Crimea and the war in the Donbass region in 2014 were still to follow this display of post-Soviet unity.
3. In her original French: 'Ils savent que nous sommes à reunion, its savent que nous discutons ce sujet. Ils veulent nous pousser jusqu'au bout. Mais que pouvons-nous faire?'
4. Different from Timbuktu and *pace* Hafstein (2018: 81–82), Bamiyan was not on the World Heritage List at the time.
5. The French-led 'Opération Serval' that recovered Malian governmental control of the north in early 2013 was nowhere in sight at this point.

Chapter 2

THE PROMISE OF WORLD HERITAGE

A year earlier at the World Heritage Committee session in the UNESCO headquarters in Paris, June 2011. A nature conservation expert working for the World Heritage Centre tells me how pleased he is with the statement that a Tanzanian government minister delivered in the session: the plan to build a highway through the Serengeti National Park has been abandoned. A major threat to the famous annual migration of the savannah animals is thus averted. One such success alone, the UNESCO official claims, compensates him for all the frustrations he must endure at the hands of uncooperative nation states, reassuring him that his work is meaningful after all.

He also alerts me how the World Heritage Committee is an ideal place to make such an announcement. Conservation NGOs such as the World Wildlife Fund had been campaigning against the project too, but by travelling all the way to Paris, the Tanzanian minister bows to UNESCO, not to Western organizations, thus avoiding the colonial aftertaste the latter might convey. Concessions to the global community, this means, bespeak noble commitments, not weakness or a loss of face. They acknowledge humanity as a co-owner of the World Heritage sites.

The 'Common Heritage of Mankind'

With all the complaints about World Heritage carrying Western bias and being appropriated by nation states, the utopian premises of the Convention tend to be overlooked. Yet the website of the World Heritage Centre maintains that 'World Heritage sites belong to all the peoples of the world, irrespective of the territory on which they are located'.[1] The World Heritage Convention establishes spaces with, ideally at least, double ownership – tracts of sovereign national territory that, at the same time, belong to 'mankind as a whole' (see below). Scholarly and popular comment on World Heritage often fails to properly appreciate how unusual an idea this is, and it is worth tracing its genesis.

The principle of the common heritage of mankind (in the male-biased language of the day) made its debut in the international law of the 1960s. Technological advances had brought spaces outside state borders – the terrestrial ones of the deep seabed and Antarctica, and the extraterrestrial ones of the moon, other celestial bodies and outer space – within the reach of conceivable economic exploitation. Some nation states, not least so the newly independent colonies, feared that potential profits would bypass them before they could catch up with the leading countries. This provoked efforts to regulate usage and the distribution of potential benefits by international law. The phrase 'common heritage of mankind' was first used in a United Nations General Assembly session on the seabed and the ocean floor in 1967, proceeding to UNGA Resolution 2749 in 1970 and further into the UN Convention on the Law of the Sea adopted in 1982 after long negotiations. Kindred formulations that likewise seek to transform legally unclaimed territory (*terra nullius*) into not just a matter open to any country's use (*res communis omnium*), but one of humanity as a whole (*res communis humanitatis*) were also employed in the 1979 Moon Treaty ('Agreement Governing the Activities of States on the Moon and Other Celestial Bodies'). No state openly opposed the concept of the common heritage, not even the United States, which did not adopt these conventions (Keyuan 1991: 177).

What exactly does the concept entail? In the Convention of the Law of the Sea,[2] the fact that 'the seabed and ocean floor and subsoil thereof, beyond the limits of national jurisdiction' (§ 1.1) and its resources 'are the common heritage of mankind' (§ 136) means that no state can claim sovereignty over them (§ 137); all rights in the resources, including archaeological and historical objects, are vested in mankind as a whole (§§ 137 and 149); and all activities in the area, including scientific research, must be carried out for the benefit of mankind as a whole (§§ 140 and 143) and for peaceful purposes only (§§ 141, 143.1 and 147.1d). The Moon Treaty[3] likewise states that 'the moon and its natural resources are the common heritage of mankind' (§ 11.1), which makes it exempt from national sovereignty and both national and non-national property claims (§ 11.2–3) and requires 'an equitable sharing by all States Parties in the benefits derived from those resources' (§ 11.7a).

Two other international treaties dealing with similar kinds of space, while not using the phrase 'common heritage', express closely related ideas. The Treaty on Principles Governing the Activities of States in the Exploration and Use of Outer Space, including the Moon and Other Celestial Bodies[4] of 1967 states that 'the exploration and use of outer space, including the moon and other celestial bodies, shall be carried out for the benefit and in the interests of all countries, irrespective of their degree of economic or scientific development, and shall be the province of all mankind' (§ 1). Free exploration, access and use for all states are foreseen, and national appropriation is ruled out (§§ 1 and 2). The Antarctic Treaty of 1959 suspends consideration of the territorial claims that seven states had already made at the time, rules out any additional ones (§ 4), allows only the peaceful use of Antarctica (§ 1), prohibits nuclear explosions (§ 5) and has a number of provisions for free access and the free exchange of plans and results (§§ 3 and 7). States from the Global South have challenged the club-like arrangement of that treaty in which decision-making powers rest with the twelve original negotiators and the seventeen further 'Consultative Parties' these twelve have admitted on condition of substantial research activity on the continent.[5] However, the actual effect of the treaty and subsequent agreements within the Antarctic Treaty System that, for example, restrict mineral resource activities come close to establishing a 'common heritage of mankind' regime on the southernmost continent (Keyuan 1991).

What all these agreements share in common is the urge to keep the non-appropriated in this condition, by way of formalizing quasi-extraterritoriality and prolonging it indefinitely (cf. Höhler 2014; Rehling and Löhr 2014; Wolfrum 2009). Success has been moderate: the United States has never signed the Convention on the Law of the Sea (Höhler 2014: 70) and the Moon Treaty has been ratified by no more than seventeen countries, none of which has a significant space programme.[6] Also, the United States in 2015 and Luxembourg in 2017 have passed national laws that allow domestic corporations – including all those with a mere branch office in the country in the case of Luxembourg – to keep the proceeds from mining the moon and other celestial bodies, despite criticism of thereby violating the international agreement (Falenczyk 2017). Obviously, all this happens at a time when such technologically challenging plans are still to be put into practice, and what the 'common heritage of mankind' concept will do in the event that this ever happens remains to be seen. Still, all these international legal measures work on the premise that for the greater good of humanity, some of the space accessible to humans is best kept removed from the national grip.

The World Heritage Convention[7] builds on the same line of thought. It claims that cultural and natural heritage 'constitutes a world heritage for whose protection it is the duty of the international community as a whole to co-operate' (§ 6.1), thus pointing to heritage in its undifferentiated entirety. This follows up on UNESCO's first legal instrument for cultural heritage, the Hague Convention for the Protection of Cultural Property in the Event of Armed Conflict of 1954, which states that 'damage to cultural property belonging to any people whatsoever means damage to the cultural heritage of all mankind'.[8] However, in its preamble, the World Heritage Convention is more selective, saying that '*parts* of the cultural or natural heritage are of outstanding interest and therefore need to be preserved as part of the world heritage of mankind as a whole' (emphasis added). For these parts, the Convention foresees a 'World Heritage List' of heritage properties 'having outstanding universal value' (§ 11.2). Yet while these choice sites qualify for global custodianship, they remain on national ground: List candidates are supposed to come from the inventories of 'property … situated in its territory' that every treaty state is encouraged to submit (§ 11.1).

What results, then, is a unique case in international law: World Heritage properties belong to 'mankind as a whole' and create obligations for it, given that 'it is incumbent on the international community as a whole to participate in the protection of the cultural and natural heritage of outstanding universal value' (Preamble). At the same time, they must be located within the borders of sovereign states. Accordingly, the possibility of listing Antarctica or the UN Headquarters building in New York was quickly dismissed in the Committee's very first session,[9] even though including Antarctica had still been regarded as conceivable during the drafting process (Batisse and Bolla 2005: 27).[10] As for the high seas, the establishment of the World Heritage Marine Programme in 2005 has spurred on discussion of bringing them within the World Heritage framework, leading to expert meetings and a 2016 report compiled by the World Heritage Centre. In addition to introducing possible candidates for listing, this report boldly reasons that the Convention and its founders cannot have meant to exclude half of the surface of the earth from consideration (UNESCO 2016: 4, 11, 46–48). It therefore contemplates options for remedying this 'historical oversight' (2016: 11), either through a liberal re-interpretation of the current Convention or the addition of an optional protocol, which States Parties would then have to adopt separately (2016: 49–51).

While any such move would amount to a major reconceptualization – probably the most innovative one since the 1990s reforms of cultural heritage (see below) – no concrete steps have been taken so far. Celestial bodies have never been discussed, meaning that for the time being, all prior applications of the 'common heritage' principle in other international law remain outside of the World Heritage purview. Instead, the Convention applies to territory that *has* been appropriated, lying within state borders, and thus creates hybrid constructs where humanity becomes a co-owner and co-custodian of portions of national territory. State spaces are pulled back into a kind of extraterritoriality, with the sovereign willingly consenting to curbing its own powers and giving an international body rights. In other contexts, nation-state enthusiasm for relinquishing control is a rare event, so how did this anomaly come about in 1972 when the Cold War was still in full swing?

The Way to the Treaty

Much as UNESCO is willing to take the credit, the World Heritage Convention would not have been adopted – or at least not so quickly and in a different form – were it not for the involvement of IUCN and the US government. IUCN had proposed a global list of national parks and nature reserves in 1958, the United States put the idea on the agenda of the UN Economic and Social Council (ECOSOC) in 1959 where it was adopted, and IUCN then worked out an initial list for the first World Parks Congress in 1961. The inclusion of cultural sites – briefly considered but then dropped – returned to the agenda in 1965 when the White House Conference on International Cooperation discussed the creation of 'A Trust for the World Heritage'. IUCN welcomed this initiative and – while hesitant to make it an own project – recommended it for adoption at the UN Conference on the Human Environment, to be held in Stockholm in 1972. In the preparatory working group, IUCN was first to put a draft convention on the table, with proposals by the United States and UNESCO only following later (Batisse and Bolla 2005: 21, 23). Into 1972, the adoption of two separate legal instruments for nature and culture and the placement of the secretariat with IUCN appeared conceivable (Batisse and Bolla 2005: 27, 31–32).

Only then were the two concerns fused and did UNESCO receive the backing of the Nixon administration for acting as the secretariat of the new convention (Batisse and Bolla 2005: 29, 70–71, 79; Stott 2011). IUCN eventually supported this move, seeing a UN-system organization as a good base for urging national commitment and counting on UNESCO's influence with the Soviet Union and the developing countries (Stott 2011: 286). The Stockholm Conference approved a revised draft of the convention submitted by UNESCO. A final drafting committee worked through no less than 128 proposed amendments, including a major impasse about the states' financial contributions, before the UNESCO General Conference adopted the treaty on 16 November 1972 (for more detailed accounts, see Batisse and Bolla (2005); Cameron and Rössler (2013); Stott (2011); Titchen (1995)).

Yet while the new convention thus landed with UNESCO, crucial elements – the phrase 'World Heritage', the List, the inclusion of both natural and cultural sites – did not originate with it. UNESCO had been working towards its own legal instrument for cultural heritage since 1966 (Cameron and Rössler 2011: 14), but in Stott's eyes, this was rather meant to establish a legal framework for its safeguarding campaigns and would have contented itself with establishing a 'Red Cross for monuments' (2011: 279, 284). The first and most famous of these much-publicized missions was the 'International Campaign to Save the Monuments of Nubia' (Allais 2013; Betts 2015; Hassan 2007; Meskell 2018: 28–58). Since its construction in 1889, the Aswan Dam had been raised twice in 1912 and 1933, and this meant that the Isis temple complex on the island of Philae was submerged by the waters of the reservoir for several months each year. In 1954, however, Nasser's Egypt decided to raise the dam yet again in the interest of electricity production, irrigation and flood and drought control, with construction starting in 1960. After completion, all the ancient monuments on the banks of the Upper Nile in the region of Nubia, straddling the border with Sudan, would be submerged by the reservoir. Egypt and Sudan therefore turned to UNESCO for help, a conveniently nonpartisan body right after the Suez Crisis of 1956 had pitted the United Kingdom, France, the United States and the Soviet Union against each other. Also, 'the conservation and protection of the world's inheritance of books, works of art and monuments of history and science' was one of the goals mentioned in the UNESCO Constitution (§ 1.2c) and had already led to the aforementioned Hague Convention and a number of official recommendations in the field of cultural heritage (Titchen 1995: 38–39, 48–51). Therefore, Director-General Vittorio Veronese launched a campaign to be coordinated by UNESCO in 1960, declaring that: 'These monuments … do not belong solely to the countries who hold them in trust. The whole world has the right to see them endure. They are part of a common heritage, which comprises Socrates' message and the Ajanta frescoes; the walls of Uxmal and Beethoven's symphonies. Treasures of universal value are entitled to universal protection' (Betts 2015: 109). That the Egyptian emergency was self-inflicted did not play a great role at a time when modernization and development were the order of the day.

Over the next two decades, donations, technical advice and on-site assistance from forty-seven countries – Egypt contributed one-third of the funds (Betts 2015: 112) – helped to cut some twenty temples and monuments into pieces, lift them to safe grounds and reassemble them there. This included Rameses II's temple at Abu Simbel, which became most strongly associated with the campaign, not least through the spectacular footage of the royal faces lifted by the cranes. In the final phase, Philae itself was also relocated. Large-scale archaeological and photographic documentation of what could not be saved was conducted alongside the salvage works, with the scope of all these operations being unprecedented. They provided a welcome opportunity for the United States to engage in Egypt: while the dam itself was constructed with Soviet support, the United States alone contributed more than three-quarters to the funding of the conservation campaign (Hassan 2007: 83), even though a good part of this consisted of writing off Egyptian debts whose repayment was uncertain to begin with (Meskell 2018: 53). Yet despite US dominance and the fact that the architectural and engineering schemes came from European corporations (2018: 33), the specialists involved on and off the sites were a very multinational mix, straddling accustomed divides in a way that amazed contemporary observers. 'There seemed to be no Cold War in the land of Kush', a *New York Times* reporter wrote (Betts 2015: 116),[11] and the archaeological surveys and excavations involved teams from more than twenty nations from around the globe, something for which coordination by a UN agency was essential (Meskell 2018: 37, 42–43).

Also due to the visually spectacular nature of the operations (Allais 2013: 28–31; Betts 2015: 115–16) and support by US celebrities (Meskell 2018: 54), European and Japanese royalty and other luminaries (Hassan 2007: 82), the campaign became globally known and turned into the signature activity of UNESCO (Batisse and Bolla 2005: 16), a source of inspiration for further such campaigns for Borobudur, Moenjodaro, Florence and Venice (Meskell 2018: 59–89) and a major factor for the organization's popular association with cultural heritage conservation. The latter made US government authorities and IUCN drop culture from their aforementioned first initiative for a UN list since culture was 'the direct province of Unesco [*sic*]' (Stott 2011: 280). Meskell argues that the campaign also shifted UNESCO attention towards the conservation of monuments and technical assistance, away from archaeological research, which ended up playing second fiddle (2018: 37–44, 57–58).

UNESCO's and the World Heritage organizations' routine way of presenting the famous safeguarding campaign as the direct forerunner of the World Heritage Convention[12] is a bit self-serving, given that UNESCO was not the dominant force in drafting the treaty and was even suspected of half-heartedly jumping the bandwagon when presenting its own draft (Batisse and Bolla 2005: 19, 26). Yet there is no doubt that the Nubian campaign was on many people's minds in those days. It naturalized international cooperation for in situ conservation and the idea that national heritage can be humanity's heirloom and responsibility at the same time (cf. e.g. Betts 2015: 110, 125).[13] Without the concrete precedent of joint international action under these premises, the World Heritage Convention would have appeared much less plausible.

Interestingly, the monuments of the Nile valley did not just provide the practical referent of the safeguarding campaign; they had also contributed to conceptualizing the common heritage of mankind in the first place. Contrary to what legal-studies overviews such as Wolfrum's (2009) suggest, it was around monuments under national or colonial rule, rather than extraterritorial spaces, that the idea first coalesced. Eighteenth- and nineteenth-century debates over international codes of conduct for warfare considered special protection for culture heritage, given that this was the common property of mankind, not just a national matter (Jokilehto 1999: 281–82). And for expanding this idea to nonviolent situations, Philae played a key role: in 1894, with the British public divided about the construction of the first Aswan Dam, which was threatening to inundate the temple complex, the Society for the Preservation of the Monuments of Ancient Egypt reasoned that the 'material evidences of Egypt's former greatness … are the common property of the cultivated in all lands' (Andersen 2011: 209), not just that of the Egyptians or their colonial overlord. In 1931, when the First International Congress of Architects and Technicians of Historic Monuments met in Athens, it debated whether sites such as Philae were the 'patrimoine commun de l'humanité' (Rehling 2014: 114). The recommendations adopted, confirmed by a League of Nations resolution the following year, confirmed 'that the testimonies of art and history constitute a common heritage for, which the possessor State is accountable to the international community' (Titchen 1995: 24–25). The debt that the World Heritage Convention owes to the Egyptian site is thus twofold. Quite fittingly, the 'Nubian Monuments from Abu Simbel to Philae' were listed as a World Heritage property early on, as part of the second annual batch of inscriptions in 1979.[14] This leads us to what happened after the World Heritage Convention was adopted and to the details of how the common heritage of humankind has actually been cared for.

Notes

1. http://whc.unesco.org/en/about (retrieved 25 August 2020).
2. http://www.un.org/depts/los/convention_agreements/texts/unclos/unclos_e.pdf (retrieved 25 August 2020).
3. https://treaties.un.org/doc/Treaties/1984/07/19840711%2001-51%20AM/Ch_XXIV_02.pdf (retrieved 25 August 2020).
4. http://www.unoosa.org/oosa/en/ourwork/spacelaw/treaties/outerspacetreaty.html (retrieved 25 August 2020).
5. Cf. https://www.ats.aq/devAS/ats_parties.aspx?lang=e&lang=e (retrieved 25 August 2020).
6. https://treaties.un.org/pages/ViewDetails.aspx?src=TREATY&mtdsg_no=XXIV-2&chapter=24&clang=_en (retrieved 25 August 2020).
7. http://whc.unesco.org/en/conventiontext (retrieved 25 August 2020).
8. http://portal.unesco.org/en/ev.php-URL_ID=13637&URL_DO=DO_TOPIC&URL_SECTION=201.html (retrieved 25 August 2020).
9. CC-77/CONF.001/9, p. 4.

10. A brief resumption of this issue in 1991 (SC-91/CONF.002/2, p. 14; SC-91/CONF.002/15, p. 20) led to no concrete steps.
11. By contrast, Allais sees a 'veritable iron curtain' running through the Nile valley, between the Soviet engineers working on the dam and the Western specialists working on the monuments (Allais 2013: 12). Betts, however, mentions teams from seven Eastern Bloc countries working on the monuments too, including one from the Soviet Union (Betts 2015: 116).
12. Cf. e.g. UNESCO 2016: 45, whc.unesco.org/en/convention (retrieved 25 August 2020).
13. However, four specially engaged countries (Italy, the Netherlands, Spain and the United States) were each allowed to take one temple home, and foreign archaeological teams were promised half of their finds (Betts 2015: 113–114, 124; Hassan 2007: 80–81, 90), calling to mind earlier colonial modes of global 'sharing'. The fact that around 100,000 people had to move as well, often under less than satisfactory conditions and complaining that '[w]e would be better looked after if we were statues' (Betts 2015: 121), is largely missing from the celebratory accounts (Betts 2015: 120–22; Hassan 2007: 85; Meskell 2018: 56).
14. Concern over what the transfer of many of the temples to new locations and the attendant damages to their fabric (Meskell 2018: 34–35) had done to their authenticity is not on record, and the property only includes Egyptian sites, not Sudanese ones.

Chapter 3

FULFILLING THE PROMISE

With adoption in 1972 and the UNESCO Director-General as the designated depositary of the World Heritage Convention, the countries of the world could start to ratify the treaty and thus declare themselves bound by it. In 1975, the twenty signatures necessary for the Convention to enter into force had been reached, and this set in motion the foreseen institutions and mechanisms. The General Assembly of States Parties congregated for the first time in 1976, electing the first member states of the Committee. This body first met in 1977 and adopted the Operational Guidelines that, continuously revised over the years, have guided its work ever since. At its second annual session in 1978, the Committee inscribed 'Galápagos Islands' (Ecuador) as the first property in the World Heritage List, and in 1979, the earthquake-stricken 'Natural and Culturo-Historical Region of Kotor' (then Yugoslavia and now Montenegro) became the first property simultaneously inscribed in the 'List of World Heritage in Danger'. Also in 1979, financial support accorded to the property 'Simien National Park' (Ethiopia) inaugurated the use of the World Heritage Fund.

The first Committee sessions had fewer than fifty participants, not the thousands assembling nowadays, and they ended after four or five days, not the almost two weeks reached later on. Secretariat duties were fulfilled by two volunteers from the UNESCO headquarter bureaucracy alongside their regular duties. With evaluating the 'outstanding universal value' of candidate properties whose nomination documents filled only a couple of pages, ICOMOS and IUCN entrusted just a single person each until the late

1980s (Cameron and Rössler 2013: 179–81, 184–85, 187–94), and site visits were not yet part of the process.[1] All this is a far cry from the present scale of operations. Yet still, the World Heritage machinery was rolling, so how did it realize the ambition of owning humanity's most valuable sites in common? By and large by expansion, the short answer is: more resources, obligations and listings as well as broadened concepts, organizational frameworks and punitive powers.

Sharing and Listing

A literal version of World Heritage extraterritoriality has never been considered: no country offers visa-free visits to its World Heritage properties, even when the sites adjoin the border or are 'transboundary' to begin with (see Chapter 7). As law scholar Carduzzi concludes, 'it is certain that legally the Convention creates no pure internationalized status and property regime for the heritage it covers' (2008: 121). Also, there is no chartered right of access for Committee emissaries, unlike, say, certain disarmament treaties that foresee controls. Conversely, no State Party has ever tried to enforce support through the treaty, such as by passing on the bills for conservation measures at World Heritage properties to the Committee.

As a more limited instrument of sharing, a World Heritage Fund benefiting the properties was included in the Convention. How to fill it became the final obstacle before adoption: many Northern countries and the Soviet Bloc were for voluntary contributions and it required a sudden turnaround of the United States to install obligatory payments, as countries from the Global South had been demanding all along (Batisse and Bolla 2005: 80–85, Cameron and Rössler 2013: 24–26). These fees were set at 1 per cent of the respective country's dues to the UNESCO budget, with the latter – as elsewhere in the UN system – calculated on the basis of GDP and population. Part of the World Heritage Fund goes to 'international assistance', and four-fifths of the projects the Committee approved under this title have been for conservation and emergency measures at World Heritage properties. All treaty states are eligible for support and 169 of them have at one point or another been granted such aid. While almost 15 per cent of the grants went to European and North American states, poorer countries enjoy priority now and the top recipients are all from the Global South. However, the distributed amounts are modest, and of the more than two thousand funded requests, only thirty-five have exceeded US$75,000, with an average of slightly more than $20,000. The annual total peaked just above $3 million in 1999, but has dropped to between one-third and one-tenth of that sum in the last decade[2] (see the Conclusion). Certainly, no Nubian campaign could be conducted from these moneys, and the main burden of conserving a World Heritage property falls on the respective country and any third-party aid it is able to enlist on its own initiative. This is in line with the Convention that – alongside its emphasis on shared responsibility – clarifies that '[e]ach State Party … recognizes that the duty of ensuring the identification, protection, conservation, presentation and transmission to future generations of the cultural and natural heritage … situated on its territory, belongs primarily to that State' (§ 4) and that '[a]s a general rule, only part of the cost of work necessary shall be borne by the international community' (§ 25). The original Operational Guidelines did include assisting member states in the protection of their World Heritage properties among the Committee's duties, but this was already deleted in the first revision of 1980 (Cameron and Rössler 2013: 162), never to return.

In place of this rather meagre carrot of financial redistribution, the World Heritage Committee could have emphasized the stick of control and sanctions to live up to its moral co-ownership, but here as well, treaty states were left undisturbed for a long time. The Convention text makes no detailed requirements and the discussion of the state of conservation of specific World Heritage properties was started only in an ad hoc fashion for the natural sites in 1983, when an IUCN representative reported on the sorry state of 'Ngorongoro Conservation Area' (Tanzania), which he happened to have just visited. ICOMOS took almost another decade to follow suit. First attempts to oblige the States Parties to regularly report about their sites in the 1980s met with resistance to this perceived assault on their sovereignty and to 'ICOMOS policing' (Cameron and Rössler 2013: 126). In 1990 and prompted by a building project near the property 'Monastery of the Hieronymites and Tower of Belém in Lisbon' (Portugal), an advance notification duty for building and infrastructural works that might impair World Heritage value was introduced. However, it took until 1994 for the monitoring of World Heritage properties to be declared an official Committee task, and a general 'Periodic Reporting' for all sites was only adopted in 1996, in the face of continued resistance from some treaty states (Cameron and Rössler 2013: 110–14, 123–25, 127–28, 131–35, 162, 183).

Yet, since its first five-year cycle started in 2000, Periodic Reporting has been a self-reporting exercise, and while the questionnaire forms distributed by the World Heritage Centre have grown increasingly sophisticated, they elicit what is the State Party view of things, not an independent assessment. Treaty states did voluntarily submit reports about site conditions already in the early period, and the results of fact-finding missions of Advisory Body personnel to World Heritage properties under threat feature in Committee session records from the early 1980s.[3] However, they did not yet play the role they were to assume later, and the Committee also acted on third-party reports and the information that uninvolved State Party delegates happened to provide, even for such matters as inscription on the List of World Heritage in Danger (Cameron and Rössler 2013: 112–13). This meant that by and large for the first two decades, global sharing stopped in front of national borders, both in terms of tangible support and control.

By contrast, sharing the treaty and its list of glory proceeded at a brisk pace. Single years saw up to ten ratifications and the World Heritage community already encompassed thirty-four states when the Committee first met in 1977, welcoming the one-hundredth country in 1988 and reaching 175 in 2002 for what was an almost complete representation of humanity, given all states with more than 12 million inhabitants today were on board. Stragglers continued to trickle in, with the current total of 193 identical with the number of states recognized by the UN and the number (but not the composition) of UNESCO member states. In terms of formal participation, the World Heritage Convention is one of the most successful international treaties.

As the list of signatories grew, so did the list of properties: the first twelve inscriptions on the World Heritage List in 1978 were followed by another forty-five the following year, and the Committee has kept adding further properties in each of its annual sessions, from seven to sixty-one – 26.6 on average – for a grand total of 1,121 properties (as of 2019). Initial forecasts had been for one hundred sites (Batisse and Bolla 2005: 78, 94) or one hundred natural and one hundred cultural sites (Cameron and Rössler 2013: 48–49). In the run-up to the adoption of the Convention, ICOMOS and IUCN urged selectivity and insisting on international significance (Cameron and Rössler 2013: 29–31), and one of the secretariat officials recommended first having the States Parties prepare their national lists from which the Convention expected them to propose candidates. But the Committee went for immediately filling the World Heritage List and also rejected a formal limit on the number of listings a country could have (Cameron and Rössler 2013: 50, 52).

The Committee did, in some cases, impose restrictions on specific types of cultural heritage. In 1979, it inscribed 'Auschwitz Concentration Camp' (Poland), but 'as a unique site and to restrict the inscription of other sites of a similar nature',[4] in an obvious attempt to protect the uplifting and celebratory character of the List. In the same session, it also listed 'Historic Centre of Warsaw' (also Poland), with the political significance of this postwar reconstruction being deemed to outweigh its lack of authenticity. However, the Bureau, a sub-body of the Committee, marked this as an exception, stating that 'there can be no question of inscribing in the future other cultural properties that have been reconstructed'.[5] And again in the same year, the

United States withdrew its nomination of 'Edison National Historic Site', Thomas Edison's residence and laboratory. The Bureau and the ICOMOS advisor had feared that accepting the bid would lead to a large number of sites associated with famous individuals, encouraging nationalism and other particularisms. Accordingly in 1980, when the Committee revised the Operational Guidelines for the first time, it modified criterion (vi) referring to the association with outstanding ideas, beliefs and events: outstanding persons – the item on which the Edison nomination had rested – were deleted from the definition (Cameron and Rössler 2013: 36–37).[6]

However, no further limitations were decided, particularly not on numbers. When fifty nominations arrived for the 1979 session (Cameron and Rössler 2013: 50), with only five of these rejected, it must have been obvious that the initial estimates would be greatly exceeded. The Committee did limit consideration to two candidates per state for 1978 (Batisse and Bolla 2005: 91), but in every session of the following decade, single countries obtained up to between four and seven inscriptions. As a consequence, List growth outpaced treaty growth so that in 1988, when the number of signatories reached 105, there were already 312 World Heritage properties, with the average number of titles per country having gone up every year.

Browsing the entries from that first decade, one finds many expected names – Machu Picchu, Versailles, the Alhambra and Venice, Yellowstone, the Serengeti and the Great Barrier Reef. One also finds sites of more limited fame, such as 'From the Great Saltworks of Salins-les-Bains to the Royal Saltworks of Arc-et-Senans, the Production of Open-pan Salt' (France), 'Castles and Town Walls of King Edward in Gwynedd' (United Kingdom), 'Castles of Augustusburg and Falkenlust at Brühl' (Germany) or 'Pirin National Park' (Bulgaria). Most of all and despite India leading the count until 1997, one finds European properties: 117 in total as of 1998. A continent with one seventh of the world's population at the time – down to a mere tenth now and never more than a third in the past – thus dominated the List. With another seventeen properties in the United States, ten in Canada, eight in Australia and three in New Zealand, the Global North occupied almost one half of it. Defending the values behind this selection as 'universal' became increasingly difficult.

Reforming Heritage Concepts

Complaints about a Western and Eurocentric vision suffusing World Heritage already arose in the 1980s. They triggered a wave of reforms that came into its own in the 1990s, making this period the 'heroic age' of World Heritage. More than ever before or after, the World Heritage arena became a site of conceptual innovation with considerable impact. Three elements stand out in particular: the new heritage category of cultural landscapes, the 'Global Strategy for a Representative, Balanced and Credible World Heritage List' and the new standards of authenticity.

The first component, cultural landscapes, was meant to bridge the two domains of culture and nature that had been united under the roof of the Convention, but then held apart in actual practice. In the first Operational Guidelines, the Committee included separate OUV criteria for the two fields (six for culture and four for nature), and separate organizations were tasked with evaluations and advice, reflective of the way in which outside North America, cultural and natural conservation are usually administered by mutually unrelated institutions and frameworks. It soon turned out that the States Parties' preference was for cultural heritage: the small number of natural candidates was already deplored in 1979[7] and as of 2018, 77 per cent of World Heritage properties are cultural. Sites excelling in both ways have been duly recognized: in 1979, the Committee listed 'Natural and Cultural Heritage of the Ohrid region' (Macedonia), 'Ngorongoro Conservation Area' (Tanzania) and 'Tikal National Park' (Guatemala) as the first of currently thirty-eight 'mixed' properties, another 3 per cent of the total.[8] However, these properties are listed for having both outstanding cultural and natural features, not for how these are interrelated.

Yet despite practical separation, there was conceptual overlap: the Convention itself includes 'the combined works of man and nature' in its specifications of a cultural heritage 'site' (§ 1), and OUV criterion (iii) for natural properties spoke of 'exceptional combinations of natural and cultural elements'. This made a French delegate raise the possibility of listing rural landscapes such as vineyards and rice terraces already in the 1984 session.[9] The Committee prevaricated and Britain's nomination of 'Lake District National Park' was postponed twice in 1987 and 1990, with IUCN doubting whether this was a natural site in the sense of the Convention,[10] which it assumed to be 'nature not modified by man'.[11] But on the recommendation of a separately held expert meeting it had commissioned, the Committee adopted the new category of 'cultural landscapes' in 1992 and annexed the definitional framework of the meeting report to the Operational Guidelines. Furthermore, it modified two cultural OUV criteria by adding 'landscape' and 'land-use', and in 1994 expunged the 'combinations of natural and cultural elements' from natural criterion (iii). Since the 2005 version, the Operational Guidelines additionally emphasize that '[c]ultural landscapes are cultural properties' (§ 47), meaning that no strong bridge across the culture-nature divide was built – IUCN left the new category to ICOMOS (Gfeller 2013: 495). To this day, however, IUCN coevaluates cultural landscape nominations by commenting on their natural features, although in my eyes with less enthusiasm than for the natural properties of the presumably pristine type.[12]

According to the annex, cultural landscapes subdivide into intentional creations such as gardens and parks and 'organically evolved' landscapes where cultural responses to the natural environment have either stopped (relict landscapes) or continue in the present, in association with a traditional way of life – rice terraces and vineyards are cases in point. Crucially, however, a third subtype are 'associative cultural landscapes' where 'religious, artistic or cultural associations' of the landscape stand in the foreground and where 'material cultural evidence … may be insignificant or even absent'. The latter was a clear opening for the sacred landscapes of indigenous peoples. Accordingly, in 1993 and 1994 and as the first cultural landscapes,[13] 'Tongariro National Park' (New Zealand) and 'Uluṟu-Kata Tjuṯa National Park' (Australia; Uluṟu is also known as Ayers Rock) – already listed for their natural features in 1990 and 1987 – were re-inscribed by the Committee, now also under cultural criteria, which reflected the spiritual importance of these sites for the local indigenous communities. Cultural landscapes have proved to be popular and with a total of 102 properties so labelled (as of 2018),[14] every sixth inscription since 1993 has been in this category.

Even broader in scope and an explicit antidote to Eurocentric bias was the Global Strategy for a Representative, Balanced and Credible World Heritage List. The Committee had already commissioned a 'Global Study' to take stock of the World Heritage List in 1988, but this exercise soon fragmented into 'thematic studies' of specific heritage types, with the great civilizations and monuments and the accustomed disciplines of art history, architectural history and archaeology retaining centre stage (Gfeller 2015: 371–73). Criticism arose and an expert meeting in 1994 recommended discarding the Global Study, making a set of recommendations, which the Committee approved later that year.[15]

The diagnosis of that meeting was quite scathing: the World Heritage List had relied almost exclusively on a monumental concept of cultural heritage and on buildings, isolating them in addition, rather than putting them within their societal context. European heritage, historical towns, religious (in particular Christian) buildings and elite architecture were overrepresented, whereas prehistory and the twentieth century were in short supply. Living and especially traditional cultures had been neglected too. Therefore, a more dynamic 'Global Strategy' was to replace the Global Study in 'a move away from a purely architectural view of the cultural heritage of humanity towards one which was much more anthropological, multi-functional, and universal'. In particular, the following topics, 'considered in their broad anthropological context through time', were found to have great potential for filling the gaps on the List:

HUMAN COEXISTENCE WITH THE LAND

– Movement of peoples (nomadism, migration)

– Settlement

– Modes of subsistence

– Technological evolution

HUMAN BEINGS IN SOCIETY

– Human interaction

– Cultural coexistence

– Spirituality and creative expression.[16]

Africa, the Pacific and the Caucasus were mentioned as the most-neglected world regions.

The meeting also recommended modifications of the cultural OUV criteria so that the latter fitted the new agenda. Consequently, in the next revision of the Operational Guidelines in 1996, 'unique artistic achievement, a masterpiece of the creative genius' was shortened to 'masterpiece of human creative genius', thereby expunging art (criterion i); 'influence' on architecture, arts, town planning and landscape design was replaced with 'interchange', a term that emphasizes mutuality rather than hierarchy (criterion ii); 'living' civilizations and cultural traditions were added to those that have disappeared (criterion iii); and technological ensembles were added to the other items illustrating significant historical stages, namely buildings, architectural ensembles and landscapes (criterion iv).[17] The urge to get away from monumentalism, elitism and a Western optic was palpable indeed.

Subsequently, the Global Strategy has stood unassailable within the World Heritage arena. It has motivated a large range of special studies, progress reports, support measures and audits, and provides the heading for a variable list of items in the formal agenda of every Committee session. The consensus that decisions must abide by the Global Strategy is undisputed, perhaps all the more so since what exactly this entails is not quite so clear: contrary to the conceptualization of cultural landscapes and the Nara Document, which were both included in the Operational Guidelines or annexed to it, there is no canonical text that spells out the Global Strategy, only the revised cultural OUV criteria and the recommendations of the 1994 expert meeting mentioned above. While the recommendations in particular are often cited, I still have the impression that most arena participants' sense of the strategy is informed most of all by the 'Representative, Balanced and Credible World Heritage List' phrase of the title and by the strategic mileage to be gained from claiming that the real List falls short of this goal, even a quarter of a century later.

The third important reform concerned the authenticity of cultural properties. The text of the Convention does not use this term,[18] but the first Operational Guidelines stipulated for cultural heritage that, in addition to fulfilling at least one OUV criterion, 'the property should meet the test of authenticity in design, materials, workmanship and setting'.[19] They further stated that 'authenticity does not limit consideration to original form and structure but includes all subsequent modifications and additions over the course of time, which in themselves possess artistic or historical values'.[20] This formulation breathes the spirit of the Venice Charter of 1964,[21] the foundational document of modern conservation adopted by a UNESCO-supported international conference of conservationists, which also decided to found ICOMOS. For restoration and archaeological excavations (§§ 9–15), the Charter prioritizes the retention of original materials, including any layers superimposed onto a building by later historical periods. For archaeological sites, it prohibits reconstruction except by anastylosis,[22] and for everything else, it allows it only in exceptional circumstances, based on solid historical documentation and when new fabric is clearly distinguished from the original one. Historical buildings are treated as 'documents embodying evidence' (Jones and Yarrow 2013: 11) rather than aesthetic wholes. This was quite a turnaround from the eighteenth- and nineteenth-century beginnings of European architectural conservation when 'restoration' was meant to realize the (presumed) original intent of the builders in as complete a fashion as possible.

Venice Charter orthodoxy works better with stone and steel monuments in Western architecture than with wooden and earthen buildings in other parts of the world. In 1992, the Committee debated the dire state of conservation of 'Kathmandu Valley', Nepal, and repair work undertaken on the wooden palaces and temples raised the question of how the replacement of material affected the authenticity of the structures. After belatedly joining the Convention, Japan presented 'Buddhist Monuments in the Horyu-ji Area' near Nara as one of its first nominations in 1993, which also demonstrated the need for a conceptual debate. This temple complex includes the oldest wooden buildings on earth, dating from the seventh and eighth centuries, but was extensively restored in the thirteenth, seventeenth and twentieth centuries. As is common with traditional wooden buildings in Japan, the temple structures were partly or wholly dismantled and then reassembled, allowing the replacement of much damaged material[23] so that, for example, the Hokkiji temple pagoda contained as little as 49 per cent of original timber already before its last restoration in 1972–75 (Gfeller 2017: 768). This led some European experts to question the authenticity of such structures. While the Hōryūji complex was listed in 1993, Japan offered to host an international workshop in Nara the following year. This meeting drafted the Nara Document on Authenticity,[24] widely cited and praised today as the World Heritage institutions' most significant contribution to the topic.

Authenticity, this text states in rather absolute terms, is 'the essential qualifying factor concerning values' (§ 10). But then it turns relativist, saying that: 'It is … not possible to base judgements of values and authenticity within fixed criteria. On the contrary, the respect due to all cultures requires that heritage properties must considered [*sic*] and judged within the cultural contexts to which they belong' (§ 11). Also, the range of aspects supporting authenticity is widened: 'authenticity judgements may be linked to the worth of a great variety of sources of information … [that] may include form and design, materials and substance, use and function, traditions and techniques, location and setting, and spirit and feeling, and other internal and external factors' (§ 13). Compared to the 'design, material, workmanship, and setting' formulation of the original Operational Guidelines, this adds quite a number of aspects, including rather ephemeral ones that cannot be read from the physical fabric alone.

Given the milestone character often ascribed to the Nara Document and the flurry of meetings it provoked (Gfeller 2017: 786–87), it is surprising how little else it specifies. Much of the text consists of general praise for heritage conservation and cultural diversity. It does not provide a definition of authenticity, nor does it say which aspects are to be given priority when assessing the authenticity of particular kinds of cultural heritage – such as that of specific world regions or historical periods – or whether there should be different standards for different materials such as, say, stone versus wood. The Nara Document claims continuity with the Venice Charter, which it 'builds on … and extends' (§ 3), and Gfeller emphasizes that it stops short of abolishing authenticity and universal standards altogether (2017: 782–84), which leaves ICOMOS and other experts as its arbiters in place. Still, I read the above paragraphs as an attempt to remove obstacles rather than pinning anything down, allowing for a very broad range of interpretations and uses. How to reconcile their cultural relativism with the 'U' in OUV is not elaborated anywhere.

A Japanese participant told me that the European and American participants at the Nara Conference appeared to her as longing for something radically different – redemption from Venice Charter constraints, so to speak – and were almost disappointed to learn through her presentation that Japanese conservation practices are not so alien after all. I am sure that expectations were influenced by the *shikinen sengū*, the ritual renewal of the most important Shinto shrine in Ise, every twenty years on one of two alternating plots of land; the previous buildings are dismantled in the process. This case is a staple when, in books on conservation, alternatives to mainstream European philosophies are introduced. Almost invariably, the practice is exaggerated then, for example, by ascribing it to Japanese shrines and temples in general when in actual fact it continues at fewer than a dozen Shinto shrines today. Also, precisely because the current iterations of these shrine buildings are so new, they are not – and cannot be – registered cultural monuments under Japanese law.

In contrast, the conservation of ancient Buddhist temples such as the Hōryūji complex differs from Venice Charter ideals only in degree, not in kind, since material replacement is kept to the necessary minimum. Also, copying damaged wooden components does not even permit conjecture, given the modular construction and sophisticated joinery of traditional Japanese architecture, which requires new pieces to fit exactly. Elsewhere, I have argued that differences to the conservation practices of, say, European cathedrals are mostly dictated by the building material, not by fundamentally different ideals of continuity (Brumann 2012b: 222–24). Accordingly, the Japanese convenors of the workshop would have been happy with minor modifications of the Operational Guidelines (cf. also Gfeller 2017: 774, 788), and it was others who pushed for the more comprehensive formulations cited above.

Integrating the Nara Document into World Heritage regulations took its time. The ICOMOS General Assembly approved it in 1999[25] and later that year, the Committee decided to include it in the revision of the Operational Guidelines,[26] where, however, it only debuted in the 2005 version, eleven years after Nara[27] (cf. also Cameron 2008: 21–22). Moreover, the formulations based on the Nara Document (§§ 81–83) continue to be immediately tempered by the admonition that 'in relation to authenticity, the reconstruction of archaeological remains or historic buildings or districts is justifiable only in exceptional circumstances. Reconstruction is acceptable only on the basis of complete and detailed documentation and to no extent on conjecture' (§ 86). To a considerable degree, the Guidelines thus try to have the cake and eat it – Torta di Venezia with Nara icing, one is tempted to say (see Chapter 7).

It will have been noted that the three conceptual innovations function as a package (see also Gfeller 2015: 378): themes emphasized in the Global Strategy such as 'settlement', 'modes of subsistence' or 'spirituality and creative expression' are present in cultural landscapes of the 'organically evolved' or 'associative' types. For confirming the authenticity of ongoing, incrementally changing cultural landscapes and of other types of heritage emphasized by the Global Strategy, the Nara Document removes the hindrances that a narrow focus on material fabric might present. Indigenous cultural accomplishments are targeted by the spiritual emphasis included in all three initiatives. And while only the Global Strategy raises an explicit challenge to Western notions of cultural monumentality, the other two moves were also expected to counter the perceived bias of World Heritage.

Interestingly, though, the reforms did not originate with the victims of Eurocentrism: the main protagonists and almost all participants of the preparatory meetings hailed from the Global North. Yet, with the exception of the French and British contributions to the cultural landscape debate, key individuals were not from 'Old Europe', but from the margins and outliers of 'the West' – Norway, Japan and the former settler colonies Canada and Australia, all countries that had not dominated heritage discourses and policies so far and were not represented in the drafting group of the first Operational Guidelines (Brumann 2014b: 2180–83; Gfeller 2013: 502–3; 2015: 367, 377–78, 2017: 761–63, 770, 784–85). The contributors who did come from 'old' heritage nations brought new specializations to the table, such as archaeology, geography and even anthropology – Romanian-born Frenchman Isac Chiva, a student of Claude Lévi-Strauss, denounced a monumentalist and elitist perspective at the Global Strategy meeting (Gfeller 2015: 376). This means that while not fully universal, the represented range was nonetheless significantly broader than, for example, that of the Venice Charter whose twenty-three drafters included nineteen Europeans, sixteen of them from Western Europe.[28] The shift is clearly illustrated by Belgian architectural historian Raymond Lemaire's role: as one of the drafters of the Venice Charter and the ICOMOS founding Secretary-General and then President, he insisted on including authenticity in the first World Heritage Operational Guidelines, fearing that 'integrity' – the alternative term then discussed – would give too much liberty to tamper with the material fabric. However, in the Nara meeting, he was superseded by the ICOMOS Secretary-General, Canadian Herb Stovel, and others, not daring to go against what he saw as a clear changing of the guard and deplored as lacking intellectual rigour (Gfeller 2017: 764, 783–84).

National interests did play a role in the reforms and not just for Japan, which was concerned for its own World Heritage prospects under Venice Charter orthodoxy. Canada and Norway as the other main backers of the Nara Document had wooden architectural heritage too, and Canada had discovered industrial heritage earlier than continental Europe (Gfeller 2017: 771–72, 776–78, 788). In Canada, Australia and Norway (with its Sami minority), indigenous cultural heritage and collaborative policies had been on the agenda (Gfeller 2015: 375–77), more so than in most European countries. Accordingly, it was two Australian majority archaeologists with experience in such projects who proposed the associative cultural landscapes, the 'human coexistence with the land' theme of the Global Strategy and the inclusion of living cultures in the OUV criteria (Gfeller 2013: 496–99; 2015: 373–74, 376). Also, Stovel – the single most influential individual (cf. Gfeller 2013: 493–94; 2015: 371–72; 2017: 781 and *passim*) – had been working to open up ICOMOS to vernacular architecture for some time (Gfeller 2017: 772–73).

However, to just see vested interests behind this 'anthropologization' (Brumann 2011: 29; 2013: 26, 2014b: 2180; Gfeller 2015) of World Heritage would unduly trivialize it. Japan did go into a listing spree over the next decade, including wooden architecture and other 'reform-type' heritage; in part, it was catching up after late ratification. In contrast, Canada had only one site with wooden architecture inscribed and Norway and Australia had none listed to which the new perspectives could be applied (see also Gfeller 2017: 788). World Heritage Centre personnel unbound by any national interests also played a crucial role, in particular for the gestation of the cultural landscape concept. To a considerable extent, the key protagonists were working out of a genuine concern for making the World Heritage List a global list, with a more comprehensive outlook that was fairer to the non-Western countries. As has already been outlined elsewhere (Brumann 2014b: 2180–83), this is all the more remarkable because modern conservation took shape during, and owes a lot to, nineteenth-century high nationalism (Larkham 1996). The fact that conservation must be in conversation with property rights, urban planning strategies, subsidies and taxation inevitably binds much heritage law and administration to a national – if not provincial or municipal – frame of reference. The typical background disciplines of heritage conservationists, such as art history, architectural history or archaeology, also often have distinct national traditions and thematic focuses, at least in Europe, and have in the past provided the scaffolding for ideas of imperial greatness that take the superiority of mostly Euro-Mediterranean 'civilizations' very much for granted. Accordingly, early international exchanges and borrowings in the field of heritage conservation were largely confined to Europe and North America or were the product of colonial imposition (Hall 2011; Swenson 2016; Swenson and Mandler 2013). This means that conservationists are not 'natural' cosmopolitans – less so than, say, anthropologists for whom the symbolism of a mud hut can be as sublime as that of a Baroque palace. The cosmopolitan impetus of the driving figures of the World Heritage reforms has to be appreciated accordingly – not as something that comes with the trade, but as a genuine effort to transcend boundaries and inbuilt biases that did not greatly concern most of their national colleagues at the time.

Applying the New Concepts

Even when formal adoption of the Nara Document lagged, the effects of the reforms already made themselves felt in the late 1990s. More inscriptions of cultural landscapes followed, including agricultural ones, such as 'Rice Terraces of the Philippine Cordilleras' (the Philippines), and landscapes that had been sources of inspiration for religion and the arts, such as 'Lushan National Park' (China). Further testimonies of indigenous lifeways were 'Laponian Area' (Sweden), a landscape shaped by Sami reindeer herding, and 'Sukur Cultural Landscape' (Nigeria), a precolonial iron-smelting centre with a distinct building and terracing style. More wooden structures both in Japan and outside as well as vernacular architecture from earth and stone were listed too.[29] With the exception of 'Ironbridge Gorge' (United Kingdom) inscribed in 1986, technical World Heritage properties had heretofore pre-dated the Industrial Revolution (cf. also Gfeller 2017: 786). Now, however, sites such as the company town 'Crespi d'Adda' (Italy), 'Völklingen Ironworks' (Germany) or 'Ir.D.F. Woudagemaal (D.F. Wouda Steam Pumping Station)' (the Netherlands) also went on the List.

As for postwar towns, 'Brasilia' – Brazil's planned capital and the most ambitious creation of architectural Modernism (Holston 1989) – had already been inscribed in the 1987 session, barely thirty years after construction began. This was due in part to a ploy for which the then UNESCO Director-General Amadou-Mahtar M'Bow takes credit: the freshly revised Operational Guidelines had ruled that 'these towns should be deferred until all the traditional historic towns, which represent the most vulnerable part of the heritage of mankind, have been entered on the World Heritage List'.[30] Yet the secretariat distributed this text only after the Brasilia decision, to the dismay of US, Canadian and Indian delegates (Cameron and Rössler 2013: 77). However, with the tail wind of the Global Strategy, no more such ruses were needed for listing 'Bauhaus and its Sites in Weimar and Dessau' (Germany) or the 'Skogskyrkogården' cemetery in Stockholm (Sweden). The trend picked up in the 2000s and culminated in the inscription of 'Sydney Opera House' (Australia) in 2007 – heritage that, having been opened in 1973, is younger than the Convention itself. Archaeological sites, rock art sites and fossil sites of human evolution likewise increased, and so did an interest for the tragic sides of history and for the 'difficult', 'dark' or 'dissonant' heritage constituted by the so-called 'sites of conscience' (Logan and Reeves 2008; Tunbridge and Ashworth 1996): after a major diplomatic balancing act, 'Hiroshima Peace Memorial (Genbaku Dome)' (Japan) was inscribed in 1996, with the United States and China only registering their reservations rather than blocking the move, and when listing 'Archaeological

Landscape of the First Coffee Plantations in the South-East of Cuba', its history of slavery was duly acknowledged.

The 1990s also saw further conceptual debate, triggered by planned or actual World Heritage nominations. All of them concerned properties lineally extended in space, and the Committee had already approved such challenges to the common person's idea of a 'site' when listing 'Hadrian's Wall' (United Kingdom) and 'The Great Wall' (China) in 1987. Now in 1993, Spain nominated 'Route of Santiago de Compostela', and the Committee, while accepting the country's offer of hosting an expert meeting on 'heritage routes' in the following year,[31] nonetheless inscribed the bid right away. This chronological order was reversed with 'heritage canals' where the expert meeting – also in 1994[32] – pre-dated the first listing, that of 'Canal du Midi' (France) in 1996. For these two special cases of cultural landscapes, basic definitions and specifications were annexed to the Operational Guidelines, just like those for cultural landscapes.[33] Finally, by nominating 'Semmering Railway' in 1996, Austria brought in mountain railways. The Bureau of the Committee deferred a decision at first, commissioning a comparative study,[34] but based on its results and a positive ICOMOS evaluation, the Committee went for inscription in 1998, with delegates hailing the arrival of yet another new category.[35]

The broad view on cultural World Heritage continued into the new millennium. The 2008 Committee session is a plausible candidate for bringing the reform spirit to a peak with its listings, both regarding new categories and old ones presented in a new key. Let us therefore inspect the list of inscriptions more closely (see Table 3.1). It includes the third mountain railway property (largely Swiss with a tiny Italian section) and an agricultural landscape on a Croatian island where grape and olive cultivation continues to follow the ancient Greek land division. Six housing estates in Berlin going back to Weimar Republic progressive housing reform and designed by the likes of Walter Gropius and Bruno Taut made the cut too, meaning that humble workers' abodes now joined the 'Palaces and Parks of Potsdam and Berlin', which had been World Heritage since 1990. By contrast, no royal residence was on the Committee agenda this time, and nor were any cathedrals, except as parts of larger urban ensembles. Instead, in terms of Christianity, the inscriptions included eight small-sized wooden churches in the Slovak Carpathians, which owe more to the local peasant world than to the great theological and aesthetic traditions of their time.

Global Strategy echoes also resound in the way in which both the Straits cities and San Miguel with the nearby church complex were interpreted as products of cultural fusion. The 'statements of OUV' adopted at inscription praised them as sites of encounter between colonial forms and local creativity, in ways that reach beyond the built fabric:

Table 3.1. Cultural World Heritage inscriptions of the 2008 Committee session.

Al-Hijr Archaeological Site (Madâin Sâlih) (Saudi Arabia)
Armenian Monastic Ensembles of Iran (Iran)
Bahá'i Holy Places in Haifa and the Western Galilee (Israel)
Berlin Modernism Housing Estates (Germany)
Chief Roi Mata's Domain (Vanuatu)
Fortifications of Vauban (France)
Fujian Tulou (China)
Historic Centre of Camagüey (Cuba)
Kuk Early Agricultural Site (Papua New Guinea)
Le Morne Cultural Landscape (Mauritius)
Mantua and Sabbioneta (Italy)
Melaka and George Town, Historic Cities of the Straits of Malacca (Malaysia)
Protective Town of San Miguel and the Sanctuary of Jesús Nazareno de Atotonilco (Mexico)
Rhaetian Railway in the Albula/Bernina Landscapes (Switzerland and Italy)
Sacred Mijikenda Kaya Forests (Kenya)
San Marino Historic Centre and Mount Titano (San Marino)
Stari Grad Plain (Croatia)
Temple of Preah Vihear (Cambodia)
Wooden Churches of the Slovak Part of the Carpathian Mountain Area (Slovakia)

Note: Cf. http://whc.unesco.org/en/list/xls/?2019.

> Melaka and George Town … reflect the coming together of cultural elements from the Malay Archipelago, India and China with those of Europe, to create a unique architecture, culture and townscape … Melaka and George Town are living testimony to the multi-cultural heritage and tradition of Asia, and European colonial influences. This multi-cultural tangible and intangible heritage is expressed in the great variety of religious buildings of different faiths, ethnic quarters, the many languages, worship and religious festivals, dances, costumes, art and music, food, and daily life.[36]

> San Miguel de Allende … acted as a melting pot where Spaniards, Creoles and Amerindians exchanged cultural influences, something reflected in the tangible and intangible heritage. The Sanctuary of Jesús Nazareno de Atotonilco

> constitutes an exceptional example of the cultural exchange between European and Latin American cultures.[37]

With their churches, mosques and temples of multiple creeds and their Sino-European shop houses (De Giosa 2016), Melaka and George Town more clearly support such a narrative than the Mexican site, a town and (Christian) church of the Spanish *conquista*. Still, the shift in emphasis is also notable when comparing the Straits cities to earlier inscriptions of colonial ports in the Indian Ocean (for a fuller analysis, see Brumann (2019a)). When listing 'Old Town of Galle and its Fortifications' (Sri Lanka) or 'Baroque Churches of the Philippines' (the Philippines) in the 1980s, for example, local adulterations of European models were still qualified in ambivalent terms, but with 'Historic Centre of Macao' (China) in 2005 and now the Straits cities, hybridity became the very point of interest.

For a different kind of global node inscribed in 2008, the Bahá'i holy places are the central pilgrimage sites of a religion known for its cosmopolitan and pan-religious leanings. In 2007, ICOMOS had found the buildings and gardens unremarkable except for the specific religious link, although it stated that 'ICOMOS does not consider that it can assess faiths',[38] and the Committee had opted for a referral. However, now that the same selection of component sites came back with a stronger emphasis on the global pilgrimage tradition, it received a green light.

Turning from the hubs to the margins of the global system, another five of the 2008 inscriptions concern the vernacular architecture or cultural landscapes of indigenous and other subaltern groups. The *tulou* property consists of forty-six stunning vernacular structures built by the Hakka and Fulao ethnic minorities in the mountains of southeastern China. The buildings stand out by their circular shape, with a thick earthen wall for defence purposes on the outside and several storeys of apartments housing as many as eight hundred people rising around an inner courtyard, facing each other in an unwitting nod to Jeremy Bentham's panopticon. The Kuk site in the New Guinean highlands encompasses a good square kilometre of swamps that have been used for agriculture for at least seven thousand years, first with cultivation mounds, then by digging ditches. This archaeological discovery (Denham et al. 2003) challenged the association of agricultural 'triumphs' with the formation of hierarchies and states as in the classical civilizations – in the highlands, egalitarianism always prevailed. The ICOMOS evaluation did not ascribe OUV to the contemporary wetland gardening by the Kawelka ethnic group, but confirmed that 'modern farming activities … do not intrude upon the archaeological features of the site'.[39] Local livelihoods are thus not affected by World Heritage fame and this must have encouraged Marilyn Strathern – ethnographer of the Mount Hagen area and one of the most prominent living anthropologists – to give her private support to the nomination initiative.[40]

As a further listing, the sacred *kaya* forests include eleven surviving patches of forest in the Kenyan coastal hinterland, which once contained fortified settlements. The Mijikenda ethnic groups had built these *kaya* since the sixteenth century, abandoning them during colonial times, but still considering them their ancestors' abode today. Little more than the layout of the villages survives, but ritual use in tandem with traditional protection of the forests continues. ICOMOS advised for a referral, just like a year earlier, but mainly because no explanation was given how the eleven component sites had been chosen from an original thirty-six. That the property as such could have OUV was not called into doubt, and the Committee chose to inscribe it right away.

The same was true for Chief Roi Mata's domain, nominated as representing the cultural landscapes of Pacific chiefly systems and the associated taboos that had conserved the sites of the last bearer of the Roi Mata title. These include his erstwhile residential compound, the beachside cave on an adjacent island on which he died, a third island where he was buried and that was thence closed to settlement and other uses, and the connecting sea space. Roi Mata is remembered for having ended a long war by introducing a system of matriclans in around 1600 and it is the link between oral history, ongoing ritual practices and restrictions and the landscape that ICOMOS found to have OUV. The evaluation did recommend a deferral, not just because of unresolved protection problems but also to include other nearby sites associated with the famous chief. Here again, however, the Committee went for immediate inscription, swayed by State Party assurances that the issues could be dealt with post hoc. Finally, significance in oral history was also crucial for acknowledging Le Morne, a striking monolith rising steeply from the Indian Ocean. Over generations, this almost inaccessible mountain provided shelter for maroons whose memory lives on in the oral and ritual traditions of Mauritian Creoles. For the latter, Le Morne is a rallying point where they celebrate the annual abolition day,[41] similar to the role that Aapravasi Ghat – the landing place of hundreds of thousands of indentured labourers, listed as World Heritage two years earlier – plays for Mauritian Indians. The tragic legend of maroons throwing themselves from the mountain rather than suffering recapture also features in Bernardin de Saint-Pierre's 'Voyages à l'Île de France' (1773), an abolitionist classic.[42]

Across these inscriptions of 2008, the attention given to the dominated and marginalized stands out. Where the properties concern the nodes of former imperial power, they feature workers' lodgings rather than palaces (Berlin) or emphasize the colonial subjects' contribution to hybrid cultural forms (the Straits cities and San Miguel). Hierarchy is only celebrated in its stateless, traditional and relatively small-scale form (Roi Mata's domain). The *tulou*, the *kaya* and Le Morne were refuges from – rather than centres of – power, and the Bahá'i adherents are a small and often persecuted religious minority wherever in the world they have spread.

Also, the corresponding lack of conventional monumentality or the dearth of physical traces as such was not held against the sites. Material evidence of Chief Roi Mata, for example, is restricted to cave images that 'could depict' him and the fact that the excavated mass burial site features a single central figure. The ICOMOS evaluation nonetheless concludes that: 'The authenticity of the nominated sites is not in doubt, including their strong associations with oral history.'[43] The statement of the nomination file that these 'oral traditions … are amongst the most remarkable feats of historical memory in the Pacific, and perhaps in the world', as they detail the exploits of more than fifty generations,[44] must have been convincing. However, the validity of these superlative claims appears not to have been questioned or closely examined.

As for Le Morne, ICOMOS could have picked on the rather thin archaeological record of human presence on the mountain – at the level of stone slabs carried into rock shelters, bones of robbed and eaten sheep, or altered soil compositions[45] – and the patchy historical record, which gives glimpses of settled communities, huts, horticulture and organized acts of resistance on Le Morne, but little evidence of their continuity over time.[46] The nomination file emphasizes the oral and ritual traditions of Creole descendants,[47] yet the appended ethnographic documentation[48] is exploratory at best, offering a somewhat haphazard collection of local testimonies of differential plausibility while noting 'numerous contradictions in the declarations of the informants'.[49] It is of course no surprise that the precarious lifestyle of isolated fugitives left only few physical, documentary and mnemonic marks, but instead of pointing the finger at the gaps, ICOMOS found the 'remains of maroon settlements' and the association of mountain and maroons authentic, not doubting their great weight in oral history either.[50] It also saw this 'symbol of slaves' fight for freedom, their suffering, and their sacrifice' as transnationally significant,[51] an assertion not even strongly made in the nomination file, which prefers to highlight the mountain's national role.

The remaining seven cultural properties inscribed in 2008 – an early colonial Cuban town, the historic centre of San Marino, twelve fortresses built by Louis XIV's military engineer Vauban, two Italian Renaissance towns, the monumental Nabatean tombs of Al-Hijr, three Armenian Christian monasteries in Iran and the ancient Khmer temple of Preah Vihear – are more conventional fare and would not have required the Global Strategy to shine; also, Camagüey is presented as a transplant of Andalusian forms and techniques to Cuba, with no prominent mention of local input or hybridity. These 'old-school' listings also managed to add two new countries

(San Marino and Saudi Arabia) to the World Heritage map. Yet they are outnumbered by inscriptions that before the 1990s reforms would have had trouble gaining World Heritage recognition, and these likewise gave first properties to Papua New Guinea and Vanuatu and a first cultural property to Malaysia. The session decisions thus clearly demonstrate that the reforms had managed to change cultural World Heritage and convert it into a more diverse and inclusive selection of human achievements, thus embracing a vision more in line with the goal of global sharing.

Organizational Growth

Alongside the roster of signatories, the World Heritage List and the cultural heritage concepts, the organizational base for administering the World Heritage Convention expanded too. The initial problem was that while the treaty designates a depositary, it does not specify who was to do the secretarial work. The two UNESCO headquarter volunteers – one each from the Culture Sector and the Natural Sciences Sector – received temporary support from additional staff. For this, however, UNESCO made the Committee pay, asking for between US$70,000 and US$260,000 per year from the World Heritage Fund in what some criticized as a misuse of this resource. However, Committee appeals for more support were regularly rejected by UNESCO. In the 1991 Committee session of Carthage, Tunisia, things came to a head, both because of disorganization and friction between the different UNESCO divisions. In response, the Spanish UNESCO Director-General Federico Mayor installed a 'World Heritage Centre' placed outside the established UNESCO sectors in 1992, with one of the first-hour volunteers, the German Bernd von Droste, as founding director under his direct supervision. The Centre started out with nine posts delegated from other UNESCO units, and the number more than doubled within the next five years, topped up further with staff dispatched and paid by volunteer States Parties. Von Droste aimed for even more, proposing a kind of think tank that would combine the secretarial functions, a World Heritage academy and a large library, and would be housed outside UNESCO precincts, but a number of countries blocked the initiative, of which ICOMOS and IUCN were also sceptical. In 1999, the next Director-General, the Japanese Matsuura Kōichirō, re-integrated the Centre into the regular UNESCO structure by placing it inside the Culture Sector. However, in terms of staff, it grew even further in the 2000s and continues to occupy its own small building on the headquarter premises (cf. also Cameron and Rössler 2013: 204–9).

In its first years and in addition to launching the *World Heritage Review* in 1996, the Centre established partnerships with several Japanese and German broadcasting stations that started long-term series of quarter- or half-hour documentaries about individual World Heritage properties, translated into multiple languages and exported globally.[52] Not only did they contribute greatly to converting World Heritage into a household name in many countries, they also brought additional revenue to the World Heritage Centre, as was also true for coffee-table books that likewise introduced individual sites and resorted to the knowledge of Centre staff during production. The Centre also launched a 'World Heritage Education Programme' in 1994, with a teacher's manual translated into more than forty languages, youth forums and camps, and seminars for educators.[53] To this initiative, Norway gave US$2 million, demonstrating a growing interest of States Parties in shaping the course of World Heritage. Money also flowed in the opposite direction: international assistance from the World Heritage Fund to the treaty states quadrupled in just four years, reaching an all-time high of US$3.148 million for 159 projects in 1999.

In response to and further encouraging these activities, public interest in World Heritage surged. The presence of the phrase 'World Heritage' in printed books began to rise more sharply after the 1990s, as already mentioned (see Table 0.1). At roughly the same time, Committee session attendance likewise increased – what had never been more than around one hundred participants rose to over two hundred in the late 1990s and anywhere between five hundred and two thousand from the mid-2000s onwards. Increasingly, the meetings took on the trappings of a global event: a veteran recalls that in Banff, Canada, in 1990, an opening ceremony and closing dinner with an indigenous cultural performance were considered sufficient social embellishment for a Committee session. She noticed a change when in Naples in 1997, participants received tickets for a Luciano Pavarotti concert in a splendid theatre and were treated to receptions and performances almost every night. As the conceptual and organizational framework had grown, so had the traction of what was fast becoming a global brand: increasingly, the world came to share an awareness of World Heritage.

Procedural Elaboration

Committee session growth meant the arrival of new players, and it is largely due to their initiative that from around 2000, the thrust of reforms moved from World Heritage substance to World Heritage procedures. Major proponents mentioned by my interlocutors were delegates from Belgium, Greece, Hungary, Lebanon, Lithuania, the Netherlands, Saint Lucia and South Africa – most of them smaller countries with relatively few World Heritage properties and/or candidates at the time. One of the key people remembers her shock when first arriving at a Committee session: 'It was the jungle!' She felt that most decisions were already taken at the meeting of the Bureau and by the Advisory Bodies in less than transparent ways, with Committee members being expected simply to rubberstamp them. Also, except for matters of World Heritage inscription, the written records of the sessions were vague, and the secretariat had much discretion as to how to implement, or else ignore, Committee decisions. Others had similar criticisms and in the course of a general strategic reflection in the run-up to the thirtieth anniversary of the Convention, the Committee changed its ways. Much of this happened at an extraordinary Committee session in Paris in 2003, which revised the Rules of Procedure for the conduct of the Committee sessions and gave the Operational Guidelines the most comprehensive overhaul in its long history of revisions.

As a result, the Bureau of the Committee was deprived of most of its powers. Heretofore, this sub-body – consisting of the chair, the rapporteur and the representatives of five other member states, usually from the five world regions distinguished by UNESCO – had met in the middle of the year in Paris to look at both the condition of listed properties and at new nominations and the Advisory Body evaluations of these. The time until the end-of-year Committee session was then used to sort out remaining problems between the Centre, the Advisory Bodies and the concerned States Parties. However, many decisions, including new inscriptions, had already been drafted in considerable detail by the Bureau and did not change much when the Committee met. Committee members not part of the Bureau and other treaty states were free to attend the Bureau sessions as observers, yet still the critics saw the practice as giving special powers to a rather limited circle of states. As a consequence, the revised Rules of Procedure limited Bureau meetings to Committee sessions, with the latter shifting to the Bureau's former mid-year position. The Bureau is now a mere sideshow, meeting in the morning before the Committee plenary for just half an hour or so and dealing with procedural details such as rearrangements of the plenary timetable; it rarely attracts more than a few dozen people. All substantive matters are now deliberated and decided by the full Committee later in the day, thus increasing transparency but also the workload, so that I heard protagonists of the reform muse that they threw out the baby with the bathwater.

The format of the plenary sessions also changed. The two main agenda items switched places so that the discussion of the state of conservation of the already-inscribed World Heritage properties now precedes that of new nominations. This not only underpinned the official priority of conservation, but also discouraged the prior pattern of many delegates leaving the session when the inscription of new World Heritage sites was over. And then the Rules of Procedure also banned advocacy, allowing States Parties to speak on their own sites only in order to answer specific questions. Here again, however, more radical proposals such as a ban on nominating properties from one's own country while serving on the Committee were rejected.

As for documentation and following the example of the UNESCO Executive Board, the former 'summary record' has been split into two documents since 2002: the 'summary record' and the 'report of the session'. The latter contains all decisions, each with its unique code number and with a couple of numbered paragraphs expressing the Committee's will, in official-sounding language and with much standardized wording for recurring content. Drafts of these decision texts prepared by the Centre and the Advisory Bodies are published ahead of the meeting and provide the basis for discussion, with many of them changing little or not at all in the session. The 'summary record', by contrast, captures the interventions made during the debate. This record – either a summary written in third-person style or a full word-for-word protocol – is sent to Committee members for corrections before it is published, ideally within three months, but often much later, and in contrast to the decisions, it is merely for the purposes of information (or demonstration, such as when delegates want to prove faithful transmission of messages to their ministry).

Just like these records, which may now run up to three hundred (the decisions) or between four and nine hundred pages (the summary record), the general paper trail of the sessions has ballooned. Much of this was as a result of the new Operational Guidelines, which swelled from 139 to 290 paragraphs when they were finally adopted in 2005, with more detailed prescriptions for nominations, monitoring and other activities. For the first time also, the Advisory Bodies described their evaluation procedures in an annex. Without this being explicitly required, the evaluations also expanded: ICOMOS produced around four or five double-column text pages per candidate site until the mid-2000s, but then doubled the text amount over the next few years, with a larger number of standardized subheadings and more specificity, such as criterion-by-criterion assessments of OUV. IUCN evaluations also grew and took on a more structured appearance around this time. And so too did the two customary documents for the state of conservation of the regular World Heritage properties and those on the World Heritage List in Danger: up until 2002, these comprised a total of fifty to seventy pages, but by the end of that decade, this had soared to the current level of around five hundred pages. The increase was due in part to a longer World Heritage List with more threatened properties, but mainly to a more elaborate structure with more detailed information about the individual cases. The budget of the Committee and its bodies also came to be presented in a clearer way.

The main protagonists behind these changes included a number of very vocal career diplomats. Heritage experts were involved as well, yet still Committee veterans who remember the earlier sessions – which implies a heritage background in almost all cases – tend to interpret the reform as part of a general diplomatic takeover that was carried in part by a mistrust against expertise. Except for the politically delicate cases, the specialists were largely among themselves until in the mid- or even late 1990s, UNESCO ambassadors and other career diplomats arrived in large numbers and asserted their presence. They often lead the delegations now and practise the routines familiar from other UNESCO meetings, such as working with draft decision texts. Critics tend to connect this with a drop in quality – the substantive discussions of earlier sessions could no longer be had with such large numbers of often unspecialized participants. 'You kill it with process', as one of the veterans phrased it to me. Given the brevity of records for the early period, it is difficult to substantiate how much nostalgia tinges such retrospective assessments. However, what is certainly true – and again deplored by some heritage veterans – is that the momentum shifted: whereas the 1990s were used for rethinking World Heritage, in particular cultural properties, the 2000s were spent rethinking the processing of World Heritage.

Using the Stick

Expansion and a heightened sense of importance must have encouraged more assertive decisions by the Committee, in particular regarding the List of World Heritage in Danger. Just like monitoring, use of the 'Danger List'[54] had a slow start: the Committee prioritized launching the regular List in 1978 (Cameron and Rössler 2013: 135–36) and while the first Danger List entry was made in 1979, as mentioned above, it was only in 1982 that the Committee asked the Advisory Bodies to prepare more detailed guidelines for this step, which were then adopted the following year (Hølleland, Hamman and Phelps 2019: 37). Danger List entries did not reach double figures until 1992. All but one of the first ten inscriptions were backed by the respective countries, which saw them as useful for what Hølleland and colleagues call the 'fire alarm' function of this sublist, namely mobilizing attention and resources for World Heritage under severe threat. The one exception, the World Heritage inscription and Danger Listing of 'Old City of Jerusalem and Its Walls' in 1982, was certainly opposed by Israel – not yet a signatory of the Convention at the time – and the territorial status of the property remains contested to this day, but the Committee has refrained from specifying the State Party of what Jordan nominated (see Chapter 5) so that there was no agreed sovereign to oppose the step.

However, in the eventful 1990s, the Committee began to consider Danger Listings that were unwelcome to the respective country. This combined the 'fire alarm' with 'naming and shaming' as what Hølleland et al. (2019) see as the other main function: putting global public pressure on a country that is negligent of its conservation duties. The 1991 Danger Listing of 'Old City of Dubrovnik' right in the middle of the Balkan War was an ambivalent case: treaty state Yugoslavia was apparently (although not too clearly) opposed to the step, yet it was its own artillery that was shelling a property that, after the Croatian declaration of independence, was no longer under its control.[55] In 1992, the Committee made four Danger List entries, which the respective States Parties had not formally requested but did not oppose either,[56] and three further such listings followed in 1996.[57] However, other cases did provoke resistance: Ecuador's vehement opposition made the Committee postpone the proposed Danger Listing of 'Galápagos Islands' in 1995. By the following year, the country had passed a new law that took care of conservation concerns so that the idea was dropped (Cameron and Rössler 2013: 139–44). The 1995 Danger Listing of 'Yellowstone National Park' on account of planned gold mining next to the park was loudly protested by the US Congress and the state of Montana. Yet the request had come from the US federal government, meaning that the State Party was in agreement with the step (Cameron and Rössler 2013: 145–47).[58]

The *cause célèbre* of the Committee's right to danger-list and the single property that most occupied it became 'Kakadu National Park' (Australia). This nature reserve had been listed in 1981, but when a conservative national government took over in 1996, plans to activate a dormant mining license for Jabiluka, an enclave surrounded by the World Heritage property, came up. The acrimonious domestic debate saw federal authorities and the mining corporation pitted against the Northern Territories government and the Mirrar traditional owners, who complained that their consent had been unfairly acquired. The involvement of indigenous rights and of the nuclear industry did little to ease the tension, and nor did the aboriginal claimants' strategy of addressing the World Heritage Committee directly, in circumvention of the national government. Australia vehemently opposed Danger Listing in the Committee sessions and spared no efforts in lobbying the member states. Therefore, in the extraordinary Committee session of 1999 – the only such meeting ever called for discussing a single property, given that construction work for the mine had begun – the Committee, after intense debate, postponed this step (Cameron and Rössler 2013: 172, 229–31; Logan 2013; Maswood 2000). However, the Committee session of 2000 – as it happened, in Cairns, Australia – drew indigenous activists from around the world, mobilizing around Kakadu National Park and calling for more involvement in World Heritage matters, and the representative of the traditional owners of Jabiluka was invited to address the Committee. Yet, Australia insisted on State Party consent to Danger Listing where it could,[59] and it required the new owner of the mine to abandon the plan.[60]

Kakadu was bound up with 'Kathmandu Valley' (Nepal), where the rampant und often illegal demolition of wooden structures in favour of concrete buildings all across the urban space – even in the seven specially protected monumental zones – and questionable restoration work had made ICOMOS recommend Danger Listing as early as 1993.[61] However, there and in subsequent sessions, Nepal was as resistant as Australia, citing its ongoing mitigation efforts. The Committee remained divided as to the limits of national sovereignty, given that general World Heritage inscription requires State Party consent according to the Convention (§ 11.3).[62] Since a study by the UNESCO legal department[63] saw it as the Committee's very own task to interpret the treaty, many hours of the 2002 regular Committee session and the aforementioned extraordinary session of 2003 were spent on discussing the issue.[64] Interestingly, the Committee's right to delete properties from the World Heritage List if they had lost their OUV was not much disputed. However, on Danger Listing, division was encouraged by the contradictory messages that can be read from the respective Convention paragraph:

> The Committee shall establish, keep up to date and publish, whenever circumstances shall so require, under the title of 'List of World Heritage in Danger', a list of the property appearing in the World Heritage List for the conservation of which major operations are necessary and for which assistance has been requested under this Convention. This list shall contain an estimate of the cost of such operations. The list may include only such property forming part of the cultural and natural heritage as is threatened by serious and specific dangers, such as the threat of disappearance caused by accelerated deterioration, large-scale public or private projects or rapid urban or tourist development projects; destruction caused by changes in the use or ownership of the land; major alterations due to unknown causes; abandonment for any reason whatsoever; the outbreak or the threat of an armed conflict; calamities and cataclysms; serious fires, earthquakes, landslides; volcanic eruptions; changes in water level, floods and tidal waves. The Committee may at any time, in case of urgent need, make a new entry in the List of World Heritage in Danger and publicize such entry immediately. (§ 11.4)

The long list of possible dangers in the middle section separates two quite distinct parts. The first two sentences outline a procedural scenario in which Danger Listing is coupled with a request for financial assistance for major safeguarding operations – a request coming from the respective State Party, as everyone understood this passage. However, the last sentence frames Danger Listing as an emergency measure, and the Committee's right of taking it 'at any time' implies swift and autonomous action without extensive consultations. Also, the State Party and its consent remain unmentioned in the entire paragraph.

Therefore, the champions of national sovereignty emphasized the prior request of assistance without which Danger Listing could not proceed in their eyes; unsurprisingly, Australia took this view, but so also did the United States, with Egypt, Zimbabwe, China and India wondering what a Danger List entry imposed on an unwilling State Party could achieve in the first place. Nevertheless, others, such as several European countries and Lebanon, emphasized the last sentence and the 'common heritage of mankind' perspective. Joining an international treaty regime meant submitting to its governing body's authority, they argued, so that Committee decisions did not require State Party permission.[65] No consensus arose and, in the end, a clearly exasperated Committee agreed to keep the old paragraphs of the Guidelines.

However, in the regular Committee session following in 2003, the tone on Kathmandu had changed:[66] shocked by a monitoring mission to Nepal that had reported large-scale destruction of the urban fabric, several delegations argued for immediate deletion from the World Heritage List; the well-conserved parts of the monumental zones could be re-nominated at a later stage. Nepal was still reluctant, but, with losing the title a real threat, found Danger Listing 'acceptable' now.[67] And so the Committee decided, asserting its right to use the stick.

The mere wielding of this instrument also prompted actions by others. Mining threats to World Heritage properties had been recurring, not just in relation to Yellowstone and Kakadu.[68] Worried about its public image, the business association of global industry leaders – the International Council on Mining and the Environment (ICME), revamped into the International Council on Mining and Metals (ICMM) in 2001 – approached IUCN and the World Heritage Centre for a dialogue, and in 2000, a joint meeting was organized at IUCN headquarters.[69] In the end, ICMM complied with the IUCN demand for making World Heritage properties 'no-go' areas and officially pledged not to mine or explore them in 2003, as also did oil giant Shell. The commitment still stands and has been joined by other resource extraction firms in the 2010s[70] (see also Cameron and Rössler 2013: 148–50).

With Kathmandu finally landing on the Danger List, two further contested cases concerning urban areas in Germany followed. The first was that of 'Cologne Cathedral', where a high-rise development across the River Rhine threatened the dominance of the building in the city's skyline. In the 2004 session, the German delegation denied any such harm;[71] however, outright opposition to Danger Listing is not on record, perhaps because the building project was only backed locally, not by the foreign ministry and the conference of the German federal state ministers of culture whose representatives led the delegation.[72] Similarly, in 2006 in relation to 'Dresden Elbe Valley' – the bridge construction controversy that opened this book – the German delegation, instead of answering the question of what it thought of Danger Listing, deplored that time had not allowed for consulting local authorities, and this again did not stop the Committee from taking the step.[73]

However, all other inscriptions on the 'Red List' during these years were clearly consensual, including that of 'Galápagos Islands' in 2007, twelve years after Ecuador had narrowly averted it. By contrast, State Party opposition in most cases meant that the Centre's and the Advisory Bodies' draft decisions foreseeing Danger Listing or conditionally suggesting it for the following session were watered down by the Committee.[74] The most intense controversy arose in 2006, when Algeria demanded that 'Tipasa' be removed from the Danger List, on which it had been since 2002. The report of a monitoring mission had acknowledged substantial progress and the dispatched expert testified in the session that some of the threats – such as illegal residential occupation of the archaeological site – no longer existed. Yet ICOMOS still wished to see the completed management plan, and several mostly Euroamerican Committee members insisted on evidence rather than promises and on working on the basis of information submitted within the statutory deadlines. Yet, the opponents – mostly from the Global South – prevailed, and the State Party had its way.[75]

Even the first deletion from the World Heritage List – that of 'Arabian Oryx Sanctuary' in 2007 – was in agreement with the State Party. A year earlier, Oman had claimed it had allowed 'nothing more than some initial research [on oil deposits], noting that there was no reason for concern about the conservation of the property'. Yet now the delegation asked for striking the reserve from the List as its OUV was gone, and IUCN confirmed that state protection had been lifted from 90 per cent of the area, with its entirety foreseen for oil drilling. Also, poaching had reduced the oryx antelopes – the property's claim to fame – to eight animals, which were facing extinction in the wild so that Oman proposed a captive breeding programme instead. So far, however, the property had not even featured on the Danger List. The Committee spent the better part of a morning session discussing the ramifications of global co-ownership – did Oman have the right to ask back what it had offered up to humanity's custody? Most delegations spoke in favour of national sovereignty and/or realism, arguing that a World Heritage property with no OUV left would impair the credibility of the Convention. Yet still, a secret ballot came out one vote short of the required two-thirds majority, so it required informal consultations over the course of the remaining day to convince the doubters. After resumption in late afternoon, it took the Committee only a couple of minutes to seal the deletion, now by consensus, and for the first time since its inception, the World Heritage List shrank.

Deleting Dresden

By striking off the Omani property, the Committee expanded the range of decision precedents yet again to include not just ignoring national wishes when formally marking the condition of a World Heritage property as threatened, but also removing the title when its foundation was gone. It remained to combine the two measures, and this happened with 'Dresden Elbe Valley'. Dresden was certainly no straightforward issue: the construction plan for the Waldschlösschen Bridge was actually mentioned several times in the nomination file that Germany submitted in 2003. Also, several witnesses confirmed that the ICOMOS representative who visited Dresden for the evaluation mission was shown the planned site (Berthod 2011: 182). However, the Committee did not raise the issue when inscribing the property in 2004, perhaps reassured by the claim in the ICOMOS evaluation that 'construction of a new bridge is foreseen 5 km down the river from the centre'.[76] This would have placed it beyond the next river bend, out of sight and close to the boundaries, but the actual location is very much at the heart of the World Heritage property, visible from the centre of Dresden and less than 3 km *up* the river.

After the Centre was informed that the locally contested project had been approved by a municipal referendum in 2005, the Committee justified the aforementioned Danger Listing of 2006 with a violation of the Operational Guidelines, as these require advance consultation of the Committee in the case of problematic construction. However, the bridge plan as such had never been hidden, and the design was still to be determined at that point. Nevertheless, the Dresden city assembly used the Danger Listing to halt the project locally and announce a second referendum, citing changed circumstances. In addition, the national Ministry of Construction indicated willingness to fund the extra costs of a tunnel alternative in 2007. Yet putting the intricacies of German federalism on stark display, the Mayor of Dresden and the superordinate authorities of the federal state of Saxony overrode the city assembly and ignored the ministry, backed by two court decisions that declared the first referendum decision to be binding. This meant that construction went ahead, with only minor modifications of the design, which ICOMOS declared insufficient in 2008. In both the 2006 and 2007 sessions, the Committee had threatened deletion for the following year without then acting on it; it repeated its threat in 2008 unless the structure would be completely removed. With construction at an advanced stage, this unsurprisingly did not happen,[77] so in the 2009 meeting, the draft decision, for the first time, foresaw immediate deletion from the World Heritage List.

Dresden was the talk of the session and when the item was opened on the third plenary day, a World Heritage Centre representative – herself a German national – summarized the long process leading up to the current impasse. An ICOMOS delegate reiterated that OUV had been irreversibly damaged and that assessment of such damage needed to take place prior to construction, not post hoc. Germany was given the floor next and its UNESCO ambassador claimed that the consultation process with the State Party had not yet been exhausted. He then handed over to the Mayor of Dresden, herself part of the delegation, who emphasized that while the city was now legally bound to building the bridge, the World Heritage property was much larger than the construction site. In the Committee debate, eight delegations argued for waiting yet another year for completion[78] and the result of another court appeal for a second referendum; however, eight other members urged immediate deletion, with Barbados emphasizing that inaction would jeopardize the credibility of the Committee and the Convention.[79]

No consensus arose and when deciding the matter by vote was proposed, confusion spread: just like over Schwetzingen three years later (see Chapter 1), delegates were uncertain in which order the different options were to be addressed, which kind of majority applied, what would happen in the event of the motion being rejected, why the vote should be secret, whether it had already started or not and so on. A spontaneous 20-minute interruption to sort out procedural details with the legal adviser and a second one for lunch delayed things further. However, after the lunch break, two secret ballots ended Dresden's short World Heritage spring: an Egyptian amendment postponing the decision was rejected with eight votes in favour and thirteen against, and the original draft decision of deletion was then adopted with fourteen votes in favour and five against. The numbers indicate that one delegation switched sides between the votes – had it not done so, the required two-thirds majority for deletion would have been missed, and one can only speculate that the limits to which this delegation was willing to comply with the bridge supporters' lobbying had been exceeded. Lamentable as the Spanish chair found the deletion and much as it signalled collective failure to her, she thus declared Dresden Elbe Valley struck from the List. All the Committee conceded was a noncommittal sentence in the decision text that 'a new nomination for the heritage of Dresden which justifies Outstanding Universal Value could be envisaged'.[80]

As already mentioned in the Introduction, this was a difficult step for many delegates: both the interventions during the debate and informal comments later on conveyed mixed feelings. However, at the same time, there was a strong sense of obligation – the Committee owed it to itself to take a tough stance after having been ignored for years.[81] In addition, budging in the face of a major country from the Global North would send the wrong message to others. Such considerations were carried by a clear sense that the World Heritage Committee should strive to be more than just the assembled countries and their vested interests. This means that at that point in time, most Committee members still agreed they were not bound by national wishes and that, if need be, the multilateral commitment required putting treaty states in their place.

World Heritage at Its Peak?

Much of what was accomplished in the World Heritage arena during the first three decades of its operations can be seen as fulfilling the promise of turning humanity's most important cultural and natural sites into a common cause. Almost all countries joined the treaty and submitted candidate properties, filling the World Heritage List far beyond initial expectations. Implicit conceptions of conventional European monumentality were duly exposed and subjected to reform, leading to a broader, more global and quasi-anthropological notion of cultural heritage, which paved the way for new types of cultural properties. In line with the rise of World Heritage to a prominent and coveted global brand, the supporting organizational framework within UNESCO headquarters expanded, and the annual World Heritage Committee sessions became well-visited global events. Responding to calls for transparency, consistency and accountability, mechanisms for monitoring conservation at World Heritage properties and for self-reporting their condition were introduced in the 1990s, and many Committee procedures underwent elaboration and systematization through the 2000s, in line with what a Foucauldian perspective on governmentality would let us expect (Foucault 1991). The produced paper trail swelled accordingly.

The Committee also flexed its muscle by ringing the official alarm bell of the 'Danger List' for troubled sites, even when the national sovereign was not in agreement. And finally in the late 2000s, the Committee went as far as withdrawing World Heritage honours when States Parties grossly violated their pledge to properly conserve what they had agreed to share with humanity. In an intergovernmental body, this required treaty states to turn against treaty states, not for any bilateral grudges, but out of a multilateral commitment to defend common interests. While the Committee was powerless to physically stop what it denounced as the destruction of World Heritage properties, it thereby did demonstrate that such measures would not pass with impunity and that expulsion from the club of superlative sites was a real possibility, thus sending a clear warning to everyone.

To be sure, the financial sharing of conservation burdens through the World Heritage framework remained modest at best, and while World Heritage titles can mobilize substantial resources, most of these have been deployed by other actors, including other intergovernmental bodies such as the International Coordinating Committee for the Safeguarding and Development of the Historic Site of Angkor (ICC-Angkor), the United Nations Development Programme (UNDP) and the World Bank. The proclaimed semi-extraterritoriality of the World Heritage properties – too important to be left to their country – remains largely symbolic: in practical political terms, they are indistinct from other national territory, and economic sanctions or armed interventions to enforce their protection are not on record. But within these constraints, World Heritage must have appeared at the peak of its powers in 2009 – grown beyond the most optimistic expectations, more widely shared and more broadly conceptualized than ever, and with a track record of actually using the disciplinary tools of which the Committee disposes.

However, as initially outlined as the central puzzle of this book, the following years saw a complete transformation of Committee operations in which the multilateral ambitions behind the Dresden decision greatly contracted. Let me next turn to how this came about.

Notes

1. IUCN started them in 1985 and ICOMOS in 1992 (Cameron and Rössler 2013: 179, 193).
2. Cf. http://whc.unesco.org/en/intassistance/action=stats for all mentioned figures (retrieved 25 August 2020).
3. Cf. the records of the respective Committee sessions, available at http://whc.unesco.org/en/sessions (retrieved 25 August 2020).
4. CC-79/CONF.003/13, p. 11. In 2008, Israel was considering the addition of other concentration camps to the World Heritage List, questioning the usefulness of this restriction (cf. https://wikileaks.org/plusd/cables/08PARIS1054_a.html (retrieved 25 August 2020)), but no such nominations have come forward.
5. CC-80/CONF.017/4, p. 4.
6. CC-79/CONF.005/6, p. 3; CC-79/CONF.005/6, Annex II, p. 7; CC-79/CONF.003/12 Rev, p. 2; CC-79/CONF.003/13, p. 9; Operational Guidelines, 1978 version, § 7; Operational Guidelines, 1980 version, § 18. The different versions of the Operational Guidelines from the initial to the current 2019 one can be found at https://whc.unesco.org/en/guidelines (retrieved 25 August 2020).
7. CC-79/CONF.005/6, p. 2.
8. Ngorongoro also inaugurated the practice of treating sites of human evolution as cultural properties, even when the timespan covered by the finds is four million years, including ancestral species whose cultural capacities must have been limited.
9. SC/84/CONF.004/9, pp. 7–8.
10. SC-85/CONF.008/9, pp. 5–6; CC-86/CONF.001/11, pp. 10–11; SC-87/CONF .004/11, pp. 9–10, 13; CC-90/CONF.003/12, p. 16; CLT-90/CONF.004/13, p. 8.
11. SC-87/CONF.004/11, p. 9.
12. For the genesis of the cultural landscapes category, cf. also Titchen (1995: 206–41) and Whitby-Last (2008).

13. This is at least the standard claim. Technically, 'cultural routes' are cultural landscapes too, so that 'Routes of Santiago de Compostela' (Spain), inscribed in 1993 (see below), shares the distinction. For another 1993 listing, 'The Sassi and the Park of the Rupestrian Churches of Matera' (Italy), landscape features are crucial too, but for some reason it was not classified as a cultural landscape. Earlier inscriptions such as 'Cliff of Bandiagara (Land of the Dogons)' (Mali) do not appear on the World Heritage Centre's list of World Heritage cultural landscapes (https://whc.unesco.org/en/culturallandscape (retrieved 25 August 2020)).
14. Cf. https://whc.unesco.org/en/culturallandscape (retrieved 25 August 2020).
15. WHC-94/CONF.003/INF.6.
16. Ibid., p. 6.
17. Cf. Operational Guidelines, 1994 and 1996 versions, § 24 (http://whc.unesco.org/en/guidelines (retrieved 25 August 2020)).
18. Authenticity only features in a technical postscript to the thirty-eight articles, in an oxymoronic way: 'Done in Paris, this twenty-third day of November 1972, in two authentic copies bearing the signature of the President of the seventeenth session of the General Conference and of the Director-General of the United Nations Educational, Scientific and Cultural Organization which shall be deposited in the archives of the United Nations Educational, Scientific and Cultural Organization.'
19. Operational Guidelines, 1978 version, p. 4 (http://whc.unesco.org/en/guidelines (retrieved 25 August 2020)).
20. Ibid.
21. www.icomos.org/charters/venice_e.pdf (retrieved 25 August 2020).
22. 'Only anastylosis, that is to say, the reassembling of existing but dismembered parts can be permitted. The material used for integration should always be recognizable and its use should be the least that will ensure the conservation of a monument and the reinstatement of its form' (Article 15).
23. http://whc.unesco.org/archive/advisory_body_evaluation/660.pdf, pp. 3–4 (retrieved 25 August 2020).
24. http://whc.unesco.org/document/9379 (retrieved 25 August 2020).
25. WHC-99/CONF.209/22, p. 39.
26. Ibid., p. 45.
27. Operational Guidelines, all versions since 2005, §§ 81–83 (http://whc.unesco.org/en/guidelines (retrieved 25 August 2020)). For more detail on this process, see Cameron (2008: 21–22).
28. Cf. https://www.icomos.org/charters/venice_e.pdf, p. 4 (retrieved 25 August 2020).
29. Examples include 'Petäjävesi Old Church' (Finland), 'Town of Luang Prabang' (Laos), 'Ancient Ksour of Ouadane, Chinguetti, Tichitt and Oualata' (Mauritania) and 'The Trulli of Alberobello' (Italy).
30. 1987 version, § 29 (http://whc.unesco.org/en/guidelines (retrieved 25 August 2020)).
31. Cf. WHC-94/CONF.003/INF.13.
32. Cf. WHC-94/CONF.003/INF.10.
33. Cf. Operational Guidelines, 2017 version, Annex 3, §§ 16–24 (http://whc.unesco.org/en/guidelines (retrieved 25 August 2020)).
34. WHC-96/CONF.201/4, p. 33.
35. WHC-98/CONF.203/18, p. 27.
36. http://whc.unesco.org/en/list/1223 (retrieved 25 August 2020).
37. http://whc.unesco.org/en/list/1274 (retrieved 25 August 2020).
38. WHC-07/31.COM/INF.8B1, p. 176.

39. http://whc.unesco.org/document/9949, p. 137 (retrieved 25 August 2020).
40. http://whc.unesco.org/uploads/nominations/887.pdf, p. iii (retrieved 25 August 2020).
41. https://whc.unesco.org/uploads/nominations/1259.pdf, pp. 33–34 (retrieved 25 August 2020).
42. Ibid., pp. 36–39.
43. WHC-08/32.COM/INF.8B1, pp. 93, 95.
44. https://whc.unesco.org/uploads/nominations/1280.pdf, p. 66 (retrieved 25 August 2020). The evaluation daintily skips over another quite proudly presented superlative, namely that the proportion of the island population buried together with Roi Mata – in an 'apparently voluntary' fashion – is a world record (ibid., p. 61).
45. Cf. http://whc.unesco.org/uploads/nominations/1259.pdf, pp. 24–28 (retrieved 25 August 2020).
46. Cf. ibid., pp. 17–24.
47. Ibid., pp. 29–39.
48. Ibid., pp. 909–1125.
49. Ibid., p. 1088.
50. WHC-08/32.COM/INF.8B1, p. 3.
51. Ibid., pp. 3–4.
52. 'Schätze der Welt' (Treasures of the World) by the German station SWR has reached 423 episodes by now (https://www.swr.de/schaetze-der-welt/denkmaeler/-/id=6837828/15uiy21/index.html (retrieved 25 August 2020)) and 'Sekai isan/The World Heritage' by the private Japanese station TBS was already up to nine hundred episodes (on more than six hundred sites) by 2015, as announced in that year's Committee session. The latter provoked a rival product by the Japanese national station NHK. By 2012, the TBS series had been exported to no less than 112 countries, as reported at the Closing Event of the Fortieth Anniversary of the World Heritage Convention in Kyoto.
53. http://whc.unesco.org/en/wheducation (retrieved 25 August 2020).
54. I use the most common informal appellation in Committee session talk even when 'In-Danger List' (Hølleland, Hamman and Phelps 2019) might be the linguistically preferable alternative.
55. Cameron and Rössler speak of a Yugoslavian representative questioning the Committee's right to decide this step against his State Party's wishes (Cameron and Rössler 2013: 227–29). The official session record only states that: 'The observer from Yugoslavia requested that the Committee should be cautious in this action' (SC-91/CONF.002/15, p. 8), however, and the legal opinion on Danger Listing presented to the 2002 Committee session also says that Dubrovnik was listed 'without the former Yugoslav government clearly opposing this inscription' (WHC-02/CONF.202/8, p. 28).
56. These included the three nature reserves 'Sangay National Park' (Ecuador), 'Mount Nimba Nature Reserve' (now 'Mount Nimba Strict Nature Reserve', Côte d'Ivoire/Guinea) and 'Manas Wildlife Sanctuary' (India), as well as 'Angkor' (Cambodia) that – with the country just emerging from its civil war – was under UN administration at the time (cf. WHC-92/CONF.002/12, pp. 8–9, 25–28, 37–38, 46). Cameron and Rössler claim that Manas was listed without State Party consent, but what they describe is better captured as State Party silence (2013: 137) and since the Centre was instructed to inform national authorities of the decision (WHC-92/CONF.002/12, p. 28), Indian representatives must have been absent from the meeting room.
57. In the run-up to the session, Tunisia had opposed the possible deletion of 'Ichkeul National Park', but not the Danger Listing, and for 'Garamba National Park' (Democratic Republic of the Congo), 'no commitment of the Zaire authorities for such listing had

been obtained' (WHC-96/CONF.201/21, p. 32). However, outright opposition of the States Parties to these or the Danger Listing of 'Simen National Park' (Ethiopia) is not on record (cf. WHC-96/CONF.201/21, pp. 28–32). The legal opinion on Danger Listing presented to the 2002 Committee session claims that the Danger List entry of Simen (now 'Simien National Park') was opposed post hoc by the State Party (WHC-02/CONF.202/8, pp. 29–30), but the 1997 session record only mentions opposition by the concerned *regional* government that had been conveyed in writing to the Committee, whereas the Ethiopian national delegate present at the session remained equivocal (cf. WHC-97/CONF.208/17, p. 12).

58. The Clinton administration then swapped the mining licence with federal land elsewhere so that the Committee could lift the alert in 2003.
59. WHC-2000/CONF.204/4, pp. 18–19.
60. In 2001, industry giant Rio Tinto, which had acquired a majority of shares of the previous owner, postponed exploitation (WHC-01/CONF.208/24, p. 30), and in 2005, it granted the traditional owners a formal veto right so that Jabiluka has remained untouched until today.
61. WHC-93/CONF.002/2, p. 24.
62. WHC-2000/CONF.204/21, pp. 27–29.
63. WHC-02/CONF.202/8.
64. Cf. WHC-02/CONF.202/INF.15, pp. 23–38; WHC-03/6 EXT.COM/INF.8, pp. 20–28, 74–85.
65. For an extensive legal comment on this issue, cf. Buzzini and Condorelli (2008: 180–90).
66. Cf. WHC-03/27.COM/INF.24, pp. 82–85.
67. Ibid., p. 84.
68. Mining plans in 'Mount Nimba Nature Reserve' (now 'Mount Nimba Strict Nature Reserve', Côte d'Ivoire/Guinea) had led to its inclusion on the Danger List in 1992. The collapse of the holding dam of a nearby mine that released toxic tailings into areas next to 'Doñana National Park' (Spain) in 1998 had also provoked controversial discussion.
69. WHC-2000/CONF.204/4, pp. 1–4.
70. Cf. https://www.icmm.com/en-gb/our-stories/some-things-you-cannot-put-a-price-on; https://www.shell.com/sustainability/environment/biodiversity/environmentally-sensitive-areas.html (retrieved 25 August 2020). Oil companies Total and Tullow Oil joined the commitment in 2013 and 2015, respectively, as did cement giant CEMEX in 2016 (https://whc.unesco.org/en/news/1101, https://whc.unesco.org/en/news/1379; https://whc.unesco.org/en/news/1490 (retrieved 25 August 2020)).
71. WHC-04/28.COM/26, p. 266.
72. After the investors had abandoned the project, of which eventually only one high-rise building was ever realized, Cologne Cathedral was taken off the Danger List in 2006 (for more detail, cf. Schmitt and Schweitzer (2007)).
73. Cf. WHC-06/30.COM/INF.19, pp. 135–40.
74. In 2008, for example, this concerned the proposed immediate Danger Listing of 'Machu Picchu' (Peru), next-year threats for 'Island of Saint Louis' (Senegal) and 'Samarkand – Crossroad of Cultures' (Uzbekistan), and a somewhat more veiled threat for 'Palace of Westminster and Westminster Abbey Including Saint Margaret's Church' (United Kingdom) (cf. WHC-08/32.COM/7B, pp. 102, 151, 198; WHC-08/32.COM/7B.Add, pp. 3, 72 as against WHC-08/32.COM/24, pp. 81, 90, 108–9, 134).
75. WHC-06/30.COM/INF.19, pp. 68–74, 88–91, 151. Cf. also Schmitt's detailed analysis of the debate he witnessed as an observer (2011: 176–95). He argues that this was a

clash of intellectual styles rather than a North-South controversy (2009: 113–19; 2011: 183–85), a question to which I will return in Chapter 9.

76. WHC-04/28.COM/INF.14A, p. 87.
77. Berthod provides an intriguing analysis of the considerable 'path dependency' of the local decision process, with multiple instances of 'goal displacement', such as when initial traffic predictions proved oversized or when the official purpose shifted from local traffic to thru-traffic and back. 'With the dikes thus dug among the participants to the debate, the bridge became an end in itself, more than a mean [*sic*]', he concludes (2011: 205).
78. Contrary to the Mayor's prediction, construction continued until 2012, and the bridge was only opened in 2013.
79. The delegations speaking in favour of deletion were Israel, Jordan, Kenya, Bahrain, Barbados, Canada, the United States and Spain; those opposing it were Madagascar, Morocco, Tunisia, Egypt, Nigeria, South Korea, Australia and Brazil (cf. http://whc.unesco.org/document/106544, pp. 92–95 (retrieved 25 August 2020) – unusually, this summary record of the 2009 session lacks an official document code).
80. WHC-09/33.COM/20, p. 44. How this might be reconciled with the diagnosed irreversible loss of the OUV of the entire cultural landscape was not explained and to my knowledge, no such initiative has been taken.
81. This at least was the general impression: most delegations, including even the German one, were at best partly aware of the actual intricacies and the fact that the bridge plan had never been hidden, given also that local bridge opponents – who actually lobbied *for* deletion at the session – had little motivation to challenge this simplified version of events.

Chapter 4

Rebellion and Peace

Transformations of institutional frameworks are often piecemeal rather than abrupt, and the post hoc insertion of breaks and turning points into continuous processes can be a somewhat artificial exercise. This is different with the transformation of the World Heritage arena: the World Heritage Committee session of 2010 not only stands out in retrospect, it already struck its participants as unprecedented while it was ongoing. Gradual processes such as the previous accumulation of dissatisfaction among many protagonists played a role, yet what they triggered in that meeting was new and widely perceived as such. In no uncertain terms, nation states resumed control of an intergovernmental framework and, more than ever before, their self-interests came to determine World Heritage outcomes.

The Rebellion of Brasilia

Brasilia in July 2010: the Brazilian capital welcomes the Committee a second time, following 1988, when the city had just been put on the World Heritage List (see above). Neither the opening ceremony with the accustomed speeches and a colourful cultural performance nor the first two session days herald anything unusual. The Committee works its way through expected items such as adopting the agenda and timetable, admitting observer participants and discussing reports about the activities of the World Heritage Centre, the Advisory Bodies, the 'Thematic Programmes' set up by the Centre for the

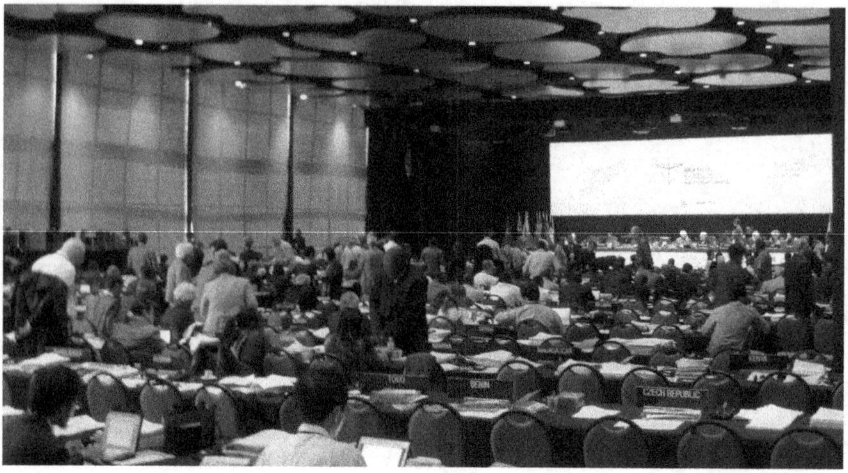

Figure 4.1. World Heritage Committee in session, Brasilia, 2010. © Christoph Brumann.

in-depth exploration of specific kinds of heritage (such as marine heritage or urban heritage) or the 'Category 2 Centres' for special World Heritage topics founded by volunteer countries with UNESCO blessing – here, most attention goes to the African World Heritage Fund set up by South Africa (see Figure 4.1). As usual, these and other introductory items give the Committee members and the occasional observer state the opportunity for general policy statements and for highlighting their own contributions, donations and special concerns. Few of the invariably small amendments that Committee states make to the prepared draft decisions encounter resistance, and a potentially explosive point on Jerusalem (see Chapter 5) is postponed to a later agenda item. The peaceful mood also holds when the Committee enters the first part of its core business, the state of conservation of World Heritage properties, beginning with those on the List of World Heritage in Danger, which are all taken up individually. Except for a vote by a show of hands on 'Medieval Monuments in Kosovo' where, due to the contested territorial status of Kosovo, substantive debate is postponed to the next session (see Chapter 5), the Committee is in agreement about leaving all sites on the Danger List where they are.

Yet consensus breaks down when the very first World Heritage inscription of 1978, 'Galápagos Islands', comes up on the third day. As already mentioned, Ecuador did not resist the Danger Listing of 2007, but now it wants the alert lifted, not least because the latter has failed to mobilize international financial support. Yet while acknowledging considerable progress, the World Heritage Centre and IUCN still see the threats of invasive species, tourism and sport fishing as insufficiently controlled, making removal from the Danger List premature in their eyes. Committee members from the same 'electoral group' (see below) rush to Ecuador's support, with the host nation Brazil taking the lead. Rather than adding new information, they evaluate the presented facts differently, arguing that the State Party's substantial efforts should be rewarded. However, other Committee members stand with the Centre and IUCN so that in the end, a vote is called. The show of hands exactly yields the two-thirds majority of fourteen votes required to remove Galápagos Islands from the Danger List. Afterwards, both a Centre official and a Western diplomat describe this as a mishap to me – had the vote been anticipated, one could have lobbied the Committee delegations, and the supporters of the strict line could have called a secret ballot, which, the official is sure, would have brought about a different result.

However, Ecuador has its way and in the following days, this precedent encourages imitation. On 'Rainforests of the Atsinanana', Switzerland and Sweden have the sense to demand a secret ballot, and the two-thirds majority required to avoid the Danger Listing proposed by the Centre and IUCN (because of massive illegal rosewood and ebony logging) is missed. But eleven votes in favour of overruling the draft decision are cast even so, despite the fact that the lobbying of Madagascar's UNESCO ambassador has been half-hearted at best.[1] Yet, where the concerned states have done their homework with the Committee members, they get what they want. And nor does anyone protest against breaches of established practice: itself on the Committee, Russia does not have the grace to let other members sponsor the desired amendments. Instead, it submits text passages for its properties 'Lake Baikal' and 'Virgin Komi Forests' itself, expunging the threat of future Danger Listing (on account of a problematic pulp mill and of gold mining, respectively) from both decisions.

With the second part of the core business – the discussion of new candidates for the World Heritage List, opened on the fifth day – Committee members drop all restraint. Unlike previous sessions, not just some but almost all sites not recommended for inscription by the Advisory Bodies are 'upgraded' – 'Central Highlands of Sri Lanka' is inscribed on the World Heritage List instead of being referred, 'Konso Cultural Landscape' (Ethiopia) and 'Fort Jesus, Mombasa' (Kenya) come off with a referral instead of a deferral, and so on. Different from previous sessions, this includes numerous sites where the experts do not just mind insufficient legal and organizational preparation, but also do not see OUV as demonstrated in the first place. Starting with 'At-Turaif District in ad-Dir'iyah' (Saudi Arabia), Committee members simply come to agree that World Heritage value is nonetheless present, resulting in five inscriptions without the required 'statements of OUV'. Since the Advisory Bodies have not prepared texts for what they could not find,

Committee adoption of these mandatory documents must eventually be postponed to the following year.

Similar deviations from customary procedures abound: the nominating State Party is given the floor early on, often repeatedly and for interventions that praise the candidate and refute Advisory Body arguments. These are clear cases of the advocacy that the Rules of Procedure prohibit – by rights, the concerned country should only speak once, at the end and for answering specific questions. Supportive interventions by Committee members barely conceal their instrumental nature and several times, speakers struggle to read out what seem to be notes drafted by others (most likely the concerned delegations). Recent visits to the discussed site that have convinced the person of its merits are a surprisingly common argument; on At-Turaif District, no fewer than three speakers claim this privilege, meaning that they have responded to clearly instrumental invitations or are simply lying. The usual order of gratification is upended when delegates argue that inscription of a given candidate site will expedite the State Party's fulfilment of the protective requirements; normally, the World Heritage title is the reward for having successfully dealt with them. In a number of cases, the Committee has no qualms to decide inscription but demand a 'State of Conservation Report' from the State Party – that is, a clear indication of trouble that should have been dealt with – right away. To the host nation's delight, 'São Francisco Square in the Town of São Cristóvão' (Brazil) is listed, even though the file is almost unchanged from two years earlier when the Committee referred the decision, against the ICOMOS advice, which was for deferral both times, as it failed to find OUV. Allowed to talk back extensively over the cultural route 'Camino Real de Tierra Adentro', a delegate from Committee member Mexico vehemently attacks the ICOMOS evaluation, notwithstanding the fact that such illicit advocacy turns him against an organization of which he himself is a vice-president. Also allowed to argue its point at length, Committee member China challenges scientific universalism over 'China Danxia', a series of karst formations in the south of the country. IUCN claims that the *danxia* category has no equivalent in internationally accepted geomorphological classifications so that pronouncing on OUV would first require preparatory research. China, however, insists that the category is properly established and the research is all there – that IUCN cannot read the Chinese-language publications is not the country's fault. In the end, the Committee inscribes the property. The trend of national self-serving is so pervasive that divergences surprise everyone: when the recommendation to defer 'Mining Sites in Wallonia' passes unchallenged, an incredulous murmur rises in the room. As in earlier such cases, Belgium has decided to accept the ICOMOS verdict and to resubmit a revised file (for eventual inscription on the Saint Petersburg session day, as will be remembered). However, to a

Western diplomat passing me by, this is a case of 'drowning when you don't fight', bespeaking weakness rather than commitment to the Convention.

Sweden, Switzerland and Estonia are the only Committee members that stand up for proper procedure and for respecting expert judgements in almost all cases. The Australian delegation – a paragon of virtue in the 2009 session – lets them down, however, proposing inscription for the Camino Real and São Cristóvão itself. Doing so does not hurt the prospects of its own candidate site 'Australian Convict Sites', which eventually gets an inscription upgrade too (from referral); one participant also points to a delegation member's candidacy for the directorship of the World Heritage Centre, which she sees as encouraging friendly behaviour.[2] New on the Committee, France could side with the other Europeans, but instead proposes 'building bridges' to the rest of the world in an internal meeting of the EU delegations, as a participant later tells me. Again, the fact that the delegation comes with two of their own nominations ('Episcopal City of Albi' and 'Pitons, Cirques and Remparts of Reunion Island') must encourage them to be nice to Committee peers, even when these bids do have Advisory Body support. Bahrain's engaged speaker from the previous meeting is now acting as the rapporteur – that is, in a neutral role. This leaves only the three European dissenters. They request secret ballots a number of times, but where the interested countries – often themselves on the Committee – have lobbied the Committee members, this is to no avail. Jubilation is often loudest and the congratulation queue longest where most pressure has been applied, and the Vietnamese delegates holding up their national flags after the inscription of 'Central Sector of the Imperial Citadel of Thang Long – Hanoi' bring an unusually nationalist tone to the room.

However, for many participants, the most blatant violation of established conduct is the circulation of signature lists. The Swiss ambassador protests it right after the nominations agenda item is opened, deploring that some members have already committed themselves to specific outcomes. As I learn over the next days, three countries – China, Vietnam and the host nation Brazil – have been urging Committee member delegations to approve amendments in favour of their sites in advance and to do so in writing on a list, which makes the reactions obvious to everyone else approached. While the signing of amendments is common in other UNESCO contexts where decision texts are developed over a longer period of time, in this case it means that minds are made up before opening discussion of the item, rendering session debate pointless. A Caribbean Committee delegate later tells me that in hindsight, the practice is indeed questionable, and the revision of the Rules of Procedure to follow in 2013 prohibits the consideration of amendments signed by more than a single Committee member (§ 23.2). But in the Brasilia session, the practice leads to the desired results – the square in

São Cristóvão and the Hanoi citadel are listed against the ICOMOS deferral recommendations, and the 'minor boundary modification' for 'Three Parallel Rivers of Yunnan Protected Areas' is approved against IUCN advice.

The latter decision in particular frustrates IUCN representatives deeply. While proposing extensions to this collection of nature reserves, China has also applied for excising a former mining area from the property in order to resume exploitation. IUCN delegates concede that the area is degraded and should never have been included in the first place in 2003; still, they fear that approval will encourage further attempts to circumvent the no-go commitment of the mining industry by conveniently redrawing World Heritage boundaries. If at all, this would require a 'significant boundary modification', subject to the same elaborate evaluation procedure as new nominations. This time, Australia stands with IUCN and the three European countries, and a secret ballot is called. Yet still, the Chinese proposal receives a solid majority, to the consternation of an IUCN representative who see this 'disaster' as another opening of the floodgates (correctly, as later sessions are to confirm).

Exceptions to the pattern of pleasing site countries only occur where States Parties – invariably those not on the Committee – have refrained from lobbying the Committee states, either for not having anticipated the new mores or as a matter of principle, as in Belgium's case above. The UK nomination of 'Darwin's Landscape Laboratory' – with, it seems, no concerted advance lobbying – leads to a long fundamental debate about the association with famous persons as a basis for OUV. Immediate inscription is in fact proposed, but after a first secret ballot brings it down, a second one upholds ICOMOS's deferral recommendation. Against the deferral for 'Dinosaur Ichnites of the Iberian Peninsula', Portugal and Spain have not mobilized, and in the case of 'Mercury and Silver Binomial' – the first transcontinental nomination, including mercury mines and related heritage in Spain, Slovenia and Mexico – an inscription upgrade of the recommended deferral is also rejected by secret ballot, just like a year earlier in Seville. Yet here, the Spanish and Slovak delegates have learnt only at the Brasilia session that the Mexican component – San Luís Potosí – also features in a different serial candidate property (the Camino Real mentioned above). That they do not push strongly for what would unite them with a partner that they feel has betrayed them, as a Slovenian delegate confirms to me, and that Mexico has a second option to bring the title to San Luís Potosí (as it turns out, already on the next day) is obvious. However, even in this case, inscription is only thwarted by a secret ballot. For 'Tajik National Park', the recommended deferral is adopted despite talk of inscription in the debate, but Tajikistan has clearly neglected the lobbying possibilities – when it is asked a question, none of its delegates is in the room. Wherever commitment is stronger, concerned states invariably get what they want.

In sum, the session awards World Heritage titles to a record twelve properties that do not have Advisory Body blessings – five upgrades from recommended referral and seven from deferral – and this more than doubles the number of new properties, as there are only nine inscriptions that have ICOMOS and IUCN support. The eight votes including five secret ballots are also at an all-time high. They signal that dissent and peer pressure (which secrecy is supposed to neutralise) are now an expected part of Committee business – the first secret ballot was only taken in 2006.[3] Contrary also to Seville in 2009, where all four secret ballots ended up supporting the expert recommendations, several votes in Brasilia overturn them, including for the crucial decisions on Galápagos and the Three Parallel Rivers.

Many Committee session regulars I talk to are appalled. Experienced Centre officials and long-term state delegation experts regard this as their worst meeting ever, with state representatives acting 'without shame'; an IUCN representative speaks of 'a disgrace'; and the head of a partner organization sees 'multilateralism at its worst', which is bound to ruin the Convention. An ICOMOS veteran of the 1990s reforms wonders why he is still tormenting himself with coming every year, and a current ICOMOS office-holder jokes to others that he hopes that the ground that the venture is now hitting consists of rubber, making World Heritage bounce back. Similar sarcasm is rampant among all those who side with expertise and established procedures, and the staged nature of much of the exchanges is also transparent to an unspecialized interpreter I talk to. However, the cheering after upgrade inscriptions shows that the decisions also please many, and in the final round of reflections, Brazil characterizes the session as 'a milestone'. Also, a Western European UNESCO ambassador I talk to welcomes the fact that previously suppressed demands and conflicts have now come out in the open, a necessary step towards joint solutions. In his view, it is the three European dissenters that with their rigidity exclude themselves from being part of these.

Comprehensive as it is, the 'Brasilia rebellion' is no carefully crafted conspiracy. The signature lists had already circulated at UNESCO headquarters in the preceding weeks, and this and the general lobbying must have signalled to everyone that States Parties would fight back against the experts' verdicts. But the session dynamic still gives the impression of contagion, with delegates emboldened by what they see others get away with, both in terms of claims and of demands. For instance, an Australian delegate tells me that after having helped Kiribati with the nomination of 'Phoenix Islands Protected Area', they had hoped for an upgrade of IUCN's deferral recommendation to a referral. But when others on the Committee proposed immediate inscription of that marine reserve, Australia did not get in the way, despite the fact that Kiribati is in no way prepared to monitor the largest

World Heritage property of all at the time. Placement within the session agenda thus plays a role – had the Konso Cultural Landscape and Fort Jesus come up at a later point, upgrades to inscription rather than just referral would have been likely, and in the final round of reflections, the Zimbabwean delegate would have had no reason to deplore a 'black day' for Africa, the only UNESCO world region passed over entirely in the inscription spree. Most participants could not know in advance that sustained resistance would only come from Sweden, Switzerland and Estonia. These are just three out of twenty-one Committee members, none of them being an international heavyweight and two having a somewhat dry style of interventions that lacks the rhetorical flourish of some of their antagonists. In the Seville meeting, by contrast, there still had been six states that upheld the strict line most of the time (in addition to Sweden, the United States, Canada, Australia, Israel and Bahrain), with Kenya and Barbados also joining them on occasion. This was more of a critical mass, resulting in only two inscription upgrades (from recommended referral) and painful decisions such as the deletion of Dresden. Now, however, the realization that the dissenters can be brushed aside with ease encourages delegations to step up their demands. The World Heritage title is up for grabs.

The Routinization of National Primacy

The Swiss ambassador's condemnation of the conduct of the session, delivered when nominations have ended, provokes a round of spontaneous applause, meaning that States Parties' opinions are divided. Also, one delegate tells me of a Russian Committee colleague who conceded to him that things had gotten out of hand, even when the Advisory Bodies, which had become 'too nosy', deserved a lesson. According to what I hear from State Party representatives and Centre officials over the following year, several countries are considering a reaction, and there is talk of a joint letter in which several Nordic states will threaten to leave the Convention if things do not change. Among veteran UNESCO diplomats too, many are shocked and angry, one of them tells me. However, nothing tangible happens and the next sessions simply pick up where Brasilia left off.

Assertions that the Committee is sovereign and not to be dictated by expert advice, and that the Advisory Bodies should engage in 'dialogue' – the word of the session in 2012, as an ICOMOS official drily observes to me – are already made on the first days, before the discussion of sites has even started. Entries into the Danger List only move forward when the country does not oppose them or by accident.[4] In most cases, even the mere consideration of Danger Listing for future sessions is expunged from

Table 4.1. World Heritage upgrade inscriptions in the Committee sessions of 2010–15.

Committee session	Total number of inscriptions	'Upgrades'	in %	Upgraded from referral	Upgraded from deferral	Upgraded from non-inscription
2010	21	12	*57.1*	7	5	-
2011	25	13	*52.0*	2	11	-
2012	26	11	*42.3*	6	4	1
2013	19	2	*10.5*	2	-	-
2014	26	10	*38.5*	2	8	-
2015	27	5	*18.5*	5	-	-

Note: Cf. http://whc.unesco.org/en/list/xls/?2019, the synoptic table on the first pages of document 8B for each of these Committee sessions and possible additional documents of the same session, such as 8B.Add, to which that table refers (all available at http://whc.unesco.org/en/sessions). ICOMOS's non-inscription recommendation for 'Birthplace of Jesus: Church of the Nativity and the Pilgrimage Route, Bethlehem' (Palestine) in 2012 concerned the special procedure chosen – inscription on an emergency basis – and did not imply the absence of OUV.

decision texts. For the candidate sites too, upgrading continues unabated, with an all-time record of thirteen such inscriptions at the next session (no fewer than eleven from recommended deferral, implying unproven OUV); Saint Petersburg in 2012 also has eleven upgrades (see Table 4.1). Resistance gradually crumbles: in 2011, Sweden still attempts a secret ballot against the inscription amendment for 'Konso Cultural Landscape' (Ethiopia), but when the latter is upheld with exactly the fourteen votes required, Switzerland, Sweden and Estonia refrain from calling any more. Their delegates begin to give up – on some overruling attempts, they remain completely silent, telling me afterwards that protest is futile. Also, the other Committee members increasingly ignore them: when Estonia and Switzerland support ICOMOS's deferral proposal for 'Residence of Bukovinian and Dalmatian Metropolitans' (Ukraine) in 2011, for example, nobody even bothers to talk back, but an inscription amendment supported by six countries is on the screen right away and is adopted in the end.

The trend of bending established Committee mores and procedures persists as well. While the chairs repeatedly invoke the rules limiting speaking time and outlawing advocacy, they only very rarely and inconsistently apply them in practice, and States Parties receive ample opportunity to speak on their own sites and talk back to the Advisory Bodies, often very early on in the debate. Clearly instrumental interventions, statements read out on others' demands, blanket assertions of OUV, feigned confusion over perfectly clear arguments and formulations, and the acceptance of unchecked last-minute information – such as a construction stop allegedly declared the very same morning[5] – also continue. Committee members remind each other that 'we inscribe sites, not files' when the Advisory Bodies miss essential preconditions for listing, and they call for encouraging States Parties, not 'punishing' them with the Danger List. States Parties insist that their ongoing building activities are only preparatory works, not the actual projects, as if gold mining operations in 'Virgin Komi Forests' (Russia) or the construction of a problematic beltway around 'Archaeological Site of Panamá Viejo and Historic District of Panamá' (Panama) could still be stopped once such preparations are under way.

Crucially too, all the sites that Advisory Body staff and critical country delegates expect to become the big fights of the session are treated according to State Party wishes: while the draft decisions foresee immediate Danger Listing for the Komi Forests (2011–13), Panamá (2012), 'Tower of London' (2011) and 'Cathedral, Alcázar and Archivo de Indias in Seville' (2012) or threaten it for the following year in the case of 'Historic Areas of Istanbul' and 'Palace of Westminster and Westminster Abbey Including Saint Margaret's Church' (both 2011) – Komi on account of mining and all the others because of disruptive construction projects – these proposals do not even get close to adoption, and resistance is weaker than some have hoped for. For the critics, there is only resignation and bitter jokes.

Continuing another Brasilia trend, procedural improvisation proliferates, and not just with postponing the 'statements of OUV' for inscriptions that the Advisory Bodies have not foreseen. The agenda is reshuffled numerous times because of specific dignitaries' time pressure, particularly in Paris (2011). I hear some grumbling, but nobody openly complains when the local mayor's return flights end up structuring World Heritage Committee business.[6] Several times in 2011 and 2012, the chairs suggest adopting unproblematic decision – such as inscriptions of properties with Advisory Body approval – entirely without debate or even en bloc, with sometimes no stronger justification than the timely start of the evening reception.[7] Upgrades from deferral to referral, which rule out a second evaluation mission in the event that the candidate site is resubmitted right away, are eagerly sought by nominating states, only to then ask for 'consultative missions' by the Advisory Bodies to still have their input. In Paris in 2011, ample use is made of interrupting agenda items and resuming them only after small drafting groups have tried to reach consensus over decisions texts, giving the proceedings a stop-and-go feel. This is somewhat less the case in Saint Petersburg in 2012, but over Schwetzingen, Committee disarray reaches unseen heights.

It is not that critical exchanges disappear and that the Advisory Bodies and the Centre are left completely in the cold. Despite going with the flow in Brasilia most of the time, Australia resists Chinese lobbying on the Three

Parallel Rivers, for example, and we already saw how France and Senegal could not bring themselves to taking Germany's last straw – the mosque in the Schwetzingen gardens – seriously. Of the new members joining the Committee in 2012 when Sweden and Australia rotate out, Algeria, Colombia, Germany, Japan and Senegal oppose the concerned States Parties' wishes at least occasionally, and this also occurs with older members France and Mexico. Over some contentious sites, hour-long debates do ensue. Yet when these delegations express support for Advisory Body reasoning, they do not insist when no majority is in sight. Also – and in contrast to Estonia and Switzerland, who remain steadfast – they join or even spearhead the parade of supportive interventions on at least some overruling attempts, if not most, demonstrating susceptibility to lobbying. For all these countries, playing along with their peers or just remaining silent is more common than supporting ICOMOS and IUCN. Concern for their own sites is a factor here: as a Western European delegate notes in 2012, Germany starts off with a critical bent, but then quietens down when the Schwetzingen debate is approaching. However, the new members also include India, and its UNESCO ambassador – a gifted and witty speaker – becomes the mouthpiece for criticizing the Advisory Bodies and their reasoning wherever possible in 2012 and 2013. Increasingly, a Caribbean delegate tells me in Paris, it is 'us against them' with ICOMOS and IUCN.

In the 2014 session in Doha, Qatar, and after Switzerland and Estonia have left the Committee, things come to a head. Not only is upgrading rampant again, but ICOMOS and IUCN are also attacked with a vehemence that several of their officials find unprecedented, an impression that critical observer States Parties share. Just when the ordeal seems to be over, with 'Other business' opened on the penultimate day, the Indian ambassador – a new face, but no less sharp than her predecessor – tables a decision proposal to set up an ad hoc working group. This is to inquire into the working methods of the Advisory Bodies, citing what she sees as their factual errors in the evaluations and a need for increased transparency and dialogue with the States Parties. The move is coordinated and receives broad backing from other Committee members, even when Switzerland and the Netherlands, as observer States Parties, object that the real problem lies with a Committee that does not follow its own rules. Ultimately, a body consisting of two states from each UNESCO electoral group is set up to meet between this session and the next in the Parisian headquarters. The Advisory Bodies are not asked for their view even once and contrary to the usual practice when discussing joint affairs, the decision text does not foresee their participation in the working group either.

At the next session in Bonn in 2015 (see Figure 4.2), ICOMOS officials tell me that their worst fears have not come to pass. The working group

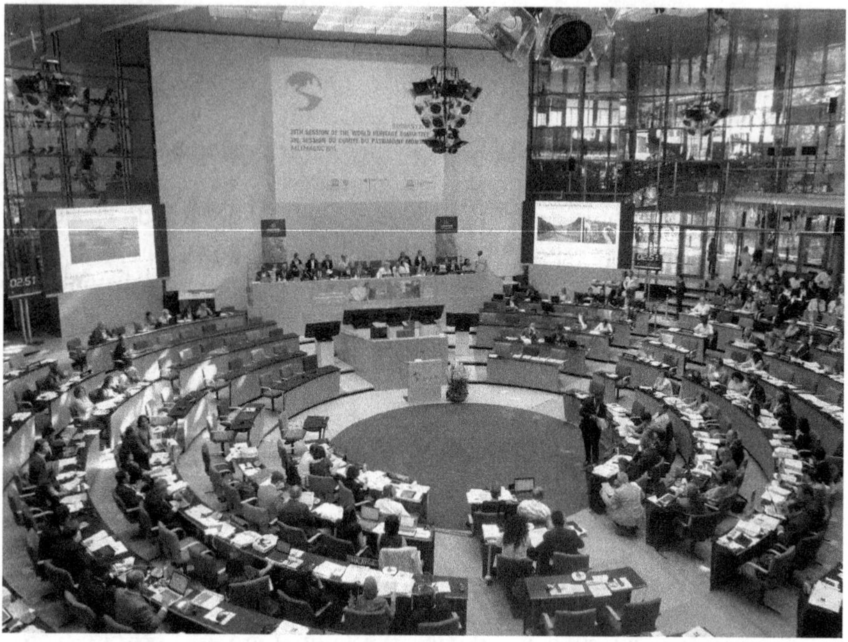

Figure 4.2. World Heritage Committee in session, Bonn, 2015. © Christoph Brumann.

did include Advisory Body representatives in their Paris meetings after a while and most of its demands[8] are acceptable, even when some – such as publishing the qualifications and CVs of the ICOMOS and IUCN World Heritage Panel members – stretch the limits. My ICOMOS informants are also positive on the demands for consultation and dialogue, which were already implemented in the evaluation process for Bonn, despite the additional workload. This meant that in a break from the past, when the Advisory Bodies only published their recommendations six weeks ahead of the session on the World Heritage Centre website, without even notifying the concerned countries, they now gave first feedback half a year in advance, and States Parties could talk through remaining issues with them in person or in video conferences. When sites end up on the List anyway, I am told, it is better to get at least some concerns addressed, so that for the ultimate goal of conservation, the new procedure is a step forward.

And indeed, the tone changes completely in the Bonn session, as already described in the Introduction. Not only do the ferocious attacks on the Advisory Bodies subside; upgrading is also comparatively restrained, with only five unsanctioned inscriptions (all from referral, that is, with OUV considered proven) out of a total of twenty-seven and with all amendments going up only 'one step', not more: four deferrals are upgraded to referrals and one noninscription to a deferral. With one exception, all decisions are made by consensus, and heated debates between States Parties hardly arise. While participants stand transfixed before the exchange of hostilities between Jordan, Israel and Palestine over Jerusalem – decided by secret ballot – and the showdown between Japan and South Korea (see Chapter 5), harmony in the Committee reigns in almost all other matters, to a degree that it has not done in a decade. The state of conservation of World Heritage properties and the candidate sites are dealt with exactly in the allotted days, to the delight of dignitaries coming only briefly to celebrate new titles. Yet there is also markedly less reason for controversy: recommending inscription for almost 60 per cent of the bids, ICOMOS (which is in charge of almost all nominations) is at its most lenient since 2004 (see Table 4.2). 'We're mean!', an ICOMOS official jokingly assures me when I connect this with the new feedback process and the personal encounters it has brought – she asserts that the quality of nominations simply fluctuates over the sessions, with this year seeing a better-than-average crop. However, in the light of what has happened since Brasilia, I have trouble believing this claim.

A New Mode of Operations

With the new peace of Bonn, the transition initiated with the rebellious outburst of Brasilia comes to a close. After the five most turbulent and quarrelsome years that the World Heritage Committee has seen, it settles into a new routine ('Brasilianism', as an Eastern European delegate jokes to me). The governing body of the World Heritage Convention finally ends up delivering what most treaty states are most interested in, that is, friendly decisions on their own World Heritage properties and candidate sites. The danger of being disappointed in this regard is now as small as it has ever been. However, those who expect the Committee to develop an independent agenda, disciplining the States Parties and defending the interests of humanity against national interests, have little to celebrate, and conservation of the World Heritage properties increasingly depends on the respective country's goodwill. Even newcomers draw their own conclusions: in Saint Petersburg, an expert from a Southern African country tells me that he had been advocating a thorough and slow-paced preparation of their current nomination projects. However, seeing what goes on at his first session, he will now tell his colleagues to go all in, sorting things out after inscription.

Delegates are not necessarily proud of what is happening: while an East African delegate approves of the recent turn when I talk to him in 2012, he

Table 4.2. ICOMOS recommended decisions for cultural and mixed World Heritage candidate properties, 2004–19.

	Total no. of candidates	Inscription	Referral	Deferral	Non-inscription	Recommended inscriptions in %
2004	32	25	0	6	1	*80.6*
2005	31	17	0	8	6	*54.8*
2006	25	10	4	7	4	*40.0*
2007	34	12	5	11	6	*35.3*
2008	31	14	8	6	3	*45.2*
2009	26	10	2	9	5	*38.4*
2010	26	9	6	8	3	*34.6*
2011	33	10	2	15	6	*30.3*
2012	33	11	6	7	9	*33.3*
2013	25	12	2	4	7	*48.0*
2014	28	9	2	14	3	*32.1*
2015	29	17	5	4	3	*58.6*
2016	24	10	0	8	6	*41.7*
2017	28	11	1	8	8	*39.3*
2018	25	12	2	4	7	*48.0*
2019	34	20	4	4	6	*58.8*

Note: Cf. http://whc.unesco.org/en/list/xls/?2019 and the synoptic table on the first pages of document 8B for each of these Committee sessions – document 14B Rev in 2004 – and any possible additional documents of the same session, such as 8B.Add, to which that table refers (all available at http://whc.unesco.org/en/sessions). I exclude nominations for mere extensions of already listed World Heritage properties from the count, as these tend to be less controversial. For the last four sessions, figures for noninscription recommendations also include the nominations withdrawn early on because of negative ICOMOS feedback (five in 2016, three in 2017, two in 2018 and three in 2019). The number of exclusively natural candidates is usually much smaller, with only three presented in Bonn, for example, so that a similar comparison for IUCN evaluations is less instructive.

still thinks that in the previous two sessions, things have gone too far. In the general debate of the 2011 General Assembly, the Bahraini chairperson quite matter-of-factly sees the Committee's reputation and credibility at stake, the US delegate finds it 'seriously off track', and many countries, including such busy nominators as Italy and Mexico, deplore 'politicization' and an overemphasis on inscriptions where OUV and conservation should have primacy. Comparing these delegation's Committee session behaviour, Mexico in particular must be in two minds if the comments are sincere, and so are other State Party delegations: when Bahrain meekly accepts the ICOMOS deferral recommendation for its candidate site in 2011, loud applause and 'bravo' cheers rise in the hall, and the chair praises the country's constructive attitude. This outburst clearly includes some of those who on their own sites feel compelled to act differently, pushing back against unwelcome recommendations with all their might. Yet however dubious the new mode of operations appears even to some of the perpetrators, nothing can stop it.

Notes

1. She tells me that she lacks clear instructions and that the decision is largely meaningless to begin with, given that the internationally recognized government that has dispatched her has lost control of much of the country to a rival government.
2. However, an Australian delegation member told me that leniency was the result of internal preparation processes, including a new expert advisor for cultural heritage, whose recommendations could not be overturned at the session.
3. https://wikileaks.org/plusd/cables/06PARIS5276_a.html (retrieved 26 August 2020).
4. In 2011, the Danger Listing of 'Tropical Rainforest Heritage of Sumatra' proposed by IUCN is adopted, despite Indonesian resistance that other Committee members support. The lack of fit between debate and decision suggests that the State Party – a marginal presence in the sessions – had failed to clearly convey its wishes to its peers.
5. For example, over 'Vat Phou and Associated Ancient Settlements within the Champasak Cultural Landscape' (Laos) in 2011.
6. One case in point is the nomination of 'West Lake Cultural Landscape of Hangzhou' (China) in 2011.
7. In Paris, the chair once suggests to decide only the inscription of a property so that the impatient minister can deliver their acceptance speech and leave; the decision text can then be dealt with later. Ultimately, this is not taken up.
8. WHC-15/39.COM/13A.

Chapter 5

The Nation State

So what explains this turn away from multilateral commitments and ambitions in the World Heritage Committee? The following chapters will take up the contributing factors one by one. As a starting point, let me address the nation state and the structural conditions that have always given it a privileged place, however world-embracing the arena may style itself.

National Actors

While many participants associate the World Heritage arena strongly with particular individuals – friends and acquaintances of long years, diplomatic colleagues, key protagonists on the podium such as the chairperson, the active and memorable speakers of the delegations – only very few individuals are personally empowered. Instead, almost everyone is answerable to an organization. The chairperson and the rapporteur are individually elected by the Committee, usually by acclamation. However, everyone else represents an institutional entity, and the chair never calls up individuals by name during the sessions, but instead gives the floor to 'Cuba', 'ICOMOS', 'the Assistant Director-General for Culture' (of UNESCO), etc. Speakers do not usually introduce themselves so that to noninitiated participants, their backgrounds (diplomat or heritage expert) and nationalities (in the case of Advisory Body and Centre personnel and also some state delegates) remain obscure. To a degree, this is intentional: ICOMOS and IUCN very much emphasize that their expertise reflects a corporate viewpoint, not a personal opinion. The organizational logic also holds for the informal interaction and negotiations between different States Parties: whoever speaks and makes promises, instead of referring to someone else on the delegation, is assumed to have the authority to do so.

Personal specifics come into view only rarely, such as when the Permanent Delegates to UNESCO – the ambassadors – exchange pleasantries ('as my esteemed colleague from Brazil just emphasized') or when a diplomat's dismissive remark about a 'professorial' viewpoint targets the university background of a heritage expert speaking earlier. The only occasion to bring out individuals more prominently is the biannual election to the Committee: the promotional material that the candidate states circulate in the run-up features the heritage experts to be included in the delegation, and for quite a few state representatives, this is a factor taken into consideration for their 'open' votes (see below). Yet even individual observers such as myself must register as representatives of an organization and in at least some years, our entries in the list of participants were by organization, not by personal name.

The most common organizations that individuals represent are the State Party delegations, of which well more than one hundred are in attendance. Sovereign states conclude and ratify the treaty, and the decision-making body consists of elected states, not named individuals. There is considerable reluctance to interfere with the internal matters of treaty states. The usual body to nominate a candidate property is the UNESCO National Commission that every UNESCO member state is held to establish, which is usually either part of the government structure or funded and closely controlled by it. The focal point for World Heritage-related communication is also determined by each national government. Whom to include in the State Party delegation, and how many, is a sovereign decision too, and there is no background check on whether Committee delegates are indeed the 'persons qualified in the field of the cultural or natural heritage' that the Convention requires (§ 9.3) or whether the specific experts promised before Committee elections are actually included. The Periodic Reporting exercise about individual World Heritage properties is self-administered by the respective treaty state, and when discussing the condition of sites, the national delegation's reports and oral interventions are usually treated as the default line. Also, submission of a State of Conservation Report by a State Party can only be requested, not enforced, and even a general refusal to comply – such as by Russia, which submitted none of the reports due on five natural properties in 2015 – does not lead to specified sanctions. Monitoring missions by the Centre and the Advisory Bodies also require an invitation by the concerned country and, again, noncompliance is usually noted with regret but not punished. Even the international Advisory Bodies' structures recognize national units: IUCN's large number of member organizations includes those of national governments, not just NGOs, and while ICOMOS sends international-level and headquarter personnel to the sessions, the organization has national committees in most countries to which all individual members belong. Also, many ICOMOS members are in their country's state service (40 per cent in 2009, according to the then president).

This means that the perception and positioning of nonstate actors too is influenced, if not crucially determined, by their nationality. Key ICOMOS officials such as the president and vice-presidents often rely on extra funding from their own government ministries or leave of absence from their ordinary tasks for fulfilling their prominent roles; it is difficult to imagine that such support comes entirely without the hope for responsiveness to national concerns. Quite a few members of the internal World Heritage Panels of both organizations – the bodies that finalize the recommendations going to the Committee – attend the Committee sessions as part of their national delegations (see Chapter 6). Individuals also move back and forth between different positions over time – Advisory Body evaluation author, World Heritage Panel member or session representative; State Party delegate; private consultant working for a specific nomination. The new ICOMOS President took this as a reason to call for confidentiality and the avoidance of conflicts of interests in his opening address to the Brasilia session, but, as already mentioned, this did not prevent one of his vice-presidents from tearing an ICOMOS evaluation of his national candidate site apart. While that Mexican delegate stepped down from his ICOMOS office soon after, in the session, many participants were not aware of this aspect, and few of those in the know seemed greatly surprised.

The most 'denationalized' actors present are the international civil servants who make up the core personnel of the World Heritage Centre, the convention secretariat in the Parisian UNESCO headquarters. Receiving an untaxed income from a UN agency in legally extraterritorial precincts, they are answerable to the UNESCO Director-General rather than any country. However, the recent financial crisis of UNESCO has brought their number down from forty-five to twenty-nine in just the two years of 2013–15, so that the Centre's extra personnel paid or seconded on a voluntary basis by specific treaty states is now almost a majority. Providing such additional staff is sometimes seen as a way for States Parties to gain influence within the Centre. Yet, the international civil servants confirm that regular staff too are often the first point of address for any requests or concerns coming from their own country. Many States Parties see a national within the Centre as a special asset and, according to a UNESCO veteran bureaucrat, some countries – he mentioned France, Italy and Japan – are quite active in pushing their citizens' careers in the headquarter bureaucracy, whereas others – such as Germany – are less engaged.[1] Sympathy of such nationals for national concerns, although by no means guaranteed according to what I witnessed, is at least hoped for and, when not forthcoming, may lead to caustic exchanges, such as between the Indian ambassador and the Director of the Centre – a compatriot – in 2012. Observations reveal a split picture: in the 2009 session, a German Centre official introduced the highly critical assessment that led to Dresden's deletion, but I also heard rumours of Centre staff actively promoting their own country's candidate sites.

Nationality also overrides organizational divisions in the case of social events: I found myself included in those of the German delegation to which Centre personnel with German passports also came, and an ICOMOS official told me that while keeping some distance, she would still attend her country delegation's celebration of a new inscription. Conversely, one delegate arguing for the deletion of Dresden found himself threatened with unspecified consequences by a German representative. The delegate represented a different country, but his German nationality appeared to provide sufficient motivation for this (futile) attempt.

It almost seems that the more transnational the environment, the less nationality and national feelings can be shed, both in actual fact and in terms of what people expect of each other. Conflict-of-interest regulations routinely take this for granted: the Committee chairperson must leave their seat and pass to a vice-chair when a national site comes up; the World Heritage Advisors drafting the ICOMOS recommendations may not be assigned to sites from their own country; and the ICOMOS World Heritage Panel members must leave the room when such sites are discussed. All this presupposes that it is the normal and natural thing to root for one's country even when humanity's heirloom is at stake.

National Self-Interests

National actors in the World Heritage arena are expected to pursue national interests. These could answer a variety of broader objectives, such as the protection of the World Heritage List from unworthy inscriptions, the proper conservation of World Heritage properties, the enactment of value-based diplomacy, the strengthening of multilateralism and global equity, the maintenance of world peace and so on. Most States Parties are indeed committed to one or more of these goals to some degree, and this is reflected in key participants' strategies and tactics. However, when they clash with the country's more narrow self-interests, they tend to step back. Almost always, the State Party's own World Heritage properties and candidate sites are the delegations' first priority and with few exceptions, the delegations strive for Committee leniency here, both in terms of judgements and demands. All other things being equal – and often there seem to be few other things – most State Party delegations want more World Heritage titles from the Committee, but with as little interference as possible.

There is the occasional Committee representative – invariably in small delegations of small countries, such as Lebanon or Israel – who believably claims to have enjoyed a completely free hand as late as in meetings of the 2000s. However, in almost all cases, for at least part of the agenda, which usually includes the country's own properties, delegates are under orders in terms of how to speak and act. These orders usually emanate from higher levels of hierarchy, such as ministry officials or even heads of state, who do not themselves participate in the sessions and may only have a limited grasp of its conditions. The pressure can then be considerable: I heard of an African UNESCO ambassador almost getting sacked when she did not secure the desired inscription of her country's candidate site; East Asian delegates claiming that they could not return home if they failed to avert the nomination quota against which their country was fighting tooth and nail; or a Latin American representative being ordered to fly back the next day after not transmitting the commanded message on a specific agenda item precisely. Even in the absence of threats and sanctions, arriving with their own nominations puts Committee member delegations under particular pressure not to make enemies. A Caribbean diplomat described such a session as nerve-racking for him, and others also admitted that they very much felt this special circumstance.

It is not always the case that a delegation's ministerial or other superiors are the root source of the pressure; they themselves may just be passing on what actors from the site regions and communities expect of them. The UK focal point for World Heritage matters told me of receiving phone calls suggesting new World Heritage candidate sites on a daily basis; her German colleague confirmed this experience and also admitted to the difficulty of implementing a consistent nomination policy in a federal system where the subsidiary units are scrambling for the limited slots. There is often no way to withhold support: a Western European UNESCO ambassador told me how he had urged the representatives of one of his country's candidate sites to withdraw their nomination after ICOMOS had recommended noninscription. As mentioned in relation to Schwetzingen in the Introduction, this averts a binding decision and is thus the least-risky path. However, the group would not hear of it, as years of preparation involving large numbers of people meant that these expected them to put up a fight. Willy-nilly, the ambassador lobbied other delegations in their favour, to his embarrassment since all other nominations with the same verdict had been quietly withdrawn. Yet still, letting down his compatriots whose interests only he could effectively defend was not an option, so he did his duty for what became a deferral decision in the end.

On the whole, however, such half-heartedness is less common than delegates fully identifying with their country's site-specific interests. I found it impressive how even otherwise level-headed participants committed to the Convention's larger goals turn into emotional defendants of national concerns, particularly in the case of candidate sites. In the Paris session of 2011, for example, the chairperson, in accordance with the rules, was replaced by a vice-chair when one such site from her own country came up. However, she then spoke from the floor for her national delegation and immediately turned into a passionate defender of that site, engaging in precisely the advocacy that, as a chair, she had at least occasionally tried to suppress. What she saw as ICOMOS's unjust treatment of the site seemed to give her licence to do so. The aforementioned Caribbean delegate who felt the pressure of being on the Committee with his own country having a candidate property has repeatedly spoken out against this conflict of interests in the meetings; nonetheless, he too minced no words about the shortcomings of the Advisory Body evaluation of that specific site in our interview. A Western European session veteran and advocate of proper procedure would nevertheless insist to me that with one of his country's nominations, ICOMOS had got it all wrong. None of these individuals was naive about the national blinkers of others; on the contrary, two of them strongly criticized what is commonly called the 'politicization' of the Committee. However, this did not prevent them from not only thinking but also feeling national and from siding with their delegations and properties. An academic occupation is no protection either: two university professors accompanying the representatives of German candidate sites shared perceptive observations with me about the Committee proceedings they followed for the first time, clearly seeing through others' vested interests. Yet, in relation to what they felt were completely mistaken Advisory Body evaluations of their own sites, they were outraged.

Interestingly, transparency has done little to discourage national commitments. In 2011, full press access and an internet livestream were decided for future sessions, where previously the mass media had been restricted to press conferences and not allowed into the meeting room. This decision on the last session day had been given little attention so that many were surprised when its implementation was announced in 2012. Proponents had counted on a restraining effect, now that the whole world was witnessing what was happening. Yet entirely to the contrary, the move put the protagonists even more under the watchful eyes of ministries and site communities, and a high-ranking Centre official was sure that two Committee delegations were 'under remote control', receiving live text messages with instructions what to say from their ministry superiors watching the stream. In unexpected ways, transparency can thus bind individuals even closer to their national agenda.

A delegate from a Pacific country recounted to me what he saw as a revealing moment: in 2012, an East Asian UNESCO ambassador had invited him to a private dinner, together with the representatives of a handful of other delegations and the Advisory Bodies. The purpose of the meeting was to explore the possibilities of bringing about reform, as the diplomat was worried about the course recent Committee sessions had taken. Only critical individuals were in attendance and the discussion was suitably high-minded. However, after two hours, the ambassador took the delegate aside to share his worries about the prospects of one of their national nominations. The informant was struck how even in the context of committed universalism, national concerns were always close by.

Lobbying and Deal-Making

I am interviewing the UNESCO ambassador of an Asian Committee member state when his phone rings and I stop the recording. An Eastern European colleague is calling, telling him that one of their candidate sites got a referral recommendation they do not agree with. The ambassador asks for English-language documentation and promises to have a look at it. After hanging up, he tells me that several countries are 'berserk' about the treatment of their nominations. What everyone calls lobbying is already in full swing.

In actual fact, the only communication the Committee requires from the States Parties is in writing, and at this point – six weeks before the Committee session, with the evaluations just uploaded on the internet – nomination files, State of Conservation Reports and the replies to any additional questions from the Advisory Bodies are already submitted and examined. In the session, the Rules of Procedure allow a country to speak on its sites in response to questions, but unlike other schemes of distributing distinctions and awards, presentations of or by the contenders are not a regular part of the proceedings. The assumption is that everything needed for conducting Committee business can already be found in the documents, to which everyone should have given due attention. Importantly too, nowhere in the rules is there mention of an *obligation* for States Parties to speak on their sites.

Nevertheless, almost always, delegations do stand ready for responding to any queries, and they often actively press for an occasion to be heard. More importantly, they propagate their wishes and points of view during the days and weeks ahead, both at the session proper and in the run-up to it. Some of this is directed at the Advisory Bodies and the World Heritage Centre, and with the latter in particular, the UNESCO diplomats who have much of their business and sometimes also their offices on headquarter precincts walk in and out rather routinely and seem to cultivate relations.[2] However, the main target are other members of the world community of nation states.

Lobbying does not come from states alone: private actors and civil society organizations are contacting the Centre, the Advisory Bodies, the Committee member states and the chairperson about their concerns all the time. This may develop into impressive campaigns: when the Great Barrier Reef was up for discussion in 2015, no fewer than 200,000 emails urging for improved protection reached the Centre.[3] Outside the Bonn meeting venue in space not under the authority of the Committee, World Wildlife Fund (WWF) activists had built a miniature reef to welcome participants. Using more conventional forms of political protests, Turkish Kurds stood with placards against the Ilisu Dam in Turkey, and a South Korean NGO kept an information stand protesting against a Japanese nomination up for decision (see below). Such lobbying continues inside the hall as well, given that NGOs can register for observer participation and also request the floor. Using this opportunity, large organizations such as WWF or Greenpeace give brief addresses over general conservation concerns in most sessions (although not always the same organizations). There are also NGOs interested only in a specific site and its destiny, and these try to lobby the delegates during the breaks and in the corridors, sometimes also seeking the floor when their site is up. In the case of Dresden, for instance, both bridge supporters and opponents were active in Seville.

The effectiveness of such nonstate lobbying varied greatly in my observation: with groups less experienced than Greenpeace, quite a few representatives were uncertain how to proceed, whom to approach and how to phrase their demands. When they were actually given the floor – invariably in the end, after the States Parties, and just once for two minutes, whereas their government delegation often spoke repeatedly – their statement usually left no strong imprint. The reader will remember the eloquent statement of the indigenous-rights NGO representative in Saint Petersburg, delivered after much uncertainty. The Kenyan ambassador felt compelled to respond, but when the other problematic sites mentioned by the activist came up, nobody referred back to his address and State Party assurances that everything was in order with the problematic sites were not questioned. What an African diplomat shared with me tells it all: serving on the Committee, they had been approached by a French NGO disagreeing with their government's conservation policy of a French site up for decision. Yet with that NGO, they interact just this one time, the diplomat said, while they have continuous business with France, meaning that they decided not to raise the NGO's concerns. While other delegations do respond to such requests if they seem warranted, direct lobbying at the Committee session appeared less effective to me than mobilizing the general public in campaigns and thereby creating indirect pressure.

Business firms involved with specific sites are rarely present in an open way, even though undercover participation by managers of Brazilian construction giant Oedebrecht in the delegation of Panama is on record (Meskell 2018: 105). I do not think that corporations such as those involved in resource extraction are disinterested in World Heritage decisions affecting their operations. However, their main channels of influence will go through sympathetic States Parties with whom they are linked in ways that often transcend the World Heritage arena and do not specifically require a Committee presence.

Instead, most lobbying activities occur between States Parties, with those currently on the Committee being the main target. A Northern European delegate found the experience extreme and likened himself to a honeypot surrounded by bees: once outside the plenary hall, he was approached by other delegates wishing to discuss their concerns immediately, and it was plain to him that the attention was all due to his function as Committee member representative, not his personal charms. Reminiscing about her Committee experience, an African delegate found the many mayors and ministers contacting her and the necessity to then update herself about the issues at hand with the help of delegation experts the greatest challenge. Even the ambassadors of the strict countries in the post-Brasilia transition would not refuse all lobbying approaches on principle: receiving concerned representatives, listening to their views and promising due consideration appears to be the non-negotiable diplomatic minimum, even when no further commitments are made. The Advisory Bodies apply this policy too, if reluctantly. One of ICOMOS's World Heritage Advisors, after receiving daily phone calls from South Korean diplomats concerned about a Japanese nomination (see below), finally decided to meet in person, if only to be done with it once and for all. Staff of IUCN's World Heritage Unit also confirm contacts by and visits of sometimes quite substantial delegations over World Heritage matters at IUCN headquarters, with China, Russia, Ecuador, India, South Korea and Canada being particularly active. When such countries sit on the IUCN Council at the same time, these encounters are difficult to avoid.

In such overtures, most countries try to win over the approached person by the polite presentation of additional information and viewpoints – that is, the power of the argument. To take such information into account and convince oneself of its validity is by no means illicit for Committee members. On the contrary, they are part of a sovereign body expected to form an independent opinion, not trusting the Advisory Bodies' judgement blindly, and in order to correct mistakes, timely information is crucial. Occasionally, however, more outright pressure is applied. In one case in the 2008 session that was independently reported to me by several targets, the Indian UNESCO ambassador personally threatened uncooperative Committee member delegates and Advisory Body personnel with reporting them to their home institutions so as to have them banned from future sessions, following through on her threat in at least one case (to no avail). This strategy did have an impact: there were considerably more speakers supporting the ambassador's view in the debate than votes in her favour in the secret ballot, where the inscription upgrade for 'Majuli Island' was roundly rejected.[4] In a later session I myself observed, an ICOMOS delegate told me about similar pressure for a Chinese candidate site.

However, playing hardball this way, although there are additional hints,[5] is not the diplomatic rule. For a start, the targeted delegation may end up supporting the specific concern because of persuasive arguments or the general will to be friendly among peers. The wish to help is also common among states seeing each other as allies, such as neighbouring countries with good relations. Yet, more important in relation to what goes on are agreements of mutual support. Some of these deals stay within the World Heritage arena: on the Saint Petersburg session day, we already saw how India's call for upgrading the Russian Kremlins nomination was immediately reciprocated by Russia over the Western Ghats bid, and after the Indian ambassador was the most vociferous supporter of the eventual inscription upgrade for Mauritius' first World Heritage nomination, 'Aapravasi Ghat', in 2006, Mauritius had an inscription amendment ready for 'Majuli Island' in 2007 before debate even opened. While such trading of favours may arise spontaneously, in most cases it is safe to assume coordination.

The substance of deals is not confined to the World Heritage arena; instead, it may include return favours in other UNESCO fora, elsewhere in the UN system and other international organizations, or even outside of them. A Western European ambassador found outright flows of money conceivable, but had no supporting evidence; if cash payments were widespread, I think I would have heard of them, and I also believe that for many protagonists, buying cooperation would cross a line, which swapping favours does not. In any event, dealings in kind do not rule out systematicity: the delegate of a Pacific country told me that he had access to a database kept by his foreign service about all deals closed in the multilateral context, both those involving his country and those between third parties they had heard of and including the agreements made when filling key positions such as that of UNESCO Director-General. Not all countries go to similar lengths, but in the trading of favours, the World Heritage Committee is a multilateral body like any other. Even the paragons of virtue on the Committee are not immune to it: Estonia and Switzerland were the first to suggest an inscription upgrade for the Swedish farmhouses in the Saint Petersburg session (see Chapter 1), much as all three countries usually made a point of sticking to the Advisory Body line. Again, the move appeared to have been coordinated.

As with the substance of deals, their actual negotiation is not confined to the World Heritage arena. While they are often struck informally at the sessions proper or at UNESCO headquarters in the preceding weeks, they can supersede that level by far, involving direct contacts between foreign ministers or even heads of state and the exchange of diplomatic notes. The aforementioned Latin American delegate ordered home, I was told, had failed to transmit what his president had promised to a colleague in office. As pointed out earlier (Brumann 2012a: 13; Meskell 2016), the US diplomatic cables published by WikiLeaks in 2010/11 – the 'Cablegate' incident – are a rich source in this regard, as they consist of Foreign Service personnel's internal reports and comments. Ranging from 2003 to 2010, they cover the country's fourth tenure on the World Heritage Committee (2005–9). Most requests for support or offers of deals documented there concern Committee elections (see below), but one also finds a letter by the Prime Minister of Cape Verde asking President Obama to support the country's first World Heritage nomination.[6]

States Parties that believably claim not to apply pressure or make deals for their own benefit do exist: in Paris in 2011, Swiss resistance stopped the attempt of five other countries to lobby other Committee members in favour of an inscription upgrade for their joint nomination of Le Corbusier works. As a chance witness told me, this came at the cost of a shouting match between ambassadors in the corridor. Without Norway's resistance against its partners' plan to push through a transboundary nomination of Viking sites against ICOMOS advice in 2015, these would now be on the List. Italy claimed not to lobby against Advisory Body recommendations and, indeed, there was no contrary evidence in the observed sessions, with referrals, deferrals and Danger Listing threats quietly accepted.[7] However, with news about the new Committee mores spreading domestically, this position had become increasingly difficult to sustain, an Italian delegate told me.

Yet while some states refrain from lobbying in their own interests, to close oneself off to *others'* lobbying appears to be more difficult while on the Committee. An Australian representative told me that prior to the aforementioned controversial 'minor' boundary modification of the Yunnan Parallel Rivers, national authorities had been visited by two Chinese delegations, and both the Prime Minister and the Governor General had received phone calls. As mentioned, the delegation did not budge on this site, but it is easy to imagine that for most actors, being given so much attention and having met the concerned people in person will make it difficult to maintain resistance. Many a government minister too will be inclined to prioritize good bilateral relations over a consistent World Heritage policy. From observing Committee interaction, it is safe to assume that much lobbying actually achieves its goals, if only for the ease with which unlobbied requests for friendly decisions are brushed aside (a matter to which I will return).

Also, even the Committee members intent on supporting ICOMOS and IUCN will find it effective to choose their fights. In one session, an Arab Committee member delegate committed to proper procedure told me that their delegation had decided to remain silent on one problematic site and focus their energies on three others that they found more troublesome. A Western European diplomat spoke of the right mixture between diplomacy and principles; his Swiss colleague had leaned too much towards the latter in Brasilia, he said, so that estranging everyone had made him ineffective.[8] From the representative of another Western European country, I heard how the votes he casts in the Committee elections do include it all: one part goes to countries for which he is ordered to vote, most likely because of deals his superiors do not tell him about, others to closely allied countries, and some to those that – also based on the promotional material – he himself regards as most qualified in terms of conservation expertise and commitment.

Responsiveness to lobbying is easiest to monitor, of course, where it concerns session behaviour. There are some limited ways of being outside the room or on the toilet 'by mistake' to avoid having to say things one does not wholeheartedly support. One delegate also recalled having privately asked the chair to overlook their name plate: after the chair had dropped the gavel and adopted the decision, she protested against the oversight and had it noted in the records, so as to placate her superiors who had instructed her to speak. However, these are exceptional moves and whether pledges of support are kept or not is usually obvious. Interestingly, promises are also honoured by and large in the secrecy of the ballot box, both in its spontaneous use over contentious issues in the sessions and in the biannual elections of Committee members. Not only could a State Party betray its deals here; the actual person filling in the form might ignore their superiors' orders in what would amount to a principal-agent problem. Delegates told me that this is indeed an option when there is no way to reconcile themselves with the prescribed course; one heritage expert also recalled her country's ambassador handing her the list with the negotiated votes for the Committee election and telling her to treat it as she saw fit.

However, the secret ballots of Brasilia and the following sessions gave little evidence of such evasion. As mentioned earlier, the votes in favour of postponing the Dresden decision in 2009 were one short of what the preceding session interventions suggested. Yet, given the lobbying efforts of Germany, more Committee members might have chosen the easy way of expressing support in the debate while showing their true colours in secret. In Saint Petersburg in 2012, many delegates were annoyed about the emergency nomination of 'Birthplace of Jesus: Church of the Nativity and the Pilgrimage Route, Bethlehem'. Following on acceptance as a full UNESCO member by a large and enthusiastic majority in the UNESCO General Conference of

2011, Palestine had joined the World Heritage Convention and submitted its first candidate site immediately. As in most cases when Israel and its neighbours are concerned, a secret ballot over the proposed inscription amendment was called right away, and while it was being prepared, I overheard one of the Palestinian delegates sitting in front of me saying to his colleagues: 'I think we've got fourteen votes.' Committee delegates had reason for defection: many felt that nominating the church as an emergency case for accelerated treatment and immediate Danger Listing took things too far – as ICOMOS had argued, nothing spoke against choosing the normal course with possible inscription the following year. But when the vote was opened, the count was indeed fourteen, exactly the two-thirds majority needed for inscription on the World Heritage List.

What motivates such loyalty to promises I can only speculate. I think that monitoring plays no more than a limited role. It is true that in the smartphone age, voters cannot be prevented from documenting their filled-in ballots. Yet while the national principal may think of asking such proof from the voting agent, a Western European ambassador told me that one's bilateral deals are usually not revealed to third parties, making the sharing of a filled-in election ballot (which in most rounds lists multiple state names on a single page) impractical. However, I do not think that insisting on proof would work well in the first place, since these exchanges have traditionally been based on honour, as an Israeli delegate told me. Delegates are aware that if bilateral promises were broken routinely, this would deprive everyone of a last resort once key national interests call for it. Also, while actual defectors cannot be identified with certainty, the suspicion does fall on them, even if it is shared. At the individual level, I assume that quite a few people will not mind abdicating responsibility over politically sensitive issues to their superiors by simply following their orders. A single defector on the Bethlehem promises, for example, would have sufficed to overturn the decision, with unpredictable political consequences, and I do not think that everyone is equally at ease with changing the course of Middle Eastern affairs. But most importantly and returning to the topic of this chapter, I am sure that in the event of doubt, most individuals at the session prefer loyalty to their own country – which in most cases is also their employer to which they have promised, or even sworn oaths of, allegiance – over the vague interests of humankind that the Committee is supposed to defend.[9]

Diplomats and Experts

One often hears that the rise of national self-interest is connected to the takeover of the World Heritage arena by career diplomats, sidelining the heritage experts who were initially dominant. The chronology varies somewhat by source – some see UNESCO ambassadors having already moved to the forefront by the end of the 1980s (Cameron and Rössler 2013: 166), while others place the transition a decade later, at the time when Committee sessions were growing into global events (Cameron and Rössler 2013: 172).[10] Two heritage specialists and session veterans from Canada and Germany remember still having been completely on their own around 2000. Nowadays, however, most Committee member delegations and also quite a few of the observer State Party delegations are headed by someone with diplomatic credentials, and in terms of what goes on at the sessions, these are clearly the dominant group.

Just like other important international organizations, UNESCO is treated as a quasi-state in international affairs so that member countries appoint a 'Permanent Delegate' to the organization, that is, a diplomat of ambassadorial rank. When countries do not sustain a large foreign service, the bilateral ambassador to France simultaneously takes on this duty and often also representation to the Organisation for Economic Co-operation and Development (OECD) (also headquartered in Paris), sometimes with the EU coming on top. However, where resources are less scarce and there is a wish to exert an influence, the UNESCO appointment is usually separate. Permanent Delegates may then still be housed in their country's embassy to France, but a sizeable number have offices in the Rue Miollis annex, half a kilometre away from the main headquarter complex at Place Fontenoy. Not just in the case of the leading countries but also for those in the second tier of international relations, such as Kenya, most delegations consist of teams of at least two officials plus a secretary. It may then be the 'number two' who is the more continuous presence at UNESCO, with the ambassador showing up only occasionally, particularly in the case of joint appointments. For smaller or less resourceful countries such as Lithuania, a continuous UNESCO presence is often a challenge, and its Permanent Delegate and other personnel may then choose to fly in only for important events such as the biannual General Conferences or the half-yearly Executive Board meetings.

Permanent Delegates from Europe and North America are often dispatched by the foreign service, whereas their colleagues from other parts of the world tend to hail from ministries entrusted with UNESCO business, such as education or culture; some appointments are intellectual or artistic luminaries, such as prominent musicians or writers. The proportion of people with multilateral experience is smaller than in other UN bodies. However, with an ambassador from the cultural field, the 'number two' is often a career diplomat, typically with a background in law or international relations and additional training by the foreign service, so that both home ministries have a foot in the UNESCO door. I heard a Western European ambassador refer to the cultural appointments as 'amateur dramatics', but, as she added immediately afterwards, influence is less closely tied to the respective country's stature than in other international settings. Unlike the foreign service staff, who typically rotate to their next assignment after three years or so, the cultural appointments and other representatives from outside the foreign service often stay in Paris for a longer time and are thus able to acquire the reputation of opinion leaders. Diplomats from smaller countries may also be less busy with writing reports to their various national ministries, giving them more time for socializing and strategizing. As a case in point, a Lebanese national paid by the tiny Caribbean country of Saint Lucia to act as its deputy Permanent Delegate was a force to be reckoned with in UNESCO affairs and widely consulted by her colleagues.

These diplomatic personnel meet one another all the time in the statutory meetings of the organization and the conventions and other activities that UNESCO is in charge of, usually in the meeting halls and elsewhere on headquarter precincts. They often also represent their country in out-of-area meetings, such as the World Heritage Committee sessions, which are usually hosted by one of the member countries. In addition, there are a large number of receptions and other social events in Paris to which they invite each other. With the use of first names and the French *tu* rather than *vous*, they cultivate the sense of an elite and cosmopolitan circle of friends, complete with the conspicuous intimacy of wide-armed hugs, cheek-kissing and expressions of delight when meeting. The style of interaction is that of equals, whatever the wealth and power differences between their countries. This bond is not to be disrupted, and one may see ambassadors socially dancing with each other at the reception after having fought intensely during the day. However, rank is important, governing the etiquette down to the seating order at dinners – first the ambassadors by seniority of UNESCO assignment and then all others, as the standard arrangement goes. The degree to which 'number twos' are taken seriously varies: while they are considered full representation among Europeans, I was told, non-European countries – especially the large countries such as China or India – often insist on ambassadors, particularly for small events such as dinners of a dozen or so invitees. This is partly due to the fact that only what the Permanent Delegate says is binding – promises made by the 'number two' can be retracted by the Permanent Delegate the next day, citing the subordinate's lack of information. An East African 'number two' told me how he resented pitching in at social functions when his boss was busy, often feeling politely ignored. Reference to him then was often by country name only, with the ambassadors not bothering to remember his personal details.

That the ambassadors and other diplomats often lead the delegations at the Committee sessions has important consequences. First, many of them lack specialized heritage expertise, being trained in other fields, and they are also on average newer to this forum than the experts. Some of the Centre and Advisory Body staff and the State Party heritage experts have known one another for a long time – not only do some of them look back to a dozen sessions or more, they also have met at the string of expert meetings on specific World Heritage topics organized around the world throughout the year; meetings and conferences of the Advisory Bodies and their national and thematic subunits and special bodies; training initiatives for site managers; evaluation, monitoring and consultative missions to the sites; or the workshops that candidate sites organize in order to prepare their nominations. Yet while the closest ties are certainly found here, these experts' occasions to meet are limited. By contrast, the UNESCO diplomats run into each other on an almost daily basis in Paris, mostly not about World Heritage, but other things from the vast UNESCO portfolio. So while their acquaintance may be measured in months rather than decades, their relations are more intense and their community is more immersive, even frantic at times. It follows that these diplomats are the ideal protagonists of lobbying and deal-making and – being permanently placed in Paris, if not on the same corridor of the Rue Miollis UNESCO headquarters annex – also the ideal targets. This is not just because of their hierarchical position and being trained negotiators, but also because they can offer the more immediate return favours from across the UNESCO universe rather than only from inside the World Heritage arena. It follows that, for example, when an East Asian UNESCO ambassador implores her Western European colleague not to call a secret ballot over one of their country's nominations, the colleague has more incentives to comply than if this would occur between people meeting only once a year or so over heritage matters. Finally, when diplomats are not under orders anyway, they are more likely to prioritize world peace, friendly intergovernmental relations or global equity than the experts and may feel that these goals are best served by fulfilling national wishes, even in the absence of lobbying.

The internal arrangements of Committee delegations at the sessions vary. Some ambassadors make a point of emphasizing value-based diplomacy and the terms of the Convention, and thus give their experts a free hand. Therefore, and except for the politically sensitive matters (see below), it was the cultural and natural experts from national ministries and agencies who spoke for the delegations of, for example, Canada, the United States or Estonia and also determined their strategies. In other cases, such as Switzerland or Sweden, ambassadors often spoke themselves, but usually to convey their experts points' of view with whom they consulted closely. However, in most delegations, it was obvious that the UNESCO ambassadors set the course, even when they sometimes had their experts speak as well or left minor decisions to their judgement, and whenever things were getting politically delicate, the turf turned diplomatic across the board. Furthermore, the average diplomat is more skilled in foreign languages and in public speaking than the average expert and acts with more self-assurance. The mutual verbal pleasantries that only occur between ambassadors also mark them off as separate and superior. Not all diplomats are domineering and some are quite unassuming in behaviour, yet still, the most active speakers almost invariably come from this group. Nobody doubts – least of all themselves – that they are in command of the Committee.

This does not mean that diplomats ignore their delegation experts' advice; often, the detailed study of Committee documents, State of Conservation Reports and nomination files is left to the latter, which may amount to weeks or even months of preparation. Some delegations such as those of Israel or Lebanon seemed to rely largely on a single individual's efforts, but with most, there are at least two delegates, invariably including cultural heritage experts and often also at least one nature specialist; the Swiss and the Estonians came with four or five experts.[11] James and Winter unpack the category of expertise for the World Heritage arena (2015): specialized knowledge and experience can refer not just to the substantive matter of heritage conservation, but also to the special history, terms and procedures of the World Heritage Committee, and the two need not coincide. An archaeologist representing an Arab member state in the observed sessions, for example, while proud to display his knowledge of ancient cultures, dreamed up the most outlandish procedural proposals at the same time. Moreover, the skills of diplomacy and experience in international organizations are expertises in their own right. And in addition, several ambassadors have heritage backgrounds themselves, such as by having practised as a conservation architect or headed a national museum before being dispatched to UNESCO. All this complicates the oft-heard call to leave World Heritage 'to the experts', which is regularly rejected in any event by States Parties insisting on their sovereignty.[12] Also, the heritage expert camp is itself divided, with culture specialists several times more numerous than nature specialists (cf. Cameron and Rössler 2013: 163–65) and both groups often knowing only little about each other's principles and practices. Crucially too, the range of sites covered by the Convention is vast for both culture and nature, with conservation architects knowing little about sacred mountains and geologists knowing little about biodiversity hotspots (an issue to which I will return).

In any event, influence in the Committee was only weakly related to in-depth knowledge of World Heritage particulars or conservation matters in the observed sessions. This leads to a curious division of labour: the larger World Heritage discourse is often dominated by specialists of some form of cultural heritage or nature conservation who, in addition, know the arena and its established practices as a result of long-term participation. They are invited to speak on such occasions as the fortieth anniversary conference in Kyoto 2012 and often populate the expert meetings held under World Heritage auspices. Many of these specialists are critical of the recent turn, and so are UNESCO bureaucrats such as the Director-General or the Director of the World Heritage Centre who also feature in such events. Yet in the Committee sessions proper, one sees the very same experts wringing their hands, forced to defer to the UNESCO ambassador leading their delegation or listen in silence. Instead, most of the speakers are diplomats and/or are measured against their efficacy for the national score. It thus happens that first-time participants such as the relatively junior member of the Brazilian foreign service who was one of the most active and aggressive speakers in 2011 or the aforementioned Indian ambassador in 2012 leave a much larger imprint on the session than any of the veterans. Clearly, they do so by virtue of representing a weighty Committee member state. While the experts may be allowed to speak truth to the power of nations on occasion, including when the World Heritage arena congregates to celebrate itself, they do so somewhat in the manner of court jesters who cannot hope to control their majesties' actions.

Being on the Committee

By now, it will not surprise the reader to be told that having a seat on the Committee is an ideal precondition for furthering national self-interests. To begin with, the twenty-one Committee member states also have a vote on their own sites. Without the ballot that Ethiopia must be assumed to have cast in favour of its own nomination, the Konso cultural landscape would not have been inscribed in 2011 (see Chapter 4). Committee members are also given the floor first and often repeatedly. In the observed sessions, they used this for advocacy and even for submitting texts on their own sites, which is prohibited by the Rules of Procedure, but is only irregularly suppressed by the chairs. Committee members are also less likely to be contradicted: States Parties will think twice about provoking them and running the risk of retaliation, particularly when they hope to have decisions for their own sites modified. Together with the UNESCO Executive Board, the World Heritage Committee is the most prestigious body in the UNESCO ambit, providing some independent motivation to serve on it. Yet still, an ICOMOS official told me that other than in order to favour their own sites, some delegation representatives she had talked to could think of no good reason to be on the Committee (cf. also Meskell 2015c: 8).

Critical delegates keep bringing up this conflict of interests and some States Parties refrain from nominating during their Committee tenure. Estonia did not present candidates and Switzerland submitted only transboundary nominations where it did not wish to hold up the numerous other involved nations. Just like other candidates, these two countries announced this policy when running for election, in the assumption that voting States Parties find such selflessness appealing. However, throughout the observed sessions, Committee members that neither promised nor practised self-restraint formed a clear majority. Making the abstention from nominating obligatory for those on the Committee has been proposed time and again from as early as 1983 (cf. Cameron and Rössler 2013: 165), also in combination with compensation mechanisms that would give Committee members extra nomination slots after stepping down. Invariably, however, Committee members who protested against this as unduly restraining their mandate and violating the States Parties' fundamental equality brought such attempts down.

Given that in addition, with currently 193 signatories, only about every tenth treaty state can be on the Committee at any given time, Committee elections – the main agenda item of the biannual General Assemblies – used to be highly contested. No fewer than 114 States Parties had run for the Committee up to 2008, but only 77 – less than half of all treaty states at the time – had ever managed to get elected.[13] And while most such states served only a single period, others did so up to five times and with back-to-back tenures, with some spending more years on the Committee than off it (see Table 5.1). Assumed competency about and commitment to World Heritage matters may have played a role in their success, but the main reason, yet again, must have been deals. In fact, it is with Committee elections that lobbying in the World Heritage arena really comes into its own. Doing a State Party a favour over one of its sites is almost cost-free for Committee members: where a solid majority is ready to provide support more or less for the asking, at least to those they regularly interact with at UNESCO, there is no need for elaborate exchanges and carefully crafted coalitions – a point that, for the post-Brasilia period, escapes the attention of Meskell and colleagues (Bertacchini et al. 2016; Bertacchini, Liuzza and Meskell 2017; Meskell 2014, 2015c). Yet at the ballot box, treaty states are forced to make choices: terms are staggered so that past elections were usually for only seven of the twenty-one seats. With mandates voluntarily shortened, this has increased to either nine or twelve open seats, depending on the year, but this still leaves around 170 countries that can throw their hat in the ring, of which up to twenty-eight (for the twelve seats available in 2009) have done so at the same time. Therefore, states run campaigns, throw expensive receptions at World Heritage meetings and even organize study trips for fellow ambassadors to ingratiate themselves with their peers. It is clear that deal-making is intense, and the two weeks of the UNESCO General Conference in which the World

Table 5.1. States Parties with the longest World Heritage Committee presence, 1977–2019.

Country	Sessions on Committee	Committee tenures	Sessions off Committee	Year of joining Convention	UNESCO electoral group
Brazil	25	5	17	1977	3
Australia	24	5	19	1974	4
France	24	5	19	1975	1
Egypt	23	4	20	1974	5b
United States	23	4	20	1973	1
Mexico	22	4	14	1984	3
Italy	21	4	21	1978	1
Lebanon	20	4	17	1983	5b
Cuba	18	4	21	1981	3
China	18	4	16	1985	4
Germany	18	4	25	1976	1
Senegal	18	4	25	1976	5a
Tunisia	18	4	25	1975	5b
Canada	17	4	26	1976	1
Colombia	16	3	21	1983	3
India	16	3	28	1977	4
Thailand	16	3	17	1987	4
Japan	14	3	14	1992	4
Peru	14	3	24	1982	3
South Korea	14	3	18	1988	4

Note: Derived from WHC/19/22.GA/INF.5B. I count ordinary Committee sessions, not years on the Committee, until 2019, meaning that I disregard the years 1976 (when there was a Committee but no session) and 2020 and 2021 (into which some tenures will reach). The number of tenures does not include those of Committee members elected in the General Assembly of 2019 (Egypt and Thailand, among others) since these will start only their mandate with the 2020 Committee session.

Heritage General Assemblies are embedded and their run-up provide ample opportunity for it. Cablegate has only three instances of site-related advocacy in the US diplomatic cables, but there are more than a dozen requests of support and offers of deals for Committee elections, and this in spite of the fact that at the time, the United States was known not to trade votes in UN bodies.[14]

As a consequence, the voting procedure is of great concern to the States Parties. Their uneven presence on the Committee and the practice of outgoing members to stand for immediate re-election has been a bone of contention from early on (Cameron and Rössler 2013: 72, 156–62), and therefore one might expect efforts for change to focus on increasing rotation. Statistics about the past Committee presence of all states are indeed distributed each time, and measures such as the voluntary reduction of mandates from six to four years in 2009[15] – a practice that everyone now follows – and the introduction of a compulsory abstention of four and subsequently (in 2014) six years[16] have indeed been adopted. However, nobody to my knowledge has ever brought up the idea of reserved seats for those who have never served. As one such seat has been given to States Parties without World Heritage properties since 2001,[17] one might wonder why.

Part of the answer lies in what the Convention does not specify. It postulates that '[e]lection of members of the Committee shall ensure an equitable representation of the different regions and cultures of the world' (§ 8.2), but says nothing about rotation at the State Party level. Accordingly, the General Assembly has focused on the issue of regional representation. Invariably, the regions in question have been taken to be the six (originally five) 'electoral groups' that UNESCO recognizes, much like other UN bodies: the Arab countries; Africa; Asia and the Pacific; Latin America and the Caribbean; Western Europe and North America; and Eastern Europe (the latter two forming a joint group until the 1990s). In fact, almost all important UNESCO bodies and the governing organs of its conventions are structured by these groups, with each of them assigned a fixed number of seats – due to its early adoption, the World Heritage Convention is the one big exception.[18] World Heritage treaty states have sometimes been mindful of this aspect in the past, with voluntary withdrawals from second or third election rounds once a group was well represented, and electoral group seats for World Heritage were discussed repeatedly (Cameron and Rössler 2013: 72, 156–62). However, in the 2007 election, all candidates from Eastern Europe were unsuccessful, yielding a Committee with no country from that group for the 2008 and 2009 sessions. This was widely regarded as a major calamity and after adopting an interim solution proposed by a working group, the General Assembly, in its first ever extraordinary session in 2014, introduced electoral groups to World Heritage too, implementing the proposal of yet another working group (cf. also Meskell, Liuzza and Brown 2015: 440–44). Now, each group must have a specific minimum number of seats[19] and if the number of group members continuing on the Committee is fewer than this, special rounds for filling up the difference must precede the general rounds for the 'free' seats. The time that the General Assembly has spent on this matter over the years greatly exceeds any other topic, including the

Table 5.2. Number of contenders in recent World Heritage Committee elections.

Year	Free seats	Candidates
2009	12	28
2011	9	22
2013	12	21
2015	9	11
2017	12	13

Note: Cf. WHC.10/17.GA/INF.10, pp. 60–61, 76–78; WHC.11/18.GA/INF.12, pp. 6–7; WHC-13/19. GA/INF.12, p. 21; WHC-15/20 GA/INF.15, pp. 21–22; WHC/17/21.GA/INF.11, pp. 7–8.

widespread dissatisfaction with the Brasilia session, the very critical external audit of the Global Strategy in 2011 or the consequences of UNESCO's financial crisis.

An equitable representation of world regions and cultures could of course be operationalized differently and in a more fine-grained way, such as by dividing up the world into smaller units. However, proposals to this effect are surprisingly rare and usually come from heritage specialists, not diplomats. By and large, the latter regard global diversity as sufficiently captured by the electoral groups, even when these are huge and unite unlikely bedfellows ('Asia and the Pacific', running the political gamut from Iran to New Zealand and including almost half of the world's population, is the most diverse). The reason for this, once again, lies in lobbying: electoral groups are used to interacting with each other, and their UNESCO diplomats have regular meetings. Most groups try to ensure at least some degree of internal rotation in the different UNESCO bodies and committees, and sometimes come up with 'clean slates', with only as many candidates as seats assigned to their group, thus guaranteeing election and saving their candidates the campaigning effort.[20] Mutual support is certainly more expected than between countries from different groups. The hope is that having a state from the same electoral group on the Committee gives one a natural contact point and ally – we already saw how group mates were often the first to intervene on behalf of dissatisfied countries in Committee debates. The larger countries and Committee veterans from one's group can then appear to be the safest bet for engaged representation of one's own interests, preferable even to doing the job oneself when a State Party is inexperienced in the arena and stretched for resources.[21] Accordingly, competition has greatly decreased since the reform of the election procedure (see Table 5.2).[22]

Yet while for many purposes, it suffices to have someone from one's group on the Committee, and while these members sometimes openly advertise

Table 5.3. World Heritage inscriptions of Committee member states and observer states compared.

Session	Total number of inscriptions	Committee states	Observer states	Mean for Committee states	Mean for observer states	Means divided
1978	12	10	2	0.67	0.11	6.00
1979	45	28	17	1.33	0.85	1.57
1980	27	10	17	0.48	0.65	0.73
1981	26	19	7	0.90	0.21	4.27
1982	24	9	15	0.43	0.39	1.09
1983	29	14	15	0.67	0.32	2.09
1984	22	9	13	0.43	0.24	1.81
1985	30	12	18	0.57	0.30	1.90
1986	29	11	18	0.52	0.28	1.89
1987	41	18	23	0.86	0.34	2.53
1988	27	18	9	0.86	0.12	7.24
1989	7	3	4	0.14	0.05	2.96
1990	16	4	12	0.19	0.14	1.37
1991	22	11	11	0.52	0.12	4.33
1992	20	8	12	0.38	0.12	3.14
1993	33	11	22	0.52	0.20	2.57
1994	29	12	17	0.57	0.15	3.83
1995	29	13	16	0.62	0.14	4.53
1996	37	18	19	0.86	0.15	5.55
1997	46	25	21	1.19	0.17	7.03
1998	30	9	21	0.43	0.16	2.63
1999	48	16	32	0.76	0.24	3.17
2000	61	17	44	0.81	0.33	2.48
2001	31	7	24	0.33	0.17	1.93
2002	9	4	5	0.19	0.03	5.52
2003	24	9	15	0.43	0.10	4.37
2004	34	12	22	0.57	0.14	4.03
2005	24	9	15	0.43	0.10	4.46
2006	18	3	15	0.14	0.09	1.51
2007	22	6	16	0.29	0.10	2.89
2008	27	7	20	0.33	0.12	2.72

Session	Total number of inscriptions	Committee states	Observer states	Mean for Committee states	Mean for observer states	Means divided
2009	13	4	9	0.19	0.06	3.45
2010	21	9	12	0.43	0.07	5.86
2011	25	9	16	0.43	0.10	4.42
2012	26	6	20	0.29	0.12	2.39
2013	19	5	14	0.24	0.08	2.86
2014	26	11	15	0.52	0.09	5.87
2015	24	6	18	0.29	0.11	2.68
2016	21	3	18	0.14	0.11	1.34
2017	21	6	15	0.29	0.09	3.26
2018	19	2	17	0.10	0.10	0.96
1978–2018	1,094	423	671	0.49	0.14	3.55
2010–18	202	57	145	0.30	0.10	3.16

Note: Derived from http://whc.unesco.org/en/list/xls/?2019 and WHC/19/22.GA/INF.5B. Transboundary properties nominated by both Committee member states and observer states were counted with the member states. For the sake of consistency with overall inscription figures, extensions that added sites in new countries to already listed properties were not counted (there have not been more than a handful of such cases, such as 'Ancient and Primeval Beech Forests of the Carpathians and Other Regions of Europe' or 'Frontiers of the Roman Empire'; see Chapter 7).

themselves as fixers for the others' concerns, such as a Brazilian representative did before the 2010 session, being there oneself brings added benefits. As Table 5.3 shows, 423 out of the 1,094 World Heritage inscriptions went to states then on the Committee, amounting to 38.6 per cent. However, already by 1981, the twenty-one members of the Committee represented a smaller percentage of treaty states, and currently the Committee has space for only 10.8 per cent of the 193 signatories. On average, Committee member states have thus realized more than 3.5 times as many inscriptions as observer states, with the ratio rising to double that value in some years, and 1980 and 2018 were the only sessions where states off the Committee were (slightly) more successful than those on it.[23]

Table 5.4 further demonstrates that a State Party tendency to schedule nominations to their Committee periods (cf. also Bertacchini and Saccone 2012: 333–34) is not the only reason, as there is manifest Committee bias in favour of its own members' bids. As recommended properties are almost always inscribed,[24] looking at the 102 cases listed against Advisory Body advice – 25 properties until 2009 and as many as 77 properties (almost

Table 5.4. Recommended and upgrade inscriptions for Committee member states and observer states compared.

year	**recommended inscriptions**	Committee member states	observer states	**upgrade inscriptions**	Committee member states	observer states	**candidates not inscribed (referral, deferral, non-inscription)**	Committee member states	observer states	**candidates not recommended for inscription (both upgraded or not)**	Committee member states	observer states	**upgrade ratio (upgrades/ total of not recommended candidates)**	Committee member states	observer states	**recommended inscription ratio (recommended inscriptions/all candidates)**	Committee member states	observer states
2003	**20**	6	14	**4**	3	1	**14**	2	12	**18**	5	13	**0.22**	0.60	0.08	**0.53**	0.55	0.52
2004	**30**	11	19	**4**	1	3	**7**	0	7	**11**	1	10	**0.36**	1.00	0.30	**0.73**	0.92	0.66
2005	**24**	9	15	**0**	0	0	**8**	1	7	**8**	1	7	**0.00**	0.00	0.00	**0.75**	0.90	0.68
2006	**12**	4	8	**6**	1	5	**15**	4	11	**21**	5	16	**0.29**	0.20	0.31	**0.36**	0.44	0.33
2007	**18**	5	13	**4**	1	3	**14**	3	11	**18**	4	14	**0.22**	0.25	0.21	**0.50**	0.56	0.48
2008	**20**	6	14	**5**	1	4	**15**	0	15	**20**	1	19	**0.25**	1.00	0.21	**0.50**	0.86	0.42
2009	**11**	3	8	**2**	1	1	**16**	5	11	**18**	6	12	**0.11**	0.17	0.08	**0.38**	0.33	0.40
2010	**9**	7	2	**12**	5	7	**9**	2	7	**21**	7	14	**0.57**	0.71	0.50	**0.30**	0.50	0.13
2011	**12**	3	9	**12**	5	7	**14**	5	9	**26**	10	16	**0.46**	0.50	0.44	**0.32**	0.23	0.36
2012	**15**	2	13	**12**	4	8	**11**	5	6	**23**	9	14	**0.52**	0.44	0.57	**0.39**	0.18	0.48
2013	**17**	5	12	**2**	0	2	**14**	4	10	**16**	4	12	**0.13**	0.00	0.17	**0.52**	0.56	0.50
2014	**15**	7	8	**11**	4	7	**10**	2	8	**21**	6	15	**0.52**	0.67	0.47	**0.42**	0.54	0.35
2015	**18**	5	13	**5**	1	4	**10**	2	8	**15**	3	12	**0.33**	0.33	0.33	**0.55**	0.63	0.52
2016	**12**	0	12	**9**	3	6	**11**	2	9	**20**	5	15	**0.45**	0.60	0.40	**0.38**	0.00	0.44
2017	**13**	2	11	**7**	2	5	**12**	2	10	**19**	4	15	**0.37**	0.50	0.33	**0.41**	0.33	0.42
2018	**12**	1	11	**7**	1	6	**11**	2	9	**18**	3	15	**0.39**	0.33	0.40	**0.40**	0.25	0.42
2003–9	**135**	44	91	**25**	8	17	**89**	15	74	**114**	23	91	**0.22**	0.35	0.19	**0.54**	0.66	0.50
2010–18	**123**	32	91	**77**	25	52	**102**	26	76	**179**	51	128	**0.43**	0.49	0.41	**0.41**	0.39	0.42
total	**258**	76	182	**102**	33	69	**191**	41	150	**293**	74	219	**0.35**	0.45	0.32	**0.47**	0.51	0.45

Note: Derived from http://whc.unesco.org/en/list/xls/?2019, WHC/19/22.GA/INF.5B, the synoptic table on the first pages of document 8B for each of these Committee sessions – document 14B Rev in 2004 – and any possible additional documents of the same session, such as 8B.Add, to which that table refers (all available at http://whc.unesco.org/en/sessions).

two-thirds of all inscriptions) since 2010 – is more instructive. And here, while observer States Parties had only 32 per cent of their unrecommended nominations upgraded to inscription, this figure is 45 per cent for the Committee states, meaning that these were more than one-third more likely to benefit from such overrulings. This is not because Committee members must make up for special Advisory Body strictness, as the nominations of members and nonmembers received inscription recommendations at roughly the same rate. The gap in inscription upgrades thus clearly confirms what was manifest in participants' reasoning about the advantages of Committee membership and in session behaviour,[25] and while I did not analyse the minor upgrades (to referral or deferral) in the same way, my observations suggest that these were subject to the same dynamic.[26]

This means that a stint on the Committee is not only widely perceived as the right time to put forward World Heritage nominations, but also brings about demonstrably more favourable results for these bids (cf. also Braun 2007: 197–201). Small wonder then that once elected, most Committee members are so averse to addressing this conflict of interests. The General Assembly would be a better place to raise it, with more states around that have no burning nomination projects on their mind, but even here, states have been absorbed by other matters – most of all ascertaining space on the Committee for one's regular cooperators, the electoral group. This means that all that is left to the critics of the status quo is voting for Committee candidates that promise self-restraint and a global rather than parochial perspective.

Within the Nation

World Heritage nominations can create substantial mobilization, drawing in large numbers of people and institutions, but there is no reason to assume nations to be automatically united over their properties and candidate sites. Instead, interests diverge over the choice between conservation and economic development, the needs of residents versus those of visitors, the distribution of costs and benefits and so on. Typically, this brings new and often external actors onto the local social scene. Tensions between different government agencies and levels can result, making for complex constellations that are difficult to generalize. As initially stated, these are largely beyond the scope of this book, even though they were central in other research conducted by myself and collaborators (Brumann and Berliner 2016b; Cheung 2019; de Giosa 2021; Marquart 2015) and there is a sophisticated literature on this subject (e.g. Bendix, Eggert and Peselmann 2012)

Instead, this section looks at how the Committee and its auxiliary organizations have dealt with domestic division. By and large – and as already described for a number of cases – the logic of national sovereignty applies here too: the internationally recognized government and whichever bodies it empowers are the default points of contact and sources of information for a country's World Heritage properties. Initially, the Operational Guidelines even discouraged publicizing List nominations domestically so as not to prejudice Committee decision-making.[27] Not before 1996 was the involvement of local communities given an unambiguously positive cast as 'essential to make them feel a shared responsibility'.[28] Since then, however, this view has become mainstream: to the 'four Cs' of credibility, conservation, capacity-building and communication – the 'strategic objectives' proclaimed at the thirtieth anniversary of the Convention in 2002 – the Committee added the 'fifth C' of community in the 2007 session, taking up an initiative forwarded by the host, New Zealand. In 2012, then, the fortieth anniversary was celebrated under the overall theme 'World Heritage and Sustainable Development: The Role of Local Communities'. Such communities are now seen as a crucial conduit for implementing the Convention and its conservation goals rather than an obstacle, and benefits for them that inscription generates are often emphasized. In actual practice, direct links between the World Heritage Centre and the Advisory Bodies and World Heritage site managers are often important and can bypass the national level to a degree, particularly when the latter is weak or disinterested.

However, whenever the national government insists on controlling interaction, the World Heritage bodies have to comply, placing the official State Party representation at an advantage over critical NGOs, disenfranchised locals and the like. This holds for authoritarian governments and what these impose on their provinces and municipalities, but also plays out in liberal democracies, particularly those with a federal system. Signing the Convention does not require adopting national implementation laws and while some states such as Australia, Canada and Italy have passed them, most do not have the formal means to enforce compliance from subsidiary government levels. Yet, practical conservation, particularly of cultural heritage, is often in the hands of provincial or local authorities, and different government levels can develop different ideas on how to deal with World Heritage properties. In numerous cases, the national government has been more conservation-minded than local authorities. In the already discussed controversies over Yellowstone, Cologne and Dresden (see Chapter 3), national governments clearly hoped to restrain development-minded subsidiary authorities by way of World Heritage, and the second and currently ongoing Danger Listing of 'Everglades National Park' in 2010 was a national agency's effort to discipline the Florida governor, as I was told. Likewise, the World Heritage listings of 'Tasmanian Wilderness' in 1982 (Rehling 2011: 429–30) and 'Wet Tropics of Queensland' (Cameron and Rössler 2013:

170) had the Australian government pitted against its own federal states and territories, up to the highest court levels and with eventual success.[29] The only two current European Danger List entries aside from the Kosovo property, 'Liverpool – Maritime Mercantile City' and 'Historic Centre of Vienna', are barely concealed parallels: while the country's delegations did not openly oppose a step that, in both cases, responded to controversial but locally supported high-rise projects, they did not stop it either, as they easily could have done.[30]

Yet while local and provincial governments are often close to business and development interests, a national/international alliance against them is not a constant. Over Kakadu National Park, the Australian national government pushed the mining plan and was furious that subsidiary levels appealed to the Committee directly (see Chapter 3). There are also dozens of other cases where national authorities privilege development over heritage concerns and opponents have a hard time resisting. Marquart (2015: 77–95) has described in detail the dashed hopes in connection with Istanbul's Metro Bridge: local opponents had built on presumably powerful UNESCO to rein in the national and municipal governments, but as already described, Turkish lobbying brought down the proposed threat of Danger Listing in 2011. The bridge was only modified slightly, to the critics' great disappointment.

National-level precedence also holds, by and large, for the way in which indigenous territories and rights have been treated in the World Heritage arena. The rise of UN institutions dealing with indigenous issues is a remarkable development: initially focused on the Working Group on Indigenous Populations meeting in Geneva (Muehlebach 2001; Siebert 1997), momentum has shifted to the UN Permanent Forum on Indigenous Issues (PFII) established in 2000, as an advisory body to the UN Economic and Social Council (ECOSOC) that meets annually at the UN headquarters and has a permanent secretariat there (Sapignoli 2017). In addition, the UN Human Rights Council has mandated an Expert Mechanism on the Rights of Indigenous Peoples (EMRIP) and a Special Rapporteur on the Rights of Indigenous Peoples. In 2007, the UN General Assembly adopted the Declaration on the Rights of Indigenous Peoples (UNDRIP) – not a binding treaty, but a fundamental text supposed to be 'mainstreamed' (i.e. implemented) in all UN agencies, resolutions and activities, comparable to the Universal Declaration of Human Rights. As Bellier (2013) and Peterson (2010) have stressed, this amounts to a major turning point in international law: for the first time, collective entities other than states have been endowed with rights, in however limited a manner. Nation states' resistance against these developments (Muehlebach 2001; Oldham and Frank 2008; Siebert 1997) illustrates the challenge that this presents to the established international system and the underlying conceptions of sovereignty.

The World Heritage arena was touched by this mobilization early on. The 1990s reforms introducing cultural landscapes and the Global Strategy had indigenous cultures in mind. As also already mentioned, when Kakadu National Park was up for debate at the 2000 session in Cairns, indigenous activists from around the world travelled to Australia and held a meeting in the foyer. Their demand for a permanent body to advise the Committee on indigenous matters was taken up in the session and eventually the Committee formed a working group to deliberate the proposal between meetings, with Canada and New Zealand on board and Australia even providing the chair. While remaining adamant on Kakadu, Australia must have been intent on showing a friendly face to indigenous matters at the same time, and Logan finds domestic division reflected in this split strategy. The mid-2001 Bureau meeting[31] discussed the working group's proposal of establishing a World Heritage Indigenous Peoples Council of Experts (WHIPCOE) as a permanent body found initial support,[32] but at the Helsinki Committee session in 2001, the proposal was torn apart by Committee members who instead suggested that indigenous peoples should participate within their national delegations. The written record of the debate[33] is sparse, but witnesses remember China and India as leading the charge in the session; prior to it, France had criticized the proposal (Meskell 2013b: 164–65).[34] In the end, the Committee refused further funding for deliberating the WHIPCOE initiative so that it was effectively aborted, to the shock of many supporters.

However, the adoption of UNDRIP shortly afterwards brought indigenous peoples back on to the World Heritage agenda, as it demands their 'free, prior and informed consent' on all matters affecting their territories. The Permanent Forum on Indigenous Issues has taken up World Heritage and in both 2011 and 2012, the Committee and UNESCO received a joint statement by more than seventy NGOs for indigenous rights that deplored insufficient UNDRIP implementation in the context of the World Heritage Convention, especially regarding consent to nominations.[35] The statements also demanded regular consultations and the resumption of the WHIPCOE idea of a permanent consultative body. PFII representatives now attend Committee meetings, and NGOs representing indigenous rights make critical statements.

At the level of official World Heritage discourse, this criticism has been taken on board. No. 62 of the official periodical *World Heritage Review* was dedicated to 'World Heritage and Indigenous Peoples' in 2012, featuring an interview with PFII chair Myrna Cunningham who voices moderately phrased criticism and introducing ICOMOS's commitment to a rights-based approach (Sinding-Larsen 2012). An expert meeting on the topic was held in the same year. Also, there is a palpable Committee romanticism for things indigenous, not just when these are highlighted in nominations, but also on other occasions. For example, a Centre official fondly remembered the expert meeting where an indigenous delegate sang for participants. In the 2012 session when, after approval of an extension to the Tasmanian Wilderness property, the Australian delegation had an Aboriginal representative speak a few words in his indigenous language, the Russian chair declared her delight about this 'gift' to the Committee. What the gift was she did not specify – given her function, she might just as well have censured use of a language other than English and French. Also on the last 2011 session day in Paris – with the few dozen tired delegates left in the hall only half-aware of what they were doing, but in the presence of China and to the instigators' great surprise – the Committee adopted a decision that 'encourages' States Parties to involve indigenous peoples and local populations in decision-making, monitoring and the evaluation of World Heritage candidate sites and properties, and to respect their rights.[36] The 'Kyoto Vision'[37] proclaimed at the closing event of the fortieth anniversary in Japan makes a point of mentioning indigenous peoples together with local communities as groups that should be more strongly involved. All these moves have committed supporters within the World Heritage Centre, the Advisory Bodies and a number of States Parties, and nowadays quite a few countries actively involve indigenous peoples in their nominations.

However, when other countries claim that all is well with their indigenous citizens, despite allegations to the contrary, this commitment recedes into the background. We already saw in the Saint Petersburg session day how the Kenyan and Indian ambassadors' assertions about the Rift Valley lakes and the Western Ghats passed unquestioned, despite the activist's statement, and how IUCN – otherwise urging indigenous involvement in its recent State of Conservation Reports and evaluations – backed the inscription of Sangha Trinational, despite indigenous concerns not being fully addressed. It would require other Committee members to step in, but reticence against interfering in other countries' domestic matters beats the commitment to UNDRIP. Accordingly, indigenous participants have taken matters into their own hands, forming the International Indigenous Peoples' Forum on World Heritage (IIPFWH) at the 2017 session in Krakow. On the initiative of Angola, the Committee hailed this as an 'important reflection platform' and the chair termed it a 'success'.[38] However, unlike the earlier WHIPCOE idea, this is an external body and whether sovereignty-conscious States Parties will heed its advice remains to be seen. When, a year later, an IIPFWH representative deplored the deleterious effect of the Gibe III Dam on indigenous livelihoods in both Ethiopia and Kenya, for example, it was the very Angolan delegation that encouraged dialogue between 'two brother countries' over the issue, effectively sidelining nongovernmental actors.[39]

Between Japan and South Korea

The national level thus reigns supreme in the Committee, even when it is challenged by subnational forces. However, what happens when the interests of two or more countries collide over a given site? Let me illustrate this with the controversy over 'Sites of Japan's Meiji Industrial Revolution: Iron and Steel, Shipbuilding and Coal Mining', a nomination that held the 2015 session in Bonn enthralled. It provides a rich example of the intricate links that the World Heritage arena forges between the global, the national and the local, the contention this can breed and everyone's dominant strategy for dealing with it.

Japan's bid was for a serial property consisting of twenty-three components in eight different places, almost all of them on the southern island Kyūshū and the adjacent part of the main island Honshū. The collection was to honour the oldest industrialization process outside the Western world: following on the listing of 'Tomioka Silk Mill and Related Sites' further north on Honshū a year earlier, the focus was now on heavy industry, with mines, charcoal kilns, iron and steel mills and furnaces, shipyards, docks, ports, sluices and pumping stations included, some of them located within currently active industrial facilities. While based on foreign models, these first steps of the Japanese economic miracle also demonstrate independent experimentation and innovation, not all of it successful. Within the context of the increasing number of industrial properties on the World Heritage List, the nomination was certainly plausible.

However, weeks before the session, the international press (Hansen 2015; Siemons 2015a) reported that a South Korean envoy was touring the capitals of Committee member states, calling for the rejection of the Japanese bid. This was because the nomination documents failed to mention that in seven component sites, Korean forced labourers had toiled during the Second World War when Japanese men were on the front. To South Korea, this represented yet another instance of Japanese historical amnesia, which did not deserve World Heritage honours. China was reported to support the campaign, as the forced labourers had also included Chinese nationals. In addition, ICOMOS, while recommending inscription in its evaluation, called for an 'interpretive strategy … which allows an understanding of the full history of each site' in its customary list of desiderata.[40]

However, Japan launched a diplomatic countercampaign, in which it qualified the forced labour issue as irrelevant for the bid. The OUV of the property rested on its role in the Bakumatsu and Meiji periods, not on what happened there more than thirty years later in the Second World War. This questionable splitting of history was undermined by Japan's own nomination file:[41] on the fifth page, this document features the morbid charm of the decaying housing compounds for the miners of Hashima, a small, largely artificial island off the coast of Nagasaki that sits above an undersea coal mine and is also known as Gunkanjima ('Battleship Island') because of its signature shape. However, these earliest high rises of Japan were only built from 1916 onwards, that is, after the Meiji period.[42] Tellingly too, the nomination file sticks to 1910 as the end of the relevant time period, although the Meiji emperor died only in 1912. Other than discouraging South Korean resistance – 1910 was the year of Korea's formal annexation by the Japanese Empire – there is little to explain this very unusual periodization.

One week before the Committee session, the foreign ministers of the two countries were reported to have reached an agreement (*Japan Times News* 2015): since both states were on the Committee and South Korea also had an ICOMOS-approved candidate up for decision, they would support each other's bids. Yet, when the Bonn session opened, it transpired that the agreement had collapsed, with both sides blaming the other. This gave rise to what was widely perceived as the drama of the meeting. South Korea continued its protest: on the second day, the delegation invited everyone to a 'side-event' during the lunch break. Interested participants were bussed to a nearby hotel and treated not only to a generous lunch, but also a symposium on 'sites of conscience'. The topic was held general and half of the panels flanking the rows of tables took up other World Heritage properties such as the Auschwitz-Birkenau concentration camp. However, the other half criticized the Japanese nomination and throughout the proceedings, a documentary on Korean forced labour was running in an endless silent loop on a screen next to the podium. Speakers included a representative of the 'Zollverein Coal Mine Industrial Complex in Essen' (Germany) – World Heritage since 2001 – who reported on their own engagement with wartime forced labour and called for full acknowledgement of the past, a former Dutch prisoner of war held in a Japanese camp in Indonesia and also a critical Japanese activist.

In addition, an information booth manned by a South Korean NGO stood opposite the entrance to the main meeting hall over several days (see Figure 5.1). The posters and placards displayed not only addressed the forced labour issue but also the inclusion of Shōka sonjuku ('Village School under the Pines') in the Japanese nomination. This private school in Hagi, Yamaguchi prefecture, was founded in 1842, with the dissemination of Western science (*rangaku*), which was barely accessible at the time, among its main objectives. As the Japanese nomination file argued, it thus gave crucial input to the subsequent industrialization. However, the school is also known for a famous student, samurai Yoshida Shōin, who became its principal in 1857. This ardent nationalist and leader of the *Sonnō jōi* ('Revere the Emperor, Expel the Barbarians') movement, who was not even thirty at the time, was executed for conspiring against the government in 1859. Crucially, he propagated the

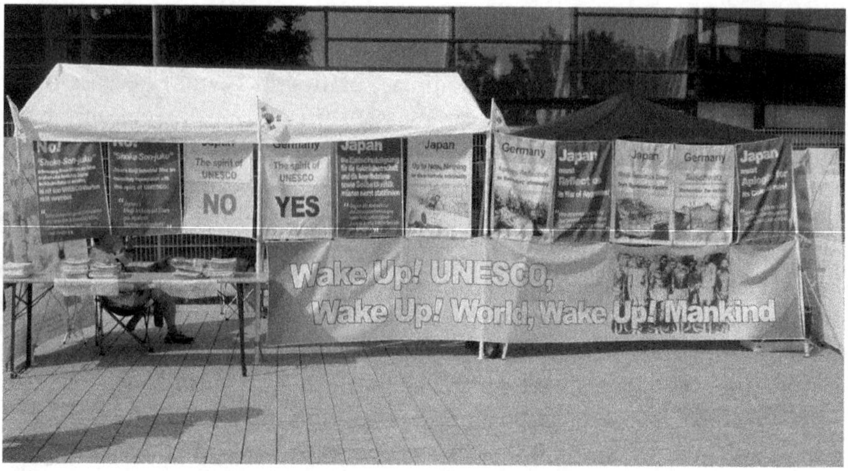

Figure 5.1. Information booth of South Korean NGO outside meeting venue, Bonn, 2015. © Christoph Brumann.

idea that Japan could only draw level with Western powers by having its own colonies and subjugating Korea. In addition, the school's numerous students who rose to eminent positions included Itō Hirobumi, who, after serving as Japanese Prime Minister, became the first General Governor of Korea in 1910. Small wonder then that South Koreans saw Shōka sonjuku as a symbol of the neighbour's imperial aggression.

During the following days and in parallel to the Committee session, confidential bilateral negotiation meetings were held on the closed upper floors of the building, with host nation Germany acting as the agreed mediator. Each Japanese or South Korean delegate descending to the foyer was immediately surrounded by the large press contingents the two countries had dispatched, but even a journalist of one of the largest Japanese dailies complained to me that they were not given any leads. Speculations flourished accordingly and the tussle was the talk of the breaks and receptions. When the South Korean property 'Baekje Historic Areas' came up on the second session day dedicated to nominations, Japan rushed to take the floor and praise the site, which was then inscribed. Despite this gesture, the negotiations dragged on and discussion of the Meiji sites was postponed twice, until the afternoon of the third and last day reserved for nominations. After the lunch break, the gallery heretofore left to independent observers such as myself was teeming with two dozen camera teams, and the general excitement was palpable.

The German chair now opened the agenda item and had ICOMOS present the nomination in the usual way, accompanied by a PowerPoint presentation. ICOMOS's most veteran World Heritage Advisor spoke and took an extra amount of time, no doubt to give no reason for criticism. She repeated the inscription recommendation and also the call to convey the 'full history'. The chair then took over again to announce that Japan and South Korea had reached an agreement, stressing how moving this turn of events was for all participants. She then explained the intricately crafted scenario on which the parties had agreed, with the series of procedural steps to be taken numbered from one to nineteen.[43] The Committee would inscribe the property on the World Heritage List, adopting the proposed draft decision in a slightly amended way in which a footnote would mention an oral Japanese statement made in the session. Japan would read this statement after inscription and South Korea would answer with its own statement. Both statements had been mutually approved and the Japanese statement – despite its presumed orality – had been distributed in writing in the hall. The chair strongly urged all Committee members to abide by this scenario – that is, not to budge.

And, indeed, nobody dared to disrupt what followed. After inscription, the Japanese UNESCO ambassador declared Japan's commitment to presenting the 'full history', as recommended by ICOMOS. She then continued:

> More specifically, Japan is prepared to take measures that allow an understanding that there were a large number of Koreans and others who were brought against their will and forced to work under harsh conditions in the 1940s at some of the sites, and that, during World War II, the Government of Japan also implemented its policy of requisition. Japan is prepared to incorporate appropriate measures into the interpretive strategy to remember the victims, such as the establishment of information centre [*sic*].[44]

A South Korean delegate then expressed his trust in Japan's promise, not without mentioning the Japanese progress report due at the 2017 session. He then said:

> Today's decision marks another important step toward remembering the pain and suffering of the victims, healing the painful wounds of history, and reaffirming that the historical truth of the unfortunate past should also be reflected in an objective manner.[45]

The chair added an own statement in which she welcomed a great victory of diplomacy and marvelled how the World Heritage Convention keeps bringing 'us all' together time and again. Until the end, the substance matter of the conflict was barely explained and chance observers of the internet livestream[46] must have had a hard time following. Yet, inside the hall, I doubt that anyone was ignorant of what was at stake.

The chair closed the item and the other delegates walked over to the two delegations to congratulate them, although the two sides did not approach each other. Within minutes, the mass media were gone, partly on their way to the Japanese delegation's press conference in a nearby hotel. However, later that afternoon, the Chinese delegation – itself not on the Committee – distributed a statement in the room that reminded participants of the two thousand Chinese forced labourers in wartime Japan, confirming in remarkable clarity that '[f]orced labour is a grave crime against humanity and a violation of human rights'.

According to my informants, the negotiations leading up to this outcome were difficult and did not take place only in Bonn, but also in the capital Berlin, to which the chair – one of Germany's deputy foreign ministers – excused herself for one-and-a-half days. They also required high-level contacts, with Chancellor Angela Merkel having a personal message conveyed to Prime Minister Abe Shinzō through a party colleague visiting Japan. In the end, this exceptional deployment of diplomatic efforts for a World Heritage candidate site hinged on a tiny detail: the Japanese had already consented to making the statement weeks ahead, realizing that this was the only way to appease the South Koreans and secure the inscription. What they did not accept was the use of the term 'forced labour', since this is the language of the Convention Concerning Forced or Compulsory Labour from 1930, adopted by the International Labour Organization (ILO), the only organization within the UN family that pre-dates the war. This convention also motivates the mention of conscription in the Japanese statement, since this is the only exception it allows for – during wartime, own nationals may be assigned to specific tasks against their will. And by Japanese reckoning, Koreans were citizens of the Japanese Empire back then – second-class citizens no doubt, paid worse than Japanese conscripts, if at all, and often maltreated so that every sixth Korean forced labourer died in Japan (Underwood 2015). But this would only constitute 'ordinary' discrimination then, not a violation of an international treaty that Japan itself had signed in 1932.

However, the fact that such reasoning implies the legitimacy of colonial rule must also have been obvious to the Japanese, so they insisted on using 'forced to work' rather than 'forced labour' in their statement, whatever the legal status of Korean colonial subjects. This is also what the foreign ministers had agreed upon in advance. Yet, when the South Korean delegation arrived in Bonn, 'forced labour' had returned to their draft statement, and the Japanese felt duped. They were also annoyed about being placed in the context of Auschwitz in South Korea's side-event and about South Korean diplomats walking over to the NGO information booth and chatting amiably with its personnel, as I myself observed. Or so the Japanese would have it[47] – South Koreans blamed an Abe confidant in the Japanese delegation for calling the agreement into question all of a sudden. The two sides engaged in shouting matches and in addition to the aforementioned mediation, it also required further direct contact between the foreign ministers to settle the issue.

Wording was not just a matter of the politics of history, but very much touched on the present: while for a long time, the countries had upheld the view that the Treaty on Basic Relations between Japan and the Republic of Korea of 1965 had settled all mutual claims, in 2012, the South Korean Supreme Court declared forced labourers' compensation claims against Japanese corporations and their successors admissible. Several South Korean courts have subsequently ordered them, in one case just days before the Committee session (Siemons 2015b; Underwood 2015). In addition, Abe cabinet members were personally affected: forced labourers had also toiled in the coal mines of Minister of Finance Asō Tarō's ancestors (not included in the property), and this former Prime Minister had great difficulty acknowledging this fact when it became publicly known in 2007 (Underwood 2015: 7).

In such a sensitive context where every detail can become juridico-political ammunition, it surprised many that the Abe government was ready to approve the above statement, even in the 'forced to work' version, and Japanese journalists jumped on this right away in the press conference. After all, Abe has been the most nationalist Prime Minister of postwar Japan, with a revisionist politics of prewar and wartime history and the amendment of the pacifist constitution as key objectives. However, the Abe government was not only intent on celebrating Japan's rise to an economic giant with the bid (to the point of reversing the schedule of Japanese World Heritage nominations where the Christian sites of the Nagasaki area were postponed to 2016). Instead, and more importantly, key figures were acting as local rather than national representatives to a large extent: Abe, Asō, Foreign Minister Kishida Fumio (head of the diplomats sent to Bonn) and Minister of Education Shimomura Hakubun (superior of the Agency for Cultural Affairs, which had prepared the nomination) all hail from the main nomination area and this is also where the latter three have their electoral districts as Members of Parliament, a base they keep even after leaving the cabinet, as Kishida and Shimomura have long done. Failure to secure the World Heritage title would have greatly disappointed local dignitaries and supporters who were looking forward to a boost in tourism, travelling to Bonn in their dozens in anticipation. Pork barrel politics of a rather conventional cast within the Japanese context (Reed 2002) justified what by Abe's usual political logic was a far-reaching concession, even though the South Koreans promised not to use the statement in the compensation lawsuits. The local will to shine on the world stage – in a country that is outdone by no other in coveting World Heritage titles (Brumann 2016b), except perhaps by its East Asian neighbours (Yan 2018) – reigned supreme.

Under these challenging circumstances, the Japanese regarded the used of 'forced to work' as a major achievement.[48] However, the South Korean Foreign Minister and both the South Korean and international press (Sankei Nyūsu 2015c, 2015f; *The Guardian* 2015) ignored the semantic subtleties, interpreting the statement as the first Japanese admission of wartime 'forced labour' in an international setting. Japanese news coverage of the foreign press reaction and its sometimes triumphant tone in turn infuriated Abe's supporters, who accused the foreign ministry of having messed up (Sankei Nyūsu 2015a). Abe had to declare himself in Parliament (Sankei Nyūsu 2015d), stressing that with the 1965 treaty covering everything, the statement was inconsequential, and Japanese diplomats complained about South Korean duplicity in the negotiations.[49] The following months saw a brief thaw when both countries reached a settlement over the so-called 'comfort women' (Jap. *jūgun ianpu*) – the mostly Korean women pressed into prostitution for the benefit of the Japanese army – but in the aftermath, relationships have soured again, with South Korean courts having taken to seize the domestic assets of Japanese corporations that refuse to pay the ordered compensation payments (*Japan Times News* 2019).[50]

As mentioned above, this highly loaded and twisted controversy about what, in the end, was a single phrase held the Committee session of 2015 in awe: unaffected participants were captivated by the high-level drama unfolding before their eyes, with quite a few appearing to relish the excitement. Committee delegates were considerably less enthusiastic, complaining how daily visits and phone calls from both sides in the lead-up had bordered on harassment – 'it's impossible!', as one of them put it to me. Participants might have taken objection to the degree to which the matter absorbed the limelight, with Japanese and South Koreans providing what felt like one-third of participants, to the extent that the Japanese delegates had been asked to take turns in actually sitting in the hall.

Yet despite all these good reasons, the tacit consensus that this was a bilateral issue requiring a bilateral solution remained unchallenged. Committee members might have instead seen the World Heritage Committee as the ideal body to take the matter *out of* the adversaries' hands: there are many other World Heritage properties with difficult pasts, including slavery and indentured labour, and the sequestration of the less presentable historical periods attempted by the Japanese nomination touches on fundamental questions of heritage interpretation. Other groups are known to ask brawlers to leave the room in order to then decide a contentious matter in their absence, and even without publicly taking sides, Committee members could have demanded a secret ballot. However, no such proposal was made. Instead, I was told that the German chair refused to even open the agenda item without an agreement. This put more pressure on the Japanese side, as their Committee term ended with the session, in contrast to South Korea, which had two years to go, and if Japan had tried to pass over the chair and insisted on opening the item, this would have been sure to backfire with the other Committee members. Combined with the ICOMOS admonition to present the 'full history', rejection by secret ballot would have been the likely result. Yet none of this was the chair's fault and, instead, her strategy stuck to a strictly bilateral logic – the two opponents were to sort things out themselves, and her and the Committee's only legitimate role was to press them to do so.

Between Other Nations

The chair herself thus ended up disempowering the multilateral body whose mandate it is to regulate World Heritage matters. However, she merely followed established practice in comparable cases in the Committee's history. Unlike the Meiji industrial sites, most of these have concerned properties on contested territory, and for the World Heritage arena, these have been the most protracted headaches.

The single most prominent issue is 'Old City of Jerusalem and Its Walls', a List entry made in 1981. The property is located in East Jerusalem, under Jordanian control since 1948, but occupied by Israel since 1967, together with the entire West Bank. Palestine claims East Jerusalem and Israel claims the entire city as its capital, despite the international status decided in the 1947 UN Partition Plan. The Temple Mount/Al-Haram Al-Sharif and its holy sites in the Old City such as the Al-Aqsa Mosque continue to be administered by the Jerusalem Islamic Waqf, an Islamic trust under Jordanian guardianship. Jordan was also Palestine's indirect mouthpiece until the latter was given observer status in 2000 and became a full State Party in 2011.

Jordan took East Jerusalem to UNESCO right away after the 1967 occupation, making the organization a site of intense Israeli–Arab confrontation. UNESCO bodies and officials repeatedly urged Israel to preserve all cultural heritage and desist from archaeological excavations, and the Director-General also appointed a special representative for Jerusalem, ICOMOS President Raymond Lemaire, whom we already encountered (see Chapter 3). When the World Heritage List opened, nomination was thus a predictable step for Jordan, and ICOMOS gave a positive evaluation. To deal with the bid, Committee members called an extraordinary session in 1981. This was the first to be dominated by diplomats rather than experts (Cameron and Rössler 2013: 167–70) and after a highly emotional debate and roll-call vote, the Committee inscribed the property. The United States protested that the Convention (in particular §§ 4 and 6) restricts State Party action to properties on their territory and that a nomination must include an effective plan for protecting and managing the site, but neither was the Old City Jordanian territory, nor could Jordan hope to implement such a plan without the de facto ruler. Still, the Committee majority decided to follow § 11.3: 'The inclusion of a property situated in a territory, sovereignty or jurisdiction over which is claimed by more than one State shall in no way prejudice the rights of the parties to the dispute.' It thus took the agnostic stance upheld ever since, namely to declare that World Heritage decisions have no bearing whatsoever on territorial claims. The Old City of Jerusalem is therefore the only World Heritage property listed under its own subheading, 'Jerusalem (Site Proposed by Jordan)', unlike all other properties, which feature under the nominating country's name.[51] In the following session of 1982, the property became the second and now longest-standing entry in the List of World Heritage in Danger.[52]

Israel was denied access to all these meetings since it had not signed the Convention. When it finally did so in 1999, it immediately nominated Mount Zion in West Jerusalem as an extension to the Old City property. In 2001, the Committee postponed consideration until such time as agreement over the status of the city was reached or a joint nomination submitted. However, Israel retains the extension on its Tentative List under the name 'Jerusalem', which draws regular Jordanian and Palestinian censure. To complicate matters further, the sand embankment and ramp leading up to the Mughrabi Gate – the only entrance to the Temple Mount for non-Muslim visitors and a possible access for Israeli security forces – collapsed in 2004. Israel's construction of a temporary bridge in 2007 provoked Muslim protests, and plans for a permanent solution submitted by both Israel and Jordan, archaeological exploration and its possible damage to adjacent Temple Mount structures, as well as a host of other points, remain hotly contested. Subsequent Committee decisions on Jerusalem have usually had two parts, one referring to the Old City and the other to the Mughrabi Ascent, which lies directly outside the World Heritage property.

Many of the arguments exchanged on these occasions, often with innocent-sounding pleas to keep politics out of things just this one time, follow a territorial and political logic, with heritage conservation playing second fiddle to one of the most jammed contemporary conflicts and its ebbs and flows. With Israel and then Palestine joining the Convention, the preferred strategy has been to open a 'side table' of confidential negotiations parallel to the main session, with agreed mediators shuttling between the adversaries' different rooms. In what most participants perceive as the good years, this leads to an agreed decision, usually distributed in the meeting room in writing, which the chair then asks all Committee members to adopt without debate, dropping the gavel within seconds. In bad years when no consensus is reached, other Committee members usually avoid a debate and call a secret ballot to vote on whichever decision proposals have been tabled. This is then followed by the adversaries' statements – in the otherwise peaceful Bonn session of 2015, these led to the most vitriolic exchange of the meeting. While Israel and the United States can be on the Committee only so many times, this body has always had Arab members, offering these a welcome and comparatively low-cost opportunity to demonstrate solidarity with Palestine.

Committee strife has distracted from quite a bit of actual cooperation on the ground, several informants told me, and UNESCO-related initiatives, also outside the World Heritage framework, appear to have played a role in these (cf. Larkin and Dumper 2009). One session veteran remembers seeing Israeli and Palestinian delegates sharing a lunch table after fighting in the session in the 2000s, showing that confrontation is partly ritualized. World Heritage has also been cherished as a communication channel still in use when everything else had broken down, as a Western European ambassador described the situation in 2013 to me.[53] However, when comparing the Committee decisions on Jerusalem over the years, it is difficult to read great impact into them: the same Israeli restoration, infrastructural and archaeological projects in and around the Old City – accused of boosting Jewish elements at the cost of Arab and Islamic elements – keep being admonished; damage caused by the various violent confrontations of recent years continues to be deplored; the Mughrabi Ascent issue is stalled; and the monitoring mission by ICOMOS and the World Heritage Centre that the Committee has been demanding since 2010 is still awaiting Israeli approval.

With Palestine becoming a State Party and the peace process deteriorating, confrontation has intensified.[54] The annual decision on Jerusalem and the Mughrabi Ascent has been leaning to the Arab viewpoint and, in the voted rather than mutually agreed decisions, habitually speaks of 'Israel, the occupying power'. Palestine repeated what it did with Bethlehem in 2012 (see above) by submitting both 'Palestine: Land of Olives and Vines – Cultural Landscape of Southern Jerusalem, Battir' ('Jerusalem' featuring once again) in 2014 and 'Hebron/Al-Khalil Old Town' in 2017 as emergency nominations, both concerning hotly disputed territory.[55] Unable to send evaluation missions because of the fast-track procedure and Israel's refusal of access, ICOMOS did not see OUV demonstrated in the nomination files. More importantly, and as with Bethlehem, it did not find the emergency nomination justified, recommending the ordinary procedure that takes one year longer. Both times after some Committee debate, a secret ballot was called, producing the required two-thirds majority for inscription and simultaneous Danger Listing, despite a considerable number of abstentions. However, in the Krakow session of 2017, exchanges were particularly aggressive,[56] and this time for once, Canada, the United States and Australia deplored the inscription of Hebron as detrimental to the peace process, while members with pro-Islamic or pro-Palestinian agendas such as Cuba and Turkey fought back. Third-party Committee members do sometimes muse about taking matters in their own hands and deciding on the basis of heritage considerations (Meskell 2018: 167–68). Yet in actual fact, members remain silent in almost all years, leaving things to the protagonists – most participants gave the impression that nothing could be further from their minds than trying to interfere. While Israel and the United States left UNESCO in 2018 because of perceived anti-Israeli bias, neither country has left the World Heritage Convention, so that World Heritage will keep the adversaries communicating, in whichever tone.[57]

As another long-term conflict, Serbia nominated 'Dečani Monastery' in 2004 and then another two monasteries and a church as an expansion to what thereby became the serial property 'Medieval Monuments in Kosovo' in 2006. Given positive ICOMOS evaluations, the serious condition of the monuments and the agreement that Committee decisions do not affect territorial status, the winners of the Balkan War did not block the inscription and simultaneous Danger Listing, with actual conservation in the hand of the then UN-administered Serbian province that declared independence in 2008. As Kosovo is neither a UNESCO member[58] nor a State Party to the Convention, the Committee has simply kept postponing debate to the next year ever since, usually after confidential negotiations; Serbian attempts to add '(Serbia)' to the property name have been refuted. The result is a curious stalemate: Serbia brings up the site, declaring its commitment to conservation and offering resources; the report of current conditions preceding the draft decision in the written document describes conservation measures taken by the de facto authorities with EU support; and the Committee refuses to discuss the property, even though it confirms its status as being 'in danger'.[59]

The most recent *cause célèbre* of bilateral strife is 'Temple of Preah Vihear', an eleventh-century Khmer sanctuary spectacularly sitting on the southern edge of a tall cliff where the Thai–Cambodian border has never been clearly delineated.[60] In 1962, the International Court of Justice (ICJ) ruled that the temple and its 'vicinity' were Cambodian territory. However, it refrained from delimiting the border itself, so that an area of 4.6 square km next to the temple remains contested. The traditional access route to the temple, coming from Thailand in the north, cuts through that area (for the historical and political background, see Grabowsky and Deth (2016); Hauser-Schäublin (2011); Hauser-Schäublin and Missling (2014); Williams (2011)).

When discussing a World Heritage nomination in 2004, both countries 'agreed at least implicitly on a joint inscription' (Grabowsky and Deth 2016: 7), but Cambodia then submitted a nomination on its own, with ICOMOS supporting inscription in its 2007 evaluation.[61] Thailand insisted on a joint bid and as a result of confidential negotiations during the 2007 session, it was agreed that 'Cambodia will propose the site for formal inscription … in 2008 with the active support of Thailand'.[62] Six weeks before that next session, the new Thai government dropped its resistance to a unilateral nomination (Williams 2011: 2). The Foreign Minister justified this with the nominated area not reaching into the disputed zone, being restricted to the temple proper now (Grabowsky and Deth 2016: 7). However, the agreement provoked a domestic outcry, particularly among 'Yellow Shirt' nationalists who took to the streets and laid siege to the parliament, and since just days before the session, a court injunction declared it invalid for lacking parliamentary approval, the Thai side revoked its consent. Yet, at this advanced stage and with OUV confirmed already in the 2007 decision, it would have required Cambodia's initiative to withdraw the bid. So in another negotiated decision adopted without debate, the Committee put Preah Vihear on the List, with the decision text including the usual disclaimer about territory, due acknowledgement of Thai reservations and gestures towards a possible binational extension of the property for the future.[63]

Not only did the listing make the Thai Foreign Minister resign immediately, it also brought troops of both sides to the area that, over the next few years, saw several clashes with casualties, displacing large numbers of civilians and wringing gunfire damage to the temple – the first (if undeclared and low-key) World Heritage war, which only ended after the ICJ ordered the withdrawal of all troops in 2011. Both countries had themselves elected into the World Heritage Committee in 2009, where they were almost entirely silent in the observed sessions, clearly absorbed by Preah Vihear, for which they achieved a negotiated decision in 2010. However, in the 2011 meeting, the 'side table' failed to reach consensus. When, against Thailand's urging, the chair did not adjourn the item, the Thai delegation – after a statement by the Minister of the Environment that carefully listed how all their grievances had been ignored – declared that it denounced the Convention, took its state name plate and walked out in a dramatic gesture, only to be back on board the following year as if nothing had happened. Interestingly, this largely ended the matter – the Committee has not discussed Preah Vihear again, normally an indication that all is well with a World Heritage site.[64]

The list of thanks delivered by the Cambodian delegate after inscription[65] give an idea how much other countries and their diplomats were drawn into the battle as mediators. When hostilities broke out on site, the UNESCO Director-General and her predecessor as envoy also made special mediation efforts.[66] But once again, this was a case in which the opponents were vastly more invested than anyone else.[67] All this suggested the usual strategy to everyone else in the Committee, namely to keep the issue as bilateral as possible and urge the contenders to find a solution instead of coming up with third-party ideas. The most telling moment in this regard followed the Thai walkout in 2011: the Committee still had to adopt the negotiated decision and was left with the single paragraph (among many) for which Cambodia and Thailand had proposed alternative texts, leading to the row (substantively, it centred on the submitted Cambodian management plan, as this included the disputed zone). The chair asked the Committee members what to do and waited for quite a while. However, nobody dared say a word, so that she simply struck out the paragraph in the end.[68]

There are more properties on the States Parties' Tentative Lists that are contested,[69] and there are cases of ambivalence about those properties that point to the historical sins of other countries: when Japan nominated 'Hiroshima Peace Memorial' in 1996, it required intense diplomatic negotiations to have the United States and also China put their reservations on record but not block the inscription, as an involved participant told me. On the listing of 'Bikini Nuclear Test Site' by the Marshall Islands in 2010, the United States did not comment in the session. In 2007, Poland had the name of 'Auschwitz Concentration Camp' changed to 'Auschwitz Birkenau: German Nazi Concentration and Extermination Camp (1940–1945)', with Germany not commenting on the move either.[70] Going back further in history multiplies the properties that speak of atrocities committed between nations, such as through warfare, colonialism or slavery, but here the past is sufficiently remote not to provoke confrontation between contemporary States Parties of the Convention.[71]

It will come as no surprise that border-crossing engagement in contemporary wars is an even more delicate issue. All World Heritage properties in Syria were put on the Danger List in 2013, all not already Danger-Listed properties in Yemen and 'Hatra' in Iraq followed in 2015, and all Libyan properties did so in 2016. War damage to World Heritage was the main theme of the 'Bonn Declaration' of 2015.[72] However, this text and many session interventions deplored 'terrorism', as if only 'Islamic State' (IS) forces had been destroying anything. They refrain from raising the involvement of the State Parties' own armies in bombing historical city centres, placing military camps in archaeological sites and so on. Instead, receipt of reports from these States Parties is duly commended and decisions appeal to all involved forces rather than naming anyone in particular. Least of all, the foreign actors in these wars, such as Russia in Syria or Saudi Arabia in Yemen, are ever challenged on their role,[73] despite the fact that the Convention requires all States Parties to abstain from 'any deliberate measures which might damage directly or indirectly the cultural and natural heritage … situated on the territory of other States Parties' (§ 6.3).

To summarize the status of binational and trinational disputes in the World Heritage arena, let us turn to the US ambassador to UNESCO's concluding comment in his report on the 2009 General Assembly, which is worth quoting in full:

> Overall, from a U.S. perspective, the results of the elections are positive and should result in a WH Committee that is more geographically balanced and that could remain politically moderate. Iran was not elected, having received 44 votes, just three votes short of capturing a seat. Both Cambodia and Thailand were elected to the WH Committee, keeping some balance, but also guaranteeing future debates on the Preah Vihear temple site, over which the two countries have had recent military skirmishes. Russia's election will not help U.S. concerns regarding the treatment of sites in Kosovo, originally nominated by Serbia. However, the election of four states that recognize Kosovo (UAE, Switzerland, France, and Estonia) will hopefully bring some balance to those discussions. The UAE and Iraq have increased the strong Arab presence now on the WH Committee, which could create difficulties for Israel, given the past disputes between Jordan and Israel concerning the reconstruction of the ascent to the Mughrabi Gate at Jerusalem's Temple Mount site.[74]

The United States was an engaged member during its Committee stint ending with those elections, leaving most interventions to the delegation experts, who tended to be motivated by conservation concerns rather than international politics. Nonetheless, the ambassador assesses the election results almost entirely for their expected effect on the Committee's most notorious border disputes. Even when assuming these to be of special interest to a freshly appointed diplomat reporting to a superpower's foreign office, this bias speaks volumes about the degree these conflicts absorb everyone's attention. I find it difficult to say whether, on balance, the World Heritage arena has helped to contain and 'civilize' these disputes, providing an outlet and diplomatic scaffolding for what otherwise might find more unstructured and violent forms, or has rather fanned the flames by giving them an intergovernmental stage. That these battles have boosted UNESCO's felt significance appears clearer in comparison.

However, what is beyond doubt is that most actors' preference for the adversaries settling things among themselves without much noise practically removes these properties from the Committee's grip. They are made extramultilateral territories, so to speak, with the Committee becoming a bystander rather than a party. This constellation is even actively sought – otherwise and even when accounting for the two countries' massive lobbying, there would be no good reason to have Cambodia and Thailand on the Committee. I am sure that States Parties were hoping for a restraining effect, voting for both opponents rather than shutting them out, but it also meant rewarding their intransigence and giving scarce seats to delegations with few other interests. Conversely, strengthening the global aspect of what has

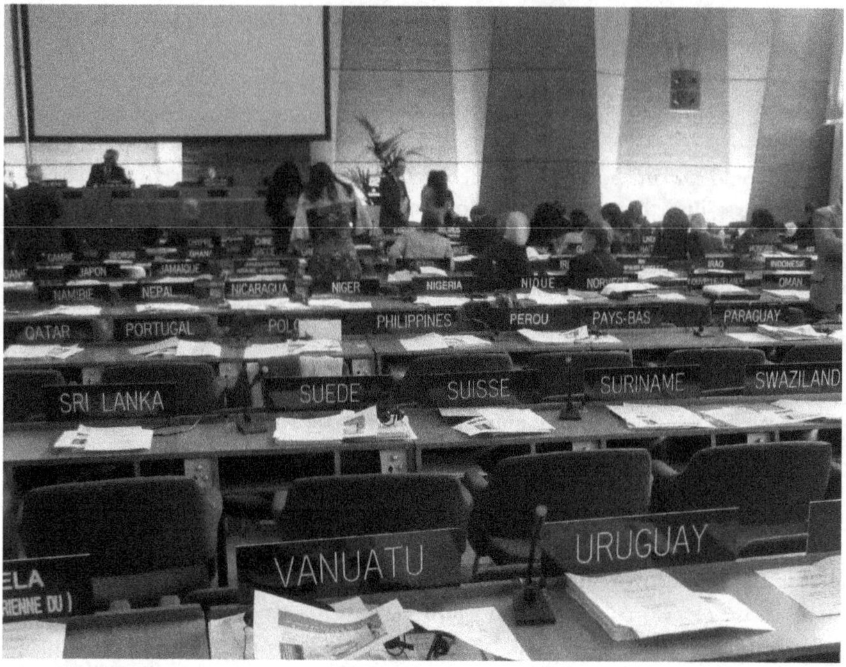

Figure 5.2. State name plates in the World Heritage General Assembly, Paris, 2009. © Christoph Brumann.

no unambiguous national colours has not been pursued in earnest, with proposals such as raising both countries' flags and that of UNESCO over Preah Vihear (Silverman 2011: 15–16) coming from external observers. This trend is of long standing, but with the general turn to prioritizing national preferences since Brasilia, it finds even less resistance than before: when two nations' strong wishes collide over a property, all others find self-denial the safest strategy.

The Intergovernmental Property of Mankind

To summarize, the first precondition for the post-Brasilia turnaround is the fact that – by virtue that it occurs in a UN setting – the 'common property of mankind', which the Convention is expected to protect, is operationalized as an intergovernmental matter, with the community of sovereign nation states standing in for humanity (see Figure 5.2). Most participants are dispatched and instructed by national organizations and agencies, and even those who represent transnational bodies can never quite shed their nationality. As national players, participants are expected to defend national interests and most often these are self-interests, not general concerns. To achieve their goals, States Parties engage in lobbying other States Parties and striking deals of mutual support, with career diplomats with no specific background or interest in heritage taking the lead.

The 21-member Committee, rather than standing above these concerns, has been a privileged place for pursuing self-interests and swapping favours, with a demonstrably better turnout in World Heritage titles for those currently on rather than off this body. Therefore, Committee elections have been a key target of lobbying and deal-making, and it is only with the belated assignment of fixed seats to electoral groups – the common framework for exchanging support at UNESCO – that this has somewhat subsided. National governments enjoy primacy where there is domestic division, with their take on things being the default even in those cases where NGOs and the like present different views. This includes indigenous peoples: fondness for their cultural expressions does not translate into taking their side when they confront national authorities. And while the Committee has challenged national governments on their properties, in those cases where two or more nations collide over territorial or symbolic ownership of a given site, all others step back, encouraging bilaterally agreed decisions with often little substantive content. The very fact of bilateral conflict might motivate the Committee to break the stalemate and find a heritage-centred solution, but in actual fact, the very opposite happens.

All this encouraged nation states to assert themselves when the opportunity offered itself in Brasilia and after. However, this would still have been more difficult if the established procedures and categories of the World Heritage bodies provided a more solid counterweight than they actually do, as will be demonstrated in the following chapters.

Notes

1. Meskell claims that Italian and Spanish nationals have dominated the World Heritage Centre (2015c: 15). US lobbying with the UNESCO Director-General for more US staff appointments at UNESCO is on record (https://wikileaks.org/plusd/cables/09PARISFR201_a.html (retrieved 27 August 2020)).
2. For example, in 2006, the US Permanent Delegation to UNESCO was pleased to hear that 'one of our Secretariat contacts said that she and her colleagues were striving to ensure a "low profile"' for an expert meeting on the effects of climate change on World Heritage properties about which the United States was particularly concerned (https://wikileaks.org/plusd/cables/06PARIS2130_a.html; cf. also https://wikileaks.org/plusd/cables/06PARIS5276_a.html (retrieved 27 August 2020)).

3. This already happened before social media caused the possibilities to balloon: a Centre official reported receiving 1,700 letters on Lyon in 2009 and 25,000 emails when the Yellowstone Danger Listing (see Chapter 4) was being considered, which resulted in her system ending up choked. More than 20,000 protest letters were received over a salt production plant that was to endanger 'Whaling Sanctuary at El Vizcaino' (Mexico) (Cameron and Rössler 2013: 218).
4. In the aftermath, the ambassador's complaints about the US delegation seconding Australia's call for the secret ballot led to US embassy staff in New Delhi being summoned to an Indian government ministry. Claiming that 'it is "common knowledge" that technical issues play only a minor role in the selection process while politics is the main determinant of which sites are inscribed', the ministry official formally expressed disappointment about the lack of US support for what would have been 'no sweat off of anyone else's back, especially not the US' (https://wikileaks.org/plusd/cables/08NEWDELHI2009_a.html (retrieved 27 August 2020)).
5. A 2009 report from the US Permanent Delegation to UNESCO has 'several experts noting that what had been subtle lobbying in the past has now become harassment and unbearable pressure' (https://wikileaks.org/plusd/cables/09PARISFR370_a.html (retrieved 27 August 2020)), without giving further details. However, I have not heard of similarly massive complaints myself.
6. https://wikileaks.org/plusd/cables/09PRAIA106_a.html (retrieved 27 August 2020). The fact that the other formal requests for nomination support in the Cablegate cables come from Burkina Faso (https://wikileaks.org/plusd/cables/09OUAGADOUGOU170_a.html (retrieved 27 August 2020)) and Saudi Arabia (https://wikileaks.org/plusd/cables/10RIYADH31_a.html (retrieved 27 August 2020)) – countries that likewise had few World Heritage properties and, in the latter case, was apparently unaware that the United States had already rotated out of the Committee by that time – suggests that the more experienced countries rely on lobbying in Paris or at the Committee sessions.
7. I will return to how such self-restraint is more easily accomplished when one is already the country with most World Heritage properties, having nominated as many as twelve for a single session. Recently, Italy has been less obedient (Hølleland and Wood 2019: 7), and throughout, it has vehemently opposed national nomination quotas whenever these were brought up.
8. In subsequent sessions, that Swiss ambassador was regularly relied on for ad hoc drafting group and even the chairmanship of a working group continuing over several years. This meant that any damage to his reputation among his peers did not last.
9. In her analyses of Committee lobbying and deal-making, Meskell has suggested that World Heritage inscriptions themselves are a kind of tradable item or currency, saying, for example, that 'inscription has become a political tool for nations to bolster their sovereign interests, using global patrimony as a pawn' (Meskell et al. 2014: 5) or that World Heritage properties 'operate as transactional devices' (Meskell 2015c: 3). However, I have trouble relating this to my own session observations: states were certainly trading favours, but World Heritage titles were the *end* of such exchanges rather than their means, and it was support that states exchanged, not titles, which, once awarded, last forever and generate revenue in many cases, but cannot be 'traded' in meaningful ways.
10. An exception was the extraordinary Committee session of 1981 in Paris over the nomination of Jerusalem, which was dominated by diplomats and government ministers (Cameron and Rössler 2013: 169).
11. I did not find the correlation between delegation size and influence in the sessions or interest and investment in World Heritage that Meskell et al. (2015: 450) and

Bertacchini et al. (2017) posit, and nor did Schmitt for the 2006 and 2007 sessions (2011: 167–69). Large teams are usually the result of including the representatives of candidate properties in the national delegation rather than having them register as independent observers, with some countries being liberal in this regard and others on the restrictive side. Of the ninety-one delegates that China sent to the 2012 session, for example – the biggest team ever, according to Meskell et al. (Meskell, Liuzza and Brown 2015: 450) – more than one-third was from the province of Inner Mongolia according to the list of participants (cf. WHC-12/36.COM/INF.2, pp. 45–56), meaning that they came to celebrate the inscription of 'Site of Xanadu' in that session. In no State Party delegation was the effective diplomatic team tasked with liaising with other delegations, lobbying and determining strategy larger than a handful of people, and rank-and-file delegation members often learned little about these core people's activities – the assumption that all members are negotiators is incorrect.

12. Cf. e.g. https://wikileaks.org/plusd/cables/08PARIS1073_a.html (retrieved 27 August 2020).
13. Cf. https://wikileaks.org/plusd/cables/08PARIS1073_a.html (retrieved 27 August 2020).
14. Before the General Assembly of 2005 when the United States (successfully) ran for a seat, the diplomatic cables record two requests of support by other candidates – Spain and Kazakhstan (https://wikileaks.org/plusd/cables/05MADRID2163_a.html, https://wikileaks.org/plusd/cables/05ALMATY3157_a.html (retrieved 27 August 2020)) – and altogether twelve reactions to US requests for support. In these, Australia, Colombia, Ireland, Costa Rica assure the country of their backing (https://wikileaks.org/plusd/cables/05CANBERRA1167_a.html, https://wikileaks.org/plusd/cables/05BOGOTA8182_a.html, https://wikileaks.org/plusd/cables/05DUBLIN1260_a.html, https://wikileaks.org/plusd/cables/05SANJOSE2413_a.html (retrieved 27 August 2020)), the Czech Republic does so too, mentioning bilateral cultural restoration projects in Iraq it would like to see supported (https://wikileaks.org/plusd/cables/05PRAGUE1026_a.html (retrieved 27 August 2020)), France promises to go by the tradition of mutual support among the 'P5' – the five permanent members of the Security Council – in UN settings (https://wikileaks.org/plusd/cables/05PARIS4802_a.html (retrieved 27 August 2020)), and the Netherlands eventually pledges support too, after some complaining about US resistance to vote-trading (https://wikileaks.org/plusd/cables/05THEHAGUE1919_a.html, https://wikileaks.org/plusd/cables/05THEHAGUE2137_a.html (retrieved 27 August 2020)). Uruguay offers support if the United States reciprocates over a Uruguayan candidacy in a different UN setting (https://wikileaks.org/plusd/cables/05ASUNCION947_a.html (retrieved 27 August 2020)) and Honduras suggests a similar deal, although it is aware that the United States is averse to these (https://wikileaks.org/plusd/cables/05TEGUCIGALPA1916_a.html, https://wikileaks.org/plusd/cables/05TEGUCIGALPA2092_a.html (retrieved 27 August 2020)). Only Bangladesh, Jamaica and Venezuela remain noncommittal (https://wikileaks.org/plusd/cables/05DHAKA3200_a.html, https://wikileaks.org/plusd/cables/05KINGSTON1698_a.html, https://wikileaks.org/plusd/cables/05ASUNCION917_a.html (retrieved 27 August 2020)). Before the General Assembly of 2007, Romania suggests mutual support (https://wikileaks.org/plusd/cables/07BUCHAREST855_a.html, https://wikileaks.org/plusd/cables/07BUCHAREST1200_a.html (retrieved 27 August 2020)), obviously an error since the United States – still on the Committee for another two years – does not stand for election. Before the 2009 elections, Mexico, Cambodia, Indonesia and Thailand ask for US support, with Thailand introducing its

delegation experts in detail (https://wikileaks.org/plusd/cables/09MEXICO741_a.html, https://wikileaks.org/plusd/cables/09PHNOMPENH505_a.html, https://wikileaks.org/plusd/cables/09PHNOMPENH689_a.html, https://wikileaks.org/plusd/cables/09JAKARTA1609_a.html, https://wikileaks.org/plusd/cables/09BANGKOK2599_a.html (retrieved 27 August 2020)). In most of these cases, the cables do not cover the entire interaction on the respective request, suggesting that these relatively formal diplomatic notes are only the tip of an iceberg of which a large part has a more informal and oral character.

15. General Assembly Rules of Procedure, 2009 version, note to § 13 (WHC-14/GA/1 Rev.3). Making this reduction mandatory would require the prohibitively impractical step of revising the Convention, which all States Parties would then have to sign again.
16. Ibid., § 13.2; General Assembly Rules of Procedure, 2014 version, § 13.2 (WHC-14/GA/1 Rev.4).
17. General Assembly Rules of Procedure, 2001 version, § 13.1 (WHC-14/GA/1 Rev.1).
18. The US ambassador marvels at the ubiquity of the electoral groups when questions of representation are discussed in 2008. She also mentions a delegate bringing up the equitable representation of cultures the Convention demands in the debate, 'a subject that others quickly dropped, as if any discussion of the subject would be like taking a walk in a diplomatic minefield' (https://wikileaks.org/plusd/cables/08PARIS1073_a.html (retrieved 27 August 2020)).
19. Africa has four seats, Asia and the Pacific three and the other four groups two, with one additional prescribed seat alternating between Asia and the Pacific and Latin America and the Caribbean. This leaves five 'free' seats for which all countries can compete.
20. Cf. also https://wikileaks.org/plusd/cables/10UNESCOPARISFR231_a.html, https://wikileaks.org/plusd/cables/06PARIS4095_a.html (retrieved 27 August 2020).
21. Other than through the electoral groups, I did not find 'regionalism' as defined by Meskell et al. (2015) an important factor in State Party interactions and patterns of support. It also does not sit well with the claim of an overlapping set of authors that the – transregional – BRICS coalition was dominating World Heritage sessions (Bertacchini, Liuzza and Meskell 2017).
22. 'Clean slates' for their assigned seats were presented by Africa in 2015 and Western Europe and North America in 2017, and only three candidates in these two elections were unsuccessful.
23. I shared this observation with arena informants and found it to be popular, seeing it appear in the already-mentioned critical external audit on the Global Strategy and the Committee session of 2011, where Estonia reported that it had been discussed at an expert meeting. It might have also inspired Cameron and Rössler's remarks (Cameron and Rössler 2013: 71–72) and the analysis by Meskell and colleagues (Meskell et al. 2014: 226–27). However, the latter lump all upgrades together, despite the very importance difference between those to inscription from those to referral and deferral. As early as 2004, Committee representatives themselves assembled similar statistics, finding a higher success rate for nominations from current members (cf. https://wikileaks.org/plusd/cables/05PARIS238_a.html (retrieved 27 August 2020)).
24. The only exceptions concerned Israeli nominations and Preah Vihear; see below.
25. The gap has actually decreased in the post-Brasilia period, as Table 5.4 demonstrates – prior to 2010, Committee members were almost twice as likely as observer states to receive one of the inscriptions by upgrade. However, the total number of cases for that period is small, concerning less than one-fifth of all inscriptions, so I hesitate to place too much weight on this finding.

26. Meskell et al. do so, but in their analyses either take 2008 as the cut-off point or lump the 2003–13 period together entirely, making comparison to my results difficult (Meskell et al. 2014: 6–7).
27. Operational Guidelines, 1994 version, § 14 (http://whc.unesco.org/en/guidelines (retrieved 27 August 2020)).
28. Operational Guidelines, 1996 version, § 16 (http://whc.unesco.org/en/guidelines (retrieved 27 August 2020)).
29. Cf. also SC-89/CONF.004/12, p. 3.
30. When asked about Danger Listing in 2011, the UK observer delegation suggested that 'a strong message should be sent to the State Party', a tacit call to go ahead that was thwarted by an inattentive Brazilian diplomat. In 2012, the delegation remained silent when the Committee took the step. Observer State Party Austria used its statement after the Danger Listing of Vienna in 2017 to put its domestic division on display in unseen fashion: a national representative pointed out that the national level had no authority over local planning decisions, passing to one city official defending the high-rise projects and another one opposing them. Not having been in Krakow myself, I can only speculate that with the session being livestreamed, this skit was meant for domestic consumption. In practice, however, the delegation's reaction amounted to tacit endorsement – committed State Party opposition would have ruled out the Danger Listing.
31. Contrary to what Meskell writes, not the twenty-fifth Committee meeting, and there also was no 2001 Bureau meeting in Winnipeg (2013b: 162, 163) – she herself correctly states that the twenty-fifth Committee session of 2001 was held in Helsinki (2013b: 163, 164) and at that time, Bureau meetings were usually held in UNESCO headquarters in Paris.
32. Cf. WHC-2001/CONF.205/10, pp. 7–10.
33. WHC-01/CONF.208/24, p. 57.
34. Meskell is surprised about postcolonial states shutting out indigenous demands (2013b: 168) but this is consistent with the initial reticence of many Asian and African countries in the aforementioned UN processes (Oldham and Frank 2008; Siebert 1997). Where there are no majority populations of European or other recently immigrated stock, identifying indigenous groups and 'first peoples' is not straightforward. National governments may, on the contrary, dread it as encouraging the separatisms that China, India and also France have been facing – *pace* Meskell (2013b: 165), the latter was probably less concerned with Bretons and Basques than with the Kanaks in French special collectivity New Caledonia in the Pacific, where the quest for independence led to an (unsuccessful) referendum in 2018. Asian and African critics have often argued that *all* their citizens are indigenous and that this is, first and foremost, an issue of the former settler colonies.
35. http://www.un.org/esa/socdev/unpfii/documents/DRIPS_en.pdf (retrieved 27 August 2020).
36. WHC-11/35.COM/20, p. 271.
37. https://whc.unesco.org/document/123339 (retrieved 27 August 2020).
38. https://iipfwh.org, WHC/17/41.COM/18, p. 12; WHC/17/41.COM.INF.18, pp. 393, 397, 420.
39. WHC/18/42 COM/INF.18, pp. 286–87. For critiques of the World Heritage arena's involvement with indigenous issues and human rights, cf. Disko and Tugendhat (2014); Larsen (2018).
40. WHC-15/39.COM/INF.8B1, p. 103. As I was told, the internal ICOMOS World Heritage Panel toned down a first draft, which was more outspoken, and it may have played a role here that, unlike in previous years, this body included a Japanese – the then ICOMOS

Vice-President and now President – but no South Korean or Chinese national (cf. http://www.icomos.org/images/DOCUMENTS/World_Heritage/ICOMOS_World_Heritage_Panel_2014.pdf (retrieved 27 August 2020)).

41. http://whc.unesco.org/uploads/nominations/1484.pdf (retrieved 27 August 2020).
42. The morbid charm of the ruins also inspired the setting of several scenes of the James Bond film *Skyfall* in 2012.
43. Cf. WHC-15/39.COM.INF.19, p. 220–21.
44. Ibid., p. 222.
45. Ibid., p. 223.
46. Cf. http://whc.unesco.org/en/sessions/39COM/records/?day=2015-07-05#tMnp-FyTHr-s11825, 0:02:00 hrs to 0:40:00 hrs (retrieved 27 August 2020).
47. Cf. the very detailed description of events on the webpage of the conservative daily *Sankei Shinbun* (Sankei Nyūsu 2015b) that reads as if spoonfed by Japanese government sources.
48. Here, the Japanese may also have been deceived by their own euphemistic translation of the statement they distributed to the press. 'Forced to work' was rendered as *ishi ni han shite hatarakasareta* – literally 'made to work against their will', sounding somewhat like detention at school. By contrast, the standard *kyōsei* for 'force' in such terms as *kyōsei rōdō* (forced labour) was studiously avoided.
49. In the Japanese press, this found some sympathy (Sankei Nyūsu 2015e), but most comments were critical of the entire affair (*Asahi Shinbun Digital* 2015; *Mainichi Shinbun* 2015).
50. Measures to convey the 'full history' at the property were still in preparation in 2018, with the World Heritage Centre and ICOMOS's State of Conservation Report urging not to restrict these to the planned information centre, and on a visit to Hashima/Gunkanjima after inscription, fellow Japanologist Aike Rots found no mention of forced labour in the available written information or in the Japanese-language guided tour (personal communication). If not on account of the Meiji sites, the forced labour issue might return when 'The Sado Complex of Heritage Mines, Primarily Gold Mines' on the Japanese Tentative List (https://whc.unesco.org/en/tentativelists/5572 (retrieved 27 August 2020)) is nominated – a leaflet distributed at the Bonn session announced Chinese resistance, given that Chinese forced labourers had worked in the mines.

    East Asian tensions about wartime history may also have affected the fate of 'Kamakura, Home of the Samurai', a nomination that received a noninscription recommendation from ICOMOS in 2013. It was duly withdrawn by Japan and is yet to be resubmitted. Founded as the seat of the shogunal warrior government, Kamakura is usually seen as the third most important historical city in Japan, not falling back behind places such as Nikkō or Hiraizumi whose sites have made it on to the World Heritage List. Even when accounting for conceptual deficiencies in the nomination, a deferral recommendation could therefore have been considered, and I was told that it was indeed discussed in the internal ICOMOS World Heritage Panel. However, the South Korean and Chinese members of a body that included no Japanese at the time insisted on rejection. Not knowing the particulars, I think that it must have played a role that for these countries, the samurai and the *bushidō* warrior ethos retrospectively ascribed to them (Benesch 2014) are not just the material of pop culture and Kurosawa movies, but also a key inspiration for Japanese colonialism and war aggression.
51. http://whc.unesco.org/en/list (retrieved 27 August 2020).
52. ICOMOS supported the step because of 'severe destruction followed by a rapid urbanization' (CLT-82/CH/CONF.015/8, p. 10); cf. also Ricca's detailed analysis (2007).

53. In 2004, the US ambassador recognized that 'UNESCO continues to be the sole UN organization where Israeli–Palestinian issues are handled with a minimum of polemics and polarization' (https://wikileaks.org/plusd/cables/05PARIS3210_a.html (retrieved 27 August 2020)), and in 2005, the Italian ambassador claimed that no other international organization engendered as much cooperation between Israel and Palestine (WHC-05/29.COM/INF.22, p. 193-194; for a similar assessment from inside UNESCO, see https://wikileaks.org/plusd/cables/08PARISFR2270_a.html (retrieved 27 August 2020)).
54. On an Egyptian initiative, the Committee had already been taking up the cultural heritage in the Palestinian territories since 2002, making it a standing agenda item in the sessions. The Committee regularly allocated money to a heritage inventory, assessment missions to several sites and specialized training, with several European countries and the United States adding their own funds, and commissioned the World Heritage Centre to report about progress (cf. also de Cesari 2014).
55. Battir is an old terraced agricultural landscape that was about to be cut apart by the wall that Israel is raising between its 1948 territory and the West Bank, separating the nominated property on the Palestinian side from the Palestinian landowners' village on the Israeli side. Unlike the rest of Hebron, the Old Town with the Tombs of the Patriarchs/Al-Ibrahimi Mosque – Abraham/Ibrahim and his family, revered by Jews, Christians and Muslims alike, as the Arab countries did not tire of emphasizing – is controlled by Israeli military for the sake of several contested Jewish settlements, in ways that impose heavy restrictions on the Palestinian majority. Also, Israel had put the Tomb of the Patriarchs on a national heritage list in 2010 (https://wikileaks.org/plusd/cables/10UNESCOPARISFR226_a.html (retrieved 27 August 2020)).
56. After a voted decision on Jerusalem, an outraged Israel called for a minute of silence for Holocaust victims in such places as nearby Auschwitz, with Cuba calling a minute of silence for Palestinian victims. These affronts resurfaced in the Hebron debate and were compounded by new ones, such as the Israeli delegate's attempt to obstruct the secret ballot by standing next to the podium and his accusing the Mayor of Hebron, who was present in the room, of having killed Israelis in the past. A Simon Wiesenthal Center representative denounced Palestine's nominations as 'war by other means'. In a way, the Committee clashed precisely because of not having discussed Jerusalem a year earlier – the attempted coup d'etat brought the 2016 session in Istanbul to a premature close, so that it could only be finished at UNESCO headquarters in Paris a few months later. Therefore, in the meantime, Arab countries tabled an 'Occupied Palestine Resolution' at the UNESCO Executive Board, where it was adopted. This document was criticized not just by Israel but also by UNESCO Director-General Irina Bokova and UN Secretary-General Ban Ki-moon for referring to the Old City and Temple Mount sites only by their Arab names.
57. This conflict also impinged on two of the only three cases where properties recommended for inscription did not make it on to the World Heritage List. The first concerned the Israeli nomination of 'Incense Route – Desert Cities in the Negev' in 2004. Egypt and Lebanon objected that the included archaeological remains of four Nabateian towns only covered a stretch of 100 km and that the nomination did not clarify the relationship with other components of the much longer historical route already listed as World Heritage – 'Petra' (Jordan) and 'Frankincense Trail' (Oman). They must also have been provoked by the map in the nomination file (http://whc.unesco.org/uploads/nominations/1107rev.pdf, p. 9 (retrieved 27 August 2020)) that depicts Israel and the occupied territories as a single, undifferentiated unit. In 2005, both the listing and the

name were no longer contested, but Oman had the name of its own site changed to 'The Land of Frankincense'. This must have been to distance itself from Israel in what – under different political circumstances – could very well have become a single, transboundary property, just like other historical routes (see Chapter 8). Also for each session from 2008 to 2011, Israel nominated 'The Triple Arch-Gate at Dan', with ICOMOS recommending inscription of this archaeological site every time. However, Arab Committee members objected that the site was not on Israeli territory and since the UN Department of Political Affairs declared itself unable to delineate the borders between Israel, Syria and Lebanon in the area with precision in 2010, a decision has been postponed until that question is solved. The Committee took care to confirm the OUV of the property each time and by the logic of World Heritage decisions having no territorial consequences, it could just as well have gone for inscription, but not even Israel pushed strongly for this.

58. Its application did not get the required two-thirds majority in the UNESCO General Conference of 2015.

59. Cf. also https://wikileaks.org/plusd/cables/09STATE63696_a.html, https://wikileaks.org/plusd/cables/10UNESCOPARISFR148_a.html (retrieved 27 August 2020). In the 2012 session with Serbia on the Committee and its ally Russia as the host, this required substantial advance diplomacy, as a Western European UNESCO ambassador told me. The Committee might see this pattern recur with 'Ancient City of Tauric Chersonese and Its Chora', nominated by Ukraine, approved by ICOMOS and with Russia as the first Committee member to speak in favour of inscription in 2013 – all just months before Russia occupied the Crimean peninsula and has since been claiming it as its own national territory, also in other UNESCO fora (Meskell 2018: 158–60). However, in the World Heritage Committee, the property has only been discussed once, with Ukraine not pushing the matter and the decision daintily appealing to 'all parties currently associated with the state of conservation of the property' (WHC/16/40.COM/19, p. 132) – not naming the parties – to fulfil their obligations in 2016.

60. France and the Kingdom of Siam agreed to go by the natural watershed in 1907, but the French map annexed to the treaty in 1908 placed that watershed further north when in actual physical fact, it is the edge of the cliff.

61. Contrary to Silverman's claim (2011: 6), I could not find evidence for a 2002 nomination.

62. WHC-07/31.COM/24, p. 154. This was the third case of a recommended inscription not being sealed by the Committee (see note 57).

63. WHC-08/32.COM/24, pp. 220–22; WHC-08/32.COM, pp. 194–98; WHC-08/32.COM/INF.8B1.Add.2. In its subsequent statement, the Thai delegation deplored the inscription, siding with the new 2008 evaluation: unlike the previous year, ICOMOS now avoided a clear pronouncement as to inscription, given that the reduced property excluded important built and landscape components and no longer justified two of the three OUV criteria. I think that implying Committee or 'UNESCO' carelessness or naivety, also in view of what followed (e.g. Silverman 2011: 14–16; Williams 2011: 3, 5–6), fails to appreciate the procedural constraints: with Cambodia insisting on something that had followed its proper bureaucratic path and Thailand objecting only at the very last moment, avoiding inscription would have required mobilizing a Committee majority in favour of suspending all established procedures, in a pre-Brasilia time when these were still more respected. It would have also meant turning against the one of the two adversaries that had remained consistent. Moreover, there is no third party, such as UNESCO bodies or officials, with formal authority over the Committee that could have ordered it to desist. I therefore go by the session chair's assessment, who told me that there was no choice, whatever one might think of the step.

64. Valid denunciation of the Convention would have required written notification, which Thailand never gave. A new ICJ ruling in 2013 (cf. Grabowsky and Deth 2016: 11; and *pace* Meskell 2018: 118) left no winner: it did make the whole promontory the 'vicinity' of the temple and thereby Cambodian, including part of an access road that has recently been built from the Cambodian side. However, much of the rest of the disputed zone outside the World Heritage property was declared Thai territory and precise delineation is still up to the two countries. Since 2014, the 'International Coordinating Committee for the Conservation and Enhancement of the Temple Preah Vihear' – modelled on ICC-Angkor and a Committee demand at inscription – has been meeting, co-presided over by China and India, including the two adversaries, but inserting Belgium, the United States, France, Japan and South Korea between them.
65. WHC-08/32.COM, pp. 195–96.
66. Based on an analysis of the WikiLeaks diplomatic cables, Meskell (2016: 118–21, 153–54; 2018) sees the United States as pulling the strings in the conflict, as it had vital geopolitical and economic interests such as offshore oil and gas deposits and restraining the Chinese advance. She suggests 'American intervention in Preah Vihear's inscription' (2018: 121), even that 'American support was instrumental' (2018: 119). As the United States was itself sitting on the Committee in 2007 and 2008, this must logically mean that it sided with Cambodia once Thailand turned to opposing the listing. However, I can find no evidence for such a strategy. It is true that in the run-up to the 2008 session, the US ambassador to Cambodia reflected about what inscription, 'if handled correctly', might do for settling other overlapping claims in the Gulf of Thailand and US access to natural resources there (2018: 120). For the suggested deal or 'emerging equation' – 'if Cambodia retained its temple, Thailand might enhance its underwater assets, and the United States might negotiate for extended contracts' – Meskell (2018: 120) cites the webpage 'Preah Vihear for Koh Kong and Natural Gas / Oil' (https://antithaksin.wordpress.com/2008/10/16/preah-vihear-for-koh-kong-and-natuaral-gasoil (retrieved 27 August 2020)). This text claims that the Thai Foreign Minister's support for inscription will bring resource concessions in the Gulf of Thailand to his former employer, the then already exiled former Prime Minister and business tycoon Thaksin Shinawatra, with Chevron waiting in the wings to exploit them. However, the webpage is maintained by Thaksin's enemies from the PAD party and itself says that '[t]his allegation may not in fact be true, but it is widely believed, and that is what counts'. Disagreeing with this view, I do not see it as sufficient proof of a major US commitment.

    Crucially too, the ambassador's assessment above was made at a time when Thailand was still *in agreement* with the Cambodian nomination. On the developments after the Thai turnaround and at the 2008 Committee session, the cables are completely silent. There is no evidence of the US favouring either of the two countries over the other or of Thai disappointment with the United States, either then or in the aftermath. Meskell herself stresses Thailand's strategic importance for the United States (2018: 121), which should have spoken for US restraint. What I gather instead from the cables is US diplomats trying to restrain the adversaries and return them to the negotiation table, being keen on maintaining friendly relations with both sides (cf. especially https://wikileaks.org/plusd/cables/08PHNOMPENH581_a.html, https://wikileaks.org/plusd/cables/08BANGKOK2428_a.html, https://wikileaks.org/plusd/cables/08PHNOMPENH679_a.html, https://wikileaks.org/plusd/cables/08PHNOMPENH694_a.html, https://wikileaks.org/plusd/cables/08PHNOMPENH701_a.html, https://wikileaks.org/plusd/cables/09PHNOMPENH101_a.html, https://wikileaks.org/plusd/cables/09BANGKOK1569_a.html and https://wikileaks.org/plusd/cables/09BANGKOK3006_a.html

(retrieved 27 August 2020)). In 2008, the United States supported bilateral talks and mediation by ASEAN rather than the Security Council debate proposed by France (https://wikileaks.org/plusd/cables/08STATE79164_a.html (retrieved 27 August 2020)) – if the United States had been intent on playing a larger role, it should have been fine with the Security Council, of which it is a permanent member. Also, while US diplomats in the cables are aware of what peace might bring to US security and business interests, they have little concern for the bone of contention as such. In fact, with one exception from 2005, which is silent on the bilateral talks about the World Heritage nomination already ongoing at the time (cf. Grabowsky and Deth 2016: 6–7), Preah Vihear does not even feature in the cables before 2008, despite it being up for decision and requiring confidential negotiations in the 2007 session when the United States was on the Committee.

67. Both nations place the glorious ancient Khmer in their ancestry; Cambodia in particular has made them and their sites central to nation-building, and Thai nationalists see Preah Vihear as yet another site taken from them in a series of colonial humiliations (Grabowsky and Deth 2016: 3, 6; Hauser-Schäublin 2011: 46–52; Hauser-Schäublin and Missling 2014: 88–89; Silverman 2011: 3, 5–6, 8–11). Current confrontation is ironic since in actual fact, this is a fringe area and historical transition zone: over centuries, the polities laying overlapping claims to it showed little practical interest in the place, and whatever borders they drew had next to no consequences for local lives and mobilities (cf. Hauser-Schäublin and Missling 2014: 89–92). Yet this changed, with the two countries even severing their diplomatic relationships over Preah Vihear in 1958 (Grabowsky and Deth 2016: 4), and in 2008, both the Cambodian government of Hun Sen and the Thai nationalist camp had special reason to drum up the issue, with Cambodian elections due only weeks after the inscription (Grabowsky and Deth 2016: 15) and with street protests and a political crisis unfolding in Thailand that ultimately led to the coup of 2014.

68. As Meskell points out (2018: 145–49), the case of 'Archaeological Site of Ani', the historical Armenian capital now part of Turkey but in viewing distance of the border, shows some parallels, as Turkey considerably downplayed the Armenian past in the nomination file for the 2016 session it itself hosted in Istanbul. ICOMOS called for 'an accurate and balanced representation of the complex history' (WHC/16/40.COM/INF.8B, p. 185), a point that – unsurprisingly perhaps by now – found favour with South Korea, which also insisted on it being retained in the decision text (Meskell 2018: 146). Meskell vividly describes massive and in the end successful Turkish pressure to overrule ICOMOS's deferral recommendation in favour of inscription. However, in contrast to the cases I discuss, Armenia welcomed the World Heritage inscription and a neighbouring country's efforts to conserve Armenian heritage, making no comments on the details of the Turkish approach (WHC/16/40.COM.INF.19, p. 145). There may have been more actual ambivalence than these words conveyed, but the State Party clearly did not contest the listing. For a US take on the early stages of Turkey's bid, see https://wikileaks.org/plusd/cables/08ANKARA2139_a.html (retrieved 27 August 2020).

69. One is 'Amami-Oshima Island, Tokunoshima Island, the northern part of Okinawa Island and Iriomote Island' (https://whc.unesco.org/en/tentativelists/6160 (retrieved 27 August 2020)). Japan stresses that OUV is restricted to the islands proper, not the surrounding seas, but this has not discouraged China from objecting to the entry as touching on the undetermined maritime borders between the two countries (Meskell 2018: 155–56). It should be noted that the islands proper – returned to Japan by the

United States in 1971 – are not disputed and are no less than 600 km away from the Chinese coast, making the Chinese maritime claim debatable.

In 2005, CNN founder Ted Turner proposed financial support for turning the demilitarized zone between the two Koreas with its undisturbed flora and fauna into a World Heritage property (https://usatoday30.usatoday.com/news/world/2005-11-18-turnerdmz_x.htm (retrieved 27 August 2020)). The idea was discussed by World Her-itage Centre personnel (https://wikileaks.org/plusd/cables/05PARIS8247_a.html, https://wikileaks.org/plusd/cables/08PARIS403_a.html (retrieved 27 August 2020)), but so far has not gone forward.

70. According to the US ambassador to UNESCO in 2006, her Israeli colleague told her that Germany felt cornered and powerless about an issue it saw as being determined between Poland and Israel (https://wikileaks.org/plusd/cables/06PARIS4363_a.html (retrieved 27 August 2020)).

71. However, border-crossing effects of contemporary action do play a role, particularly with water bodies where the chain of causation is most obvious. The most prominent recent case is that of 'Lake Turkana National Parks' (Kenya), with that lake fed by the River Omo on which 300 km further upstream, Ethiopia has now completed the enormous Gilgel Gibe III Dam. The reservoir and a linked sugar cultivation project reduce both the amount and the seasonality of the water intake. A small part of the lake is located in Ethiopia, just like the prehistoric fossil property 'Lower Valley of the Omo' that is also affected and has been raised repeatedly by the Committee, but the Lake Turkana World Heritage property lies entirely within Kenya. Committee decisions included a Danger Listing threat as early as 2011, but Kenya only came around to supporting this step in 2018, with observer country Ethiopia resisting, but the Committee (including its African members) putting the property on the Danger List. Previously, the Kenyan delegation had to hold back because of a prime ministers' agreement to prioritize neighbourly relations and to give Kenya a share of the electricity, and both countries dragged their feet over the environmental and heritage impact assessments demanded by the Committee. Therefore, the 2018 turnaround came too late for the heavy impact on the lake ecosystem already confirmed by IUCN. Kenya itself has been advancing several controversial development projects in the lake area, to which the Danger Listing decision also refers. However, once again, other Committee members only proceeded to Danger Listing when the State Party of the property asked for it, whereas before, they built on a bilateral rather than multilateral solution.

Also, until 2009, the Committee repeatedly discussed Ukrainian work on a deep-water connection between the Danube and the Black Sea that threatened to affect the natural property 'Danube Delta' (Romania), although the main battles were fought in other bodies (cf. https://wikileaks.org/plusd/cables/08BUCHAREST511_a.html (retrieved 27 August 2020)). The already-mentioned Kurdish protests outside the Bonn meeting hall in 2015 targeted the Ilisu Dam that Turkey has been building on the Tigris. In addition to the likely inundation of Kurdish sites in Turkey not on the List, they targeted the effects on the already-nominated candidate site 'The Ahwar of Southern Iraq: Refuge of Biodiversity and the Relict Landscape of the Mesopotamian Cities' (Iraq), as the included wetland marshes are partly fed by the river. IUCN did mention the Ilisu Dam and a further weir on the Iraqi–Irani border as a threat in its evaluation for the 2016 session when the Iraqi property was inscribed, and the most recent 2018 decision urges the three countries to reach a sustainable water-sharing agreement. However, none of these cases got anywhere near Danger Listing and the States Parties did not fight about them in the Committee session.

72. WHC-15/39.COM.INF.19, pp. 312–15.
73. This strategy of avoidance extends to recently ended conflicts: lost sections of the long-ruined Bagrati Cathedral in Kutaisi, Georgia – part of the property 'Bagrati Cathedral and Gelati Monastery' – were controversially reconstructed from 2010, in violation of Venice Charter standards. However, in the 2012 session in Saint Petersburg, there was broad consensus among diplomats that the item should be exceptionally adjourned to next year, given that the session was chaired by a country that had recently been at war with Georgia. The works completed later in 2012 thus went uncommented upon, despite ICOMOS regarding them as irrevocably destroying OUV.

    Countries condoning illegal trade that threatens World Heritage properties in other countries are not openly criticized either: just before the Atsinanana rainforests in Madagascar were inscribed on the Danger List in 2009 (see Chapter 4), IUCN reported that 95 per cent of the illegally logged rosewood was going to China. Itself on the Committee, China condemned illegal trade in its intervention. However, it did not comment on the allegation, and nor did other Committee members take it up, with Australia's ironic praise for what 'our colleague from China has so eloquently outlined' about smuggling remaining the most open jibe.
74. https://wikileaks.org/plusd/cables/09PARISFR1458_a.html (retrieved 27 August 2020).

Chapter 6

Procedures

The World Heritage arena has developed into a hothouse of textual production. Not only are ever more documents being produced, but those required for the operations of the Committee have also grown in size, in most cases many times over, as already described (see Chapter 3). The Operational Guidelines have 290 paragraphs now, not just twenty-eight as in 1977, and the record of Committee decisions between 240 and 290 pages in the last decade, not a mere thirteen as in the first two years. Advisory Body evaluations of candidate sites and the World Heritage Centre and the Advisory Bodies' draft decision for listed properties with their summaries of conservation issues have also ballooned (cf. also Schmutz and Elliott 2017). Some nomination dossiers for candidate properties run into the thousands of pages these days and even when subtracting the extensive supplementary material often included (such as legal texts), the core part usually is in the three figures. Organizations empowered by the Committee are held to submit regular reports about their operations – annual in the case of the Centre and the Advisory Bodies – and all special activities commissioned by the Committee, such as expert meetings on particular topics, invariably generate a report too. New kinds of texts have been added, such as the summary record of all session interventions, and previously unpublished texts are now put online for everyone's consumption, including most of the conservation reports sent by States Parties on the Committee's request; reports of the Centre and the Advisory Bodies' monitoring missions to World Heritage properties; nomination dossiers (at least in some years); the factual error letters that States Parties send when they find any in the Advisory Body evaluations; and even the draft decisions prepared by the Centre and the Advisory Bodies, allowing for comparison with what the Committee makes of them in the sessions. On the World Heritage Centre website, one also finds extensive supplementary material, such as the 140-page nomination manual, the monthly *World Heritage Review*, or the 'thematic studies' of particular kinds of heritage, which are supposed to guide future nominations and evaluations, often in the *World Heritage Papers* series. And then there are also meta-texts, such as the audit reports about the World Heritage Centre, the Global Strategy or the Advisory Bodies' evaluation procedures.

This discursive efflorescence reflects the elaboration and systematization of World Heritage procedures over the years. Almost everything that is being done in the name of the Committee now follows a much more tightly regulated course than in the past, with more detailed requirements, a more intricate schedule and a wider circle of involved individuals. By and large, this development is not due to the Centre or the Advisory Bodies feeling driven to continuously refine their procedures on their own initiative; rather, it responds to dissatisfied States Parties' demands for more transparency, objectivity, rationality and/or predictability. Growth ends up making the World Heritage machinery more opaque for those who lack the commitment or resources to work their way through it, but this does not discourage the treaty states from calling for more, with nothing ever really shrinking in return.

However, textual and operational sophistication has not produced widespread satisfaction with either the processes or their results and, crucially, it has not prevented the Brasilia rebellion and its aftermath. In this chapter, I will show how contingency and illicit influences continue to influence preparatory procedures. However, this is less momentous overall than the one element that appears to have lost rather than gained sophistication and consistency, namely the decision-making of the Committee. Not only is this due to directly concerned nation states being strongly invested, there is also a surprising degree of improvisation and much less reference to precedent than might be expected. But let me follow the procedural sequence and first look at the nomination and monitoring processes, that is, the preparatory stages of Committee decisions.

Nominations

In order to obtain a World Heritage title for a site on its territory, a treaty state to the Convention must first put it on the national Tentative List, which all but nine of the 193 treaty states have by now.[1] These lists must be formally approved by the Committee and then appear on the World Heritage Centre website,[2] complete with brief descriptions of the properties. The entries and the descriptions are unexamined, remaining the 'sole responsibility … of the States Parties', as a disclaimer introduced in 2017 now emphasizes.[3] How much to include in these lists, how often to update them – some have not been changed in twenty years – and whether to treat them as binding nomination commitments are all up to the respective country. When there are as many as fifty-six (Iran), sixty (China) or even seventy-eight entries (Turkey), it would take the better part of this century to actually work through them if the current quotas persist. However, where the Tentative List is more exclusive, it often amounts to an actual nomination plan for the next decade or so. Some of the eager nominators organize their own internal processes of nomination and selection, inviting constituent units such as federal states or provinces to submit candidates, and then screening these through the input of expert panels and the like – when visiting Görlitz in 2014, for example, I was surprised to see banners publicizing the city's candidacy not for World Heritage status, but for the German Tentative List. Yet, despite such procedures, there is a total of 1,716 entries on the Tentative Lists (as of 2019) – almost ten per country and more than on the World Heritage List proper, suggesting that the Committee will not run out of business.

World Heritage bids have become major undertakings. The official manual suggests a year for collecting the respective data and another for actually writing the dossier (UNESCO 2011: 6), but this is an optimistic estimate. I have heard of cases where this took a decade or more and even in well-resourced countries, the preparatory steps of mobilizing local support, improving the conservation and management conditions, passing legal regulations and doing the necessary background research can run to similar lengths. Some of the writing work may be outsourced to external consultants, such as the Italian I talked to who had drafted half-a-dozen dossiers in the Arab countries, and the nomination manual suggests forming an expert group including international specialists to assist the process (UNESCO 2011: 69–70). However, the intellectual requirements of a World Heritage nomination are not beyond trained academic professionals in the respective disciplines, provided that they have a grasp of World Heritage categories and discursive habits, such as through attending the Committee sessions. Differences in the cost estimates I heard, with one ICOMOS official giving an average of US$500,000 and another just half that sum, may depend on whether state employees' work time is costed or not. Much larger preparatory expenses are sometimes claimed – no less than US$200 million for 'China Danxia' (see Chapter 4), for example.[4] This must mean that major infrastructural measures were included, but whenever there is more to do than just preparing a document, the ICOMOS estimates appear modest to me. Also and even in low-wage countries, expenses are clearly beyond the maximum of

US$20,000 of 'preparatory assistance' from the World Heritage Fund, which States Parties can apply for.

Once a nomination dossier covering all the mandatory points in the Operational Guidelines is ready, the required number of copies and an electronic version must be submitted to the World Heritage Centre by the end of January prior to the year in which Committee approval is sought and four months earlier if the Centre's facultative precheck is sought. This schedule makes for about one and a half years of processing time and also means that at each session, the list of next year's new candidates is already publicized (as document INF.8B3). Eager States Parties sometimes turn submission of hardbound gold-lettered tomes into a ceremony, but Advisory Body personnel assured me that they do need more than a PDF file for most purposes. The Centre is held to reject dossiers that lack required elements and in almost every round, this makes for one or two victims, mainly from among the less experienced and resourceful countries, but sometimes also the likes of France. However, there is no provision for putting nominations on hold for other reasons, such as for collecting a number of similar ones over a couple of years and then choosing only the worthiest. Once a dossier is in and complete, it must be processed as an individual case.

For this purpose, nominations are passed on to two of the three Advisory Bodies for an evaluation. Their role is statutory, given that the World Heritage Convention establishes their right to participate in Committee deliberations, and encourages the Committee and the secretariat to make use of their services. The oldest and largest body founded in 1948, IUCN, is arguably the first 'GONGO' at the international level, since the almost 1,300 organizational members of this umbrella organization are both governmental and nongovernmental. IUCN is led by a president and an elected council and, with almost 1,000 staff in the headquarters in Gland near Geneva and in forty-five offices worldwide, is closer in size to UNESCO in its entirety than to the other two Advisory Bodies. Whereas IUCN deals with natural properties, cultural properties are addressed by ICOMOS, which was founded in 1965 as a result of the Venice Conference, which also produced the aforementioned Venice Charter. ICOMOS is nongovernmental and mainly a membership organization of 10,500 individuals in 151 countries – 107 of these with national ICOMOS committees – even though it has 270 institutional members too.[5] A triennial General Assembly elects a president and the other office-holders who make up a twenty-member board. Headquarters in Paris have a staff of ten and while they recently moved to new and larger premises on the outskirts, the old office near UNESCO headquarters gave the impression of a small-sized university department in the humanities rather than the seat of a multinational that IUCN's Swiss premises suggest. The third Advisory Body, the International Centre for the

Study of the Preservation and Restoration of Cultural Property (ICCROM) in Rome, which was founded in 1956, is itself an intergovernmental organization with 135 member states, a General Assembly, which elects a Council of named experts, and a director-general. It has almost forty staff, among them the 'Sites Unit', which deals with World Heritage. ICCROM takes charge of World Heritage training programmes and capacity-building. As it is usually not involved in nominations and only rarely with monitoring, its team at and input into the Committee sessions are smaller than those of the other two bodies.

While cultural properties go to ICOMOS and natural properties go to IUCN, mixed properties are fully evaluated by both, and for the cultural landscapes evaluated by ICOMOS, IUCN gives desktop-only comments on the natural aspects. IUCN has a standing 'World Heritage Unit' with a staff of twelve, which takes care of drafting the evaluations. Several of ICOMOS's headquarters personnel have likewise dealt with World Heritage operations for many years, but ICOMOS outsources the drafting work, assigning each nomination to one of between six and nine 'World Heritage Advisors'. These are consultants paid in proportion to their amount of work, usually as a sideline to their other sources of income, even though the core people have been doing this work for close to a decade or more. The Advisors also present the nominations in the sessions and are used for purposes such as expert meetings and special publications so that many of them are familiar faces to Committee regulars.

The evaluation includes both a 'desk review' of the nomination dossier and an 'evaluation mission'. For the desk review, both organizations contact recognized experts, both internally – ICOMOS from its members and IUCN from the experts registered in its World Commission on Protected Areas – and externally, including from academic institutions. Reviews are supposed to comment on the nomination and its OUV on condition of confidentiality. ICOMOS representatives gave an average of forty contacted individuals for each bid, but total numbers of received desk reviews – 140 for twenty-seven cultural nominations in 2017; 170 for twenty-two cultural nominations and 80 for nine natural nominations in 2018[6] – suggest modest response rates. The process resembles academic peer review, down to the lack of remuneration.

The evaluation mission is a preannounced inspection visit to the property, coordinated with and hosted by the State Party, which only became compulsory in 1993.[7] Its limited mandate is to examine the current state of conservation and management, given that for non-emergency World Heritage inscriptions, these must be up to scratch. State Party perception of the evaluation process often hinges on the site visit and the dispatched individuals – normally two in the case of IUCN and one in the case of ICOMOS. As this used to be the only personal encounter in the evaluation process until recently, the treaty states and site representatives have set great store in pleasing the emissaries. No matter how often the limited mandate is being emphasized, an ICOMOS office-holder told me, the red carpet is still being rolled out: the experts are shown the property in its Sunday best, sometimes cleared from all other visitors, and are treated to stimulating encounters with a whole range of specialists. Then it is often impossible not to fall in love with the site, the office-holder said, and mission experts can hardly resist the temptation of becoming advocates, giving advice to their hosts (even in press interviews, which ICOMOS asks them to avoid), commenting on OUV or even breaking the rules by sharing their report with the hosts. That report should instead restrict itself to the aforementioned practical matters and go to the Advisor in charge, never to be published. While remuneration is rather modest (cf. also Sheppard and Wijesuriya 2018: 58), going on evaluation missions is eagerly sought, with many ICOMOS members ready to sell their first-born child for it, as the office-holder joked, and bitterly complaining when they are not picked.

State Party grievances often centre on the mission, with disappointing evaluation outcomes being linked to the dispatched expert's lack of preparation, understanding or interest, or appearing as entirely mysterious after what appeared as a pleasant encounter.[8] Lack of expertise in the specific kind of property is a common accusation in the case of ICOMOS, and States Parties have pressed for improvements and the use of two experts rather than one in recent sessions. To some degree, I think that they are barking up the wrong tree, focusing on what they see of a process in which in ICOMOS's case, the World Heritage Advisor plays a larger role. One Advisor expressed continuous surprise at the leeway they have in conceptualizing the evaluations, and whether or not a candidate site is then assigned to the colleague with the special knack for making cultural landscapes shine can make a perceptible difference. Also, an ICOMOS official told me that having few desk reviews must not be a problem, as one or two good ones or a single useful sentence in an otherwise weak text can do a lot to clarify the case. This again presupposes selectivity and a large role for the Advisor who is piecing it all together. Compared to the average academic journal editor, the reviewers' and mission experts' input appears to be less binding.

Contrary to an earlier time when a single individual was responsible for all evaluations – like someone playing god, as an IUCN official phrased it – and these were not even put down in writing initially (Cameron and Rössler 2013: 179–81, 185–86, 187–94), the draft evaluations are now screened by an internal 'World Heritage Panel' in both Advisory Bodies in November or December. The ICOMOS World Heritage Advisors present the nominations, much like they would in the Committee session, but have no vote, and of IUCN's World Heritage Programme, only the director is included as the likewise nonvoting panel chair. Mission experts too are not usually part of the Panel in the case of ICOMOS, and IUCN tries to avoid this too. In principle then, these Panels bring in fresh eyes. Both World Heritage Panels reserve around five seats to a number of the larger organization's subunits. In the case of ICOMOS, these are its international scientific committees on such topics as rock art, cultural landscapes or historic towns and 'affinity organizations' such as the International Committee for the Conservation of the Industrial Heritage (TICCIH), depending on what is on the agenda. IUCN invites staff from its Global Protected Areas Programme and individuals appointed by its commissions on protected areas and on species survival, for example. Both panels also include experts invited in their personal capacity, five to nine in the case of ICOMOS and up to seven in that of IUCN; the latter recently formalized appointments by introducing fixed terms and an application process. Whereas ICOMOS takes care to spread nationalities for Panel members and mission experts, with most countries represented by only a single individual in each group, IUCN does not give members' nationalities.[9] The biggest difference between the two is in involving the organization's larger hierarchy: while this is represented only by a single, nonvoting appointee of the Director General in the case of IUCN, the ICOMOS panel includes half-a-dozen or more voting members of the elected officials from the ICOMOS Board, usually including at least three or four of the key office-holders (president, the five vice-presidents, treasurer and secretary general) who most often also chair the proceedings.

The ICOMOS World Heritage Panel can spend an entire day on a single bid, I was told. However, ICOMOS, with its far larger number of nominations, is usually more pressed for time, ideally moving on after 45 minutes, but nonetheless forced to continue its work past midnight on some days. An ICOMOS World Heritage Advisor told me that the Panel can be daunting, with members having no qualms about criticizing and radically revising the Advisors' drafts, so that an evaluation recommending deferral may end up advocating inscription. In quite a few cases, the Panel session generates additional questions, which are sent to the States Parties, giving them one month for a reply. With this input, a second session of the World Heritage Panels in March or April – partly virtual in IUCN's case – finalizes the texts. At this stage, the Panel also looks at resubmitted nominations referred in earlier Committee sessions – given that these are supposed to require only minor additions and amendments, they can be submitted a year later than the new nominations. The evaluations are then put online six weeks ahead of the Committee session, usually by May or June, and include recommended decisions for every bid. The contents prompt some States Parties to write factual error letters, thus completing the input that the Committee receives for its decisions. Missions cost about US$25,000 to 30,000 per candidate property and Advisory Body, and just like the monitoring missions the Committee requests (see below), they are paid from the World Heritage Fund. While costs are modest compared to nomination expenses, having nominating countries pay them has been proposed at times, but never adopted.

All contributors to the evaluations have to pledge impartiality and confidentiality. However, participants in both World Heritage Panels found illicit influence on outcomes conceivable and one ICOMOS informant spoke of active lobbying similar to the Committee sessions proper. Panel members must leave the room when their national candidates are up, but are not completely blocked from the meeting as such, a fact that several informants found problematic. There are also no abstention clauses so that Panel members can – and many do – join their national delegation in the following Committee session, even when the country is currently on the Committee. They may thus end up being judge and jury, and all of them must fight the temptation to pass on information to their national delegation – whose public service is often their employer – and to other friends and allies. An Australian delegate claimed that his compatriots on the ICOMOS Panel had never leaked anything to him; nevertheless, the insider knowledge that other delegations shared with him usually turned out to be correct.

Whatever the extent of such breaches, they do not protect active nominators such as France, Germany and Italy – countries with many ICOMOS members and one or two nationals on every ICOMOS World Heritage Panel – from negative judgements. As already mentioned, there is also no statistical evidence for Advisory Body bias in favour of Committee member bids. I was told of highly suspect cases in the more remote past, with recommendations turned around spontaneously in the Committee session.[10] However, ICOMOS officials believably argue that such things can no longer happen, given that their session delegation is now bound by the written recommendations. Yet, for the recent verdicts too, there are cases where other than by lobbying and the influence of personal connections, I have difficulty explaining complete turnarounds in subsequent evaluations of the same candidate property and glaring discrepancies between critical substantive content and friendly conclusion (a point to which I shall return). For another potential conflict of interests, some ICOMOS World Heritage Advisors work as consultants whom interested States Parties can employ for preparing nominations – an added motivation to take on a position whose financial compensation does not match the required qualifications and workload. Such Advisors must also leave the room when their properties are up, but it is difficult to imagine that all other Panel members manage to shut out this aspect completely when passing judgement.

As already mentioned, complaints about the Advisory Bodies have led to recent changes, allowing States Parties further input into the evaluation process. During the first World Heritage Panel meeting, nominating countries can now meet Panel members for feedback and questions. Also and on top of the additional questions addressed to some countries, all of them receive an interim report by the end of January, that is, about half a year ahead of the session, over which they can again consult the Advisory Bodies. This has led to hopeless bids being withdrawn at that stage rather than later and fewer noninscription recommendations being challenged at the session. It still strikes me that ICOMOS has recommended 50 per cent of all cultural nominations (70 out of 140) for inscription in the five sessions since these options were introduced in 2015 – in the five preceding sessions, that proportion was only 35 per cent (51 out of 145; cf. Table 4.2). While this may reflect genuine last-minute improvements and perhaps some resigned accommodation to post-Brasilia Committee mores on the experts' part, I still think that increased interaction with States Parties and the rapport this builds cannot help but encourage a more lenient position.

Another double bind arises from increasing efforts to draw the Advisory Bodies and the World Heritage Centre into the preparation of nominations. Dissatisfaction with unsuccessful bids led to the instalment of the so-called 'Upstream Process' in 2010, where ten nomination pilot projects – two from each world region (i.e. the electoral groups, with the two European groups treated as a single region) – were chosen for special expert support and advice from an early stage.[11] The idea as such has been widely hailed as an improvement. Four of these projects have since led to inscription.[12] This output that has been considered a less than staggering result in most Committee discussions, but any further increase might come at the cost of evaluation standards. Still, an 'Upstream Process Request Format' for help in updating the Tentative List or preparing concrete nomination projects has recently been added to the World Heritage Centre's regular portfolio.[13] The nomination projects for the so far largest cultural routes, 'Qhapaq Ñan, Andean Road System' in six South American countries and the Silk Roads, which involved a dozen states, have been under parallel preparation for over a decade, with one Centre official acting as coordinator for each and extensive input from ICOMOS. Arena participants qualified them as 'too big to fail', too symbolically charged, extensively celebrated in UNESCO outreach and bound up with ICOMOS's own advice so that ICOMOS could not recommend anything else than inscription for Qhapaq Ñan and for at least one of the first two Silk Roads segments, 'Silk Roads: The Routes Network of Chang'an-Tianshan Corridor' (China, Kyrgyzstan and Kazakhstan), when the three were officially nominated for the 2014 session. For more ICOMOS involvement in nomination processes, the post-Brasilia period has seen impatient States Parties go for a 'referral with mission' rather than the deferral recommended by ICOMOS and IUCN. By rights, a referral is only for minor amendments and signals the presence of OUV, so there is no need for a new evaluation and evaluation mission, and the candidate can already be resubmitted to the following session. Yet since the Advisory Bodies' deferral recommendation usually implies the need for major work, their advice is still being sought. Therefore, in some cases, the Committee recommends and the State Party pays for an 'advisory' or 'consultative' rather than an evaluation mission. This practice is criticized as undermining standard procedure and also by the Advisory Bodies themselves, but they have played along at least a number of times so as to add at least some improvement to what they cannot stop.

There are thus numerous regards in which evaluations are not the detached process envisioned and where in the interest of impartial judgement, contact between the treaty states and those called to assess their nominations is minimized. Rather, such contact has expanded recently, with States Parties pressing for earlier information and opportunities to respond, usually in the name of transparency and dialogue. Yet transparency also means exposure for the contributors and is therefore considered deeply problematic by the Advisory Bodies, which see themselves under pressure. Also, it leads to perceived conflicts of interests, as numerous Committee discussions of this aspect demonstrate. ICOMOS and IUCN keep stressing that they cannot be authors and judges of nomination dossiers at the same time. And while some delegates of Committee member states brush this concern aside, arguing that the Committee is not bound by Advisory Body recommendations, others are indeed worried that the Upstream Process might produce bias in favour of the respective properties.

Monitoring

Compared to the treatment of nominations, which is the subject of much attention and concern, leading to detailed explanations by the Advisory Bodies and regular modifications of the process, the procedures for the already listed properties arouse less controversy overall and appear more loosely regulated. For most properties, treaty state obligations are already met by submitting the periodic report demanded every six years, and if all is considered well, the site might never return to the Committee's agenda after inscription. However, sites in a potentially problematic condition are subjected to what is summarily called 'reactive monitoring'. In fact, the Committee and its auxiliary bodies are proactive here, rather than responding to State Party initiatives as with the nominations – in the vast majority of cases, the World Heritage Centre and the Advisory Bodies cast the first stone. With their worldwide networks of members and registered professionals and with the Centre's website being the first internet search hit for 'World Heritage', they fulfil a watchdog function, receiving a steady stream of alerts, and sometimes it takes a Centre official's accidental reading about high-rise plans in Vienna in the newspaper to kick off the process. The Centre is held to check the accuracy of worrying information, with the World Heritage 'focal point' determined by the treaty state. For anything that might affect OUV, such as major construction and infrastructural works, States Parties are obliged to contact the Centre on their own initiative and well in advance in any case, and some indeed do. The Centre and the Advisory Bodies claim,[14] and I also heard from others, that in numerous cases, consultations and the exchange of materials and sometimes a State Party-paid visit already suffice to settle the concerns. Then, the respective property stays off the session agenda and the Committee may never learn of the matter.

However, where problems remain in the eyes of the Centre and the Advisory Bodies, they can include the respective properties under agenda item 7B, 'State of Conservation of World Heritage Properties'. Then, the draft decision usually requests a 'State of Conservation report' from the State Party. With upgrading becoming rampant in recent years, it is now often the Committee itself that asks for such a report at the time of inscription, implicitly acknowledging ICOMOS's and IUCN's counter-arguments. Nevertheless, the vast majority of items under 7B are follow-up cases from previous sessions where the Committee must react to the reports demanded back then and the measures taken in response. Properties currently on the Danger List always require a decision and go first under the separate agenda item 7A, 'State of Conservation of the Properties Inscribed on the List of World Heritage in Danger'. Given the two-year reporting cycle (see below) and the decision rhythm for 7B properties, the figures amount to around 250 properties under State of Conservation procedures at any given point (see Table 6.1), that is, almost one-quarter of List entries. The requested State Party reports are due at a specified deadline around half a year before the session, giving the Centre and the Advisory Bodies time to evaluate them and prepare their draft decisions. However, in many cases, reports are late or do not arrive at all, delaying the completion of the drafts.

Draft texts under these agenda items usually have two parts. The first is a summary of the situation prepared jointly by the Centre and the Advisory Bodies, describing the threats the property faces and any new developments since the last decision; confusingly, this is also called a 'State of Conservation Report' although only these 'SOC reports' are official session documents, not those from the States Parties on which they build. The second part of the

Table 6.1. State of Conservation Reports up for Committee decision in 2015–19.

Committee session	Danger List entries	Other follow-up cases from previous sessions	New cases	*… after new inscription*	*… added by Centre and Advisory Bodies*	Total
2015	46	73	22	*7*	*15*	141
2016	48	82	26	*17*	*9*	156
2017	55	79	20	*11*	*9*	154
2018	54	84	19	*16*	*3*	157
2019	54	91	21	*5*	*16*	166

Note: Cf. WHC-15/39.COM/7, p. 1; WHC/16/40.COM/7, p. 1; WHC/17/41.COM/7, p. 2; WHC/18/42 .COM/7, p. 2; WHC/19/43.COM/7, p. 2. For the prior years, no such statistics were prepared.

text is the draft decision proper and it is only the latter that is adopted by the Committee, either unchanged or in an amended form. The decision consists of from around half-a-dozen to well over a dozen numbered paragraphs, all with the initial 'The World Heritage Committee' as the grammatical subject. The first paragraphs give the context, with the Committee 'having examined' the draft decision and 'recalling' the relevant past decisions, the number codes of which are then enumerated. The text then moves on to assessing recent developments, with the Committee 'regretting' unapproved measures, 'noting with concern' persisting threats, 'acknowledging' or 'welcoming' corrective action taken and so on, with the respective verb – of which there are many customary variants – underlined. Then, attention turns to the future, with the Committee 'requesting' or 'urging' specific measures such as the provision of information, more dialogue with World Heritage authorities, a construction stop, protective action and the like. In serious cases, the Committee 'requests' the State Party to invite what is then called a 'monitoring' mission by the Centre and the concerned Advisory Body to the property. The text closes with the next formal steps, 'requesting' an updated report from the State Party for a specific future session – one year ahead for most Danger-Listed properties and two years ahead for most others since 2011, as this leaves more time for implementation than an annual rhythm.[15] If the property is in serious trouble, the Committee can 'inscribe' it on the Danger List; for all properties already placed there in previous sessions, it either 'retains' or 'removes' that entry. A further customary option is mentioning Danger Listing or even deletion from the World Heritage List as a possibility for the next session if no progress has been achieved by that time. This is a nonbinding and largely symbolic declaration of intent, but is nonetheless considered a substantial warning, marked by being printed in bold, and, as already described, it often unleashes the same strenuous resistance as the actual Danger Listing or deletion. By contrast, when conditions of a threatened property have sufficiently improved, the draft text requests no follow-up and, if adopted by the Committee, makes the property rejoin the ranks of those that the Committee honours by *not* taking them up.

Monitoring missions are not just ICOMOS's and IUCN's preserve, but also include a World Heritage Centre official in most cases. The justification is that in what are often tense constellations in a process that might arouse little enthusiasm in the first place, the UNESCO bureaucrat can manage interaction with State Party representatives so that the Advisory Body expert is free to focus on the property and its problems. The Centre has currently almost thirty permanent posts for UNESCO international civil servants and an least equal number of secondments from States Parties and temporary consultants, with a director, a deputy director and currently seven 'units' – one each for the five world regions, one for natural properties and one for policy and organizing the meetings, each with a chief – and a small 'nomination team'. While part of this personnel rotates in and out from and to other UNESCO units, there is a core of veterans, and the current director has worked at the Centre since its start in 1992. A large part of Centre staff attend the Committee sessions and the General Assemblies where their work for introducing the 'SOCs' and other reports and for finalizing the decision texts is indispensable. Many have considerable knowledge about World Heritage properties and their histories with the Committee and some were also trained in related disciplines, meaning that they are specialists in their own right, although mostly of cultural properties; there have rarely been more than two or three for natural sites. As they are usually more experienced with mission situations – with a former director going on as many as eleven a year, as I was told – some rivalry with the Advisory Body expert may arise. However, the mission report is a joint one and usually it is put online as a document ahead of the session.

The World Heritage Centre is on UNESCO headquarters precincts, in a separate building that, while much smaller than the main building, still underlines the special status of the Convention. This setting means that the director in particular is part of the diplomatic microcosm and its events and receptions, and can, for example, be found filling in the US ambassador about the political pitfalls of an upcoming Committee session in Cablegate.[16] Rank-and-file staff can stay aloof, but member state diplomats walk in and out of the Centre rather routinely to raise their concerns. The Advisory Bodies are less easily approached and this may be a reason why, in several ICOMOS representatives' assessments, the Centre often tends to be on the soft rather than the strict side whenever they disagree over State of Conservation decisions. However, most of the time, the ICOMOS texts are just cut and pasted unchanged into the draft decision finalized by the Centre, and almost always in the observed sessions, Centre and Advisory Bodies were pulling in the same direction. IUCN and in particular ICOMOS are more often targeted by States Parties, but this is because the Centre is not involved with evaluating nominations. By contrast, when national delegates are not happy with what is suggested for the already inscribed sites, they very much bundle Centre and Advisory Bodies into a single opponent.

The majority of postponed or withdrawn candidate sites are revised and resubmitted for a subsequent session, and may then go through a second evaluation and inspection mission, making for a recursive dynamic in which States Parties respond to the initial outcome and the Committee and the Advisory Bodies then respond to the response. When listed properties stay under reactive monitoring or on the Danger List for a decade or more, this cycle has many more iterations and the generally low turnover already indicates that many sites with issues occupy the Committee for a long time (cf. Table 6.1). This means that proposed decisions become more strategic: not only are they negotiated between the Centre and the Advisory Bodies, they also build on the experience with the State Party and site managers in prior communication, the sometimes long string of monitoring missions, reactions to prior decisions and advice, and the strategic needs of the State Party, which, while most often striving for leniency, may on the contrary prefer stern wording when struggling with domestic opponents such as recalcitrant provinces and municipalities (cf. Chapter 3). I therefore think that it is even more difficult to be consistent across cases than with nomination recommendations and see this confirmed in the actual draft decisions, a point to which I shall return. In addition, there is on the whole surprisingly little resistance to the process as such; on the contrary, State of Conservation procedures are universally praised as being at the heart of the Convention and the Committee's most essential purpose, something also underlined by their repositioning before rather than after the new nominations in 2003. I also cannot recall challenges to the right of the Centre and the Advisory Bodies to put a property on the session agenda. If treaty states found this objectionable, they could make this conditional on the consent of the chair who is in office all year or the UNESCO Director-General, the depositary of the Convention, who has the right to initiate emergency procedures. So far, however, this has not been proposed. Partly at least, this reflects everyone's increasing certainty that in the Committee sessions, there are multiple ways to turn unwanted decisions into something more in line with the national agenda.

The Committee

With as many as 1,907 participants from 127 countries registered in Bonn in 2015, World Heritage Committee sessions are a lot of things at the same time. Outside the plenary session, dozens of side-events and receptions are being organized, often in parallel slots, alerting everyone to the importance of specific topics, initiatives, organizations and concerns, such as community participation in World Heritage sites, the Centre's partnerships with business corporations, the heritage of space travel, a specific university's World Heritage studies programme or a national candidacy for the Committee elections. By sponsoring all of this, the respective States Parties and other players demonstrate their ambitions, commitment and generosity. Well over one hundred academics, graduate students, consultants and NGO representatives in the heritage field update themselves about the new developments, showing their faces and sometimes advertising their services alongside. Friends, acquaintances and lovers reconvene, some of them only on this occasion. Venues, food, drinks and cultural performances are often stimulating, and in addition to the excursion on the break day used for finalizing the decision texts, many participants add extra vacation days in a country they may be visiting for the first time. Sociality is greatly accelerated and people talk endlessly – someone always talks, even when the formal proceedings enforce silence from all others. But first and foremost, Committee sessions are the occasions to take World Heritage decisions: almost all of these must be approved in the plenary session that fills about seven hours – sometimes up to ten – of almost all days. As just outlined, the Committee has discussed the Centre's and the Advisory Bodies' operations and the nomination and monitoring processes tirelessly, commissioning a stream of meetings, reports and audits, and introducing innovations, which make them ever more intricate. In comparison, it is striking how little time is spent on self-reflection, with the Committee's own procedures of decision-making remaining largely unexamined. There are complaints about politicization, time pressure, slow progress or rule-bending, but these remain at a general level and little is done to address them by structured measures.

Experienced session participants might disagree. After all, there is a fixed overall structure, with many agenda items continuing with the same name and number since 2005 so that, for example, item 8A is always 'Tentative Lists of States Parties Submitted in Conformity with the Operational Guidelines'. Also since 2003, the Committee has been working on the basis of draft decisions prepared by the World Heritage Centre and the Advisory Bodies, and while the Committee may freely amend these, the outcome is still a formal decision text, not just a loose record open to interpretation. The self-imposed Rules of Procedure[17] determine the general order of sessions, the Committee's organs and office-holders and their rights, the conduct of business and the voting and reporting modalities. Further fixtures are set by custom rather than formal rules, such as the seating order on the podium, the time limit for interventions counted down on a large clock visible to everyone – three minutes for Committee members and two minutes for all others – or the typical arrangement for dealing with an agenda item, namely having the Centre and/or the Advisory Bodies introduce the subject matter, debating it among the Committee members and then adopting the decision, either in its entirety or paragraph by paragraph, with more discussion following at this stage if the proposed amendments are contested. The language of decisions is formal, even stilted, and much of the time, many participants are only mildly interested, if their chatting with neighbours and use of electronic devices is any hint, or escape to the foyer for refreshments.

Moreover, a considerable number of draft decisions see only minimal changes or are not discussed at all. Many reports of particular organizations and activities, after being read out or summarized, are often only 'taken note of' or 'commended' by the Committee and do not provoke more than a couple of supportive comments from the members, but no controversial discussion. In particular, a large number of decisions regarding the state of conservation of specific World Heritage properties are adopted without debate. I found this strange in the beginning and so did a Centre official who told me that, nonetheless, this is their preferred option: since their and the Advisory Bodies' drafts are usually on the strict side, adopting them unamended will serve their conservation objectives. Properties on the Danger List and those proposed for that list must be discussed, but for all others, it requires either the Centre and the Advisory Bodies or a State Party to open the debate. Recently, this has been made conditional on two weeks' notice and a stated reason, raising the threshold even further. There is thus a curious discrepancy: the Committee threatened 'Archaeological Areas of Pompei, Herculaneum and Torre Annunziata' (Italy) with a next-year Danger Listing in 2009, but I believe that most session participants did not even notice – other than being read out once in the long list of State of Conservation draft decisions adopted without debate, the property did not feature.

However, my claim of unstructured and unexamined procedures refers to the moment when the Committee members start their debate. Beyond the time limit and the inconsistently enforced restrictions for States Parties speaking on behalf of their own properties, Committee members can say pretty much anything with impunity, and the sequence is only determined by the order in which they ask for the floor. Thus, while the Operational Guidelines prohibit consideration of unexamined last-minute information, nobody interrupts delegates when they bring it in and others when they take it at face value. Conservation experts have conducted site visits, but their assessments are summarized in dry reports and evaluations, which many delegations have not cared to properly read, making it easy for unspecialized diplomats to counter this with interventions describing their own impressions in alleged recent visits, with others declaring themselves persuaded. Everything is supposed to have been thoroughly examined, but when this has been difficult for one reason or another, delegates plead for trusting each other or for inscribing sites, not nomination files. And while delegates can claim to have been won over by State Party assertions or by the supreme value of a candidate property, they do not need to specify their reasons – whatever Committee members consider as legitimate and valid ends up being so.

If this were a law court, opponents might object to a specific point raised as irrelevant or against the rules and the judge might then grant the objection, removing the point from consideration. But this is not what Committee chairs do, and it also presupposes same-level adversaries who dispose of the same procedural options. However, the World Heritage Centre and the Advisory Bodies are not the equals of Committee members. They may only respond to their interventions in the end, not directly, and often must sit in silence through half-a-dozen delegations or more confirming to each other that ICOMOS and IUCN got it all wrong yet again, with the State Party being unfairly treated. When their turn finally comes, they are often the first to emphasize that they only give advice, with decisions being the Committee's prerogative, and Committee members themselves also remind each other that 'States Parties are the bosses!', as a Kenyan delegate put it in 2012. Moreover, the Advisory Bodies also stick to a studiously polite, measured and factual tone even when State Party speakers wax emotional and aggressive. The two British veterans in ICOMOS's and IUCN's teams in particular struck me as excelling in the art of verbally cushioning their retorts to allegations they clearly considered unfounded or absurd. There is also no habit of countering claims with facts: one might think of calling mission experts to the witness stand or putting the photos showing mining preparations in the Komi Forests denied by Russia on the screens, instead of the Centre merely verbally claiming to have such evidence (in the 2012 session). And when ICOMOS and Egypt were at odds about the existence of a problematic building project at 'Abu Mena' in the same session, a cameraperson at the property and a live broadcast could have clarified the matter. If the Committee were a law court, one would certainly expect it to make stronger efforts to establish certainty and to honour the results of the elaborate fact-finding procedure commissioned by it. However, in this respect, it leans more towards a parliament, with expert input receding to the background once the elected members start their debate.

To be sure, there are lots of possible sources for digression and confusion in a session routine that might be more suited for a small number of decisions with a large share of already well-prepared and prenegotiated items, such as in the World Heritage General Assemblies or in the UNESCO General Conferences. But despite as many as 290 decisions on its agenda, the Committee still takes these up one by one in plenary rather than breaking out into smaller groups and dividing the task. Delegates speak one at a time and about anything on their mind, and when half an hour or more passes before a speaker's turn is up, it is inevitable that the discussion fragments into several parallel strands, with arguments half-forgotten when the riposte finally arrives. Prior interventions sometimes do not register on the following speakers and simple misunderstandings persist, since those able to clarify them are still waiting for their turn. Debate may slow down to a snail's pace on matters large and small, and the Committee may start to go in circles at any time about the details of wording – when finally taking up Timbuktu on the Saint Petersburg session day, for example, considerable time was spent on whether 'destruction', 'damage' or 'deterioration' accurately captured what was happening at the monuments and whether sending a monitoring mission 'at the earliest possible time' meant that it would have to depart right away, irrespective of the security situation (for a parallel, see Hafstein (2018: 76)). When proposed amendments abound, they are usually displayed in track-changes mode on the screens for everyone's orientation, to be continuously updated by the typists (see Figure 6.1). But this may encourage further confusion over what has already been approved, modified, deleted or just scrolled out of sight, which paragraph is up when the automatic numbering has changed, why the English version in the left column does not match the French version in the right column and so on. The legal advisor is consulted frequently, even on elementary questions, and the UNESCO officials chosen for this task vary greatly in terms of the lucidity of their explanations. As we saw with Schwetzingen (see Chapter 1), significantly more time than in other formal proceedings is spent on clarifying the rules of the game and on getting on the same page in terms of the currently discussed point, rather than on substantive debate proper, and quite a few participants find this tiring and annoying.

The key figure for controlling this, everyone agrees, is the chairperson of the session or the vice-chair pitching in when – as in Brasilia 2010 or in Paris 2011 – the chair is absent much of the time. Those who in the past managed to keep the time, apply the rules and order the debate acquire almost legendary status in participants' recollections. The chairs I witnessed did not rise to the same level of general appreciation, with particularly the one in Seville 2009 irking many by her lose time-keeping and her tolerance of repetitive interventions – night sessions and an unprecedented ninety-one hours of

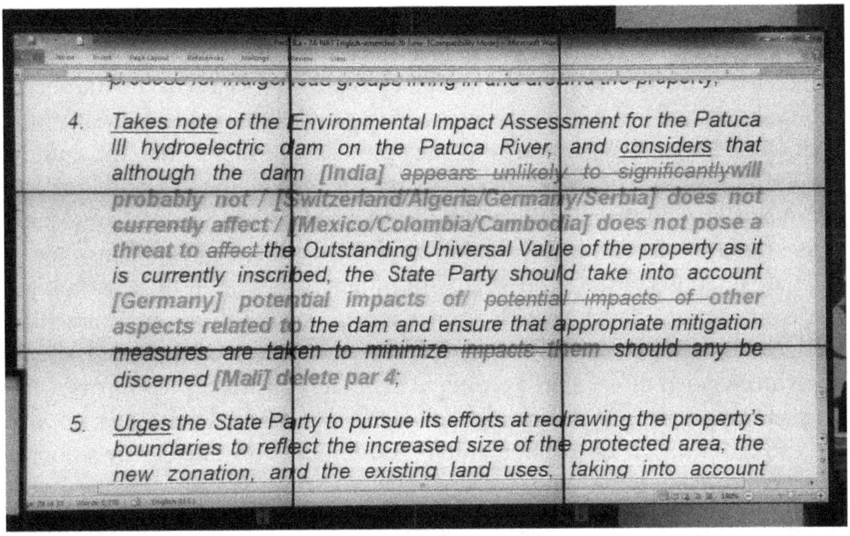

Figure 6.1. Micro-editing decision texts live on screen, Saint Petersburg, 2012. © Christoph Brumann.

plenary time were the result. However, there is little that delegates can do about this – open challenges to a chairperson's session conduct appear to be out of the question. Chairs did actively try to speed up matters on occasion, such as by suggesting to withhold repetitive or congratulatory statements, identifying a majority opinion within the interventions in order to speed up things or asking three or four delegations to form a drafting group when consensus was particularly elusive. Yet some past chairs were described to me as more active in this regard and I hardly ever saw anyone go beyond these steps. Such measures as identifying distinct issues when the debate is getting confused and then settling these one after the other, inviting delegates who wish to comment on the last intervention rather than on a different point to jump the queue, or proceeding from the key question towards the more peripheral ones – common strategies in other public discussions and working meetings I have attended – were never tried. In the state of conservation decisions, for example, the crucial element of Danger List inscription or the threat thereof features only at the end, but chairpersons would still work their way from the first paragraphs most of the time, even when their substance depended on what was still to come further down.

There are also no attempts to structure items beyond the general agenda, particularly not within the largest portions of Committee work, items 7A (state of conservation of properties on the Danger List), 7B (state of conservation of other properties on the World Heritage List) and 8B (nominations), which together take up almost a week. Instead, each property is a separate numbered sub-item (such as '7B.135') in what is a long parade of individual sites. Within each larger item, the natural properties precede the mixed and the cultural properties, and each of these blocks is further divided by the five world regions. Yet this still produces an accidental order where any links between subsequent properties are only accidental. Taking up those with similar problems (say, all the state of conservation properties with mining issues or all those with high-rise construction plans) or with similar recommended decisions (immediate Danger Listing, conditional Danger Listing next year, recommended referral or recommended deferral) as a set might facilitate consistency across cases, but no such proposals were ever made. Following the standard pattern in popular representation, such as in photographic exhibitions, on the Centre webpage or in coffee-table books and TV documentaries, the World Heritage List is instead broken up into its constituents – the single entries – with their mutual connections becoming invisible.

In this regard, the Committee swings towards the law court again, hearing cases as they arise rather than, say, postponing the robberies since this month, fraud cases are up. However, courts can usually rely on a detailed legal code, whereas in the Committee sessions, citing specific parts of the Operational Guidelines for or against a particular decision almost never occurs, given that these rules lack the required detail. One might expect this to be replaced by arguing with precedents, as legal professionals in a common law framework would do, often with considerable sophistication and tenacity in locating the cases that fit their purpose. Nevertheless, doing so is surprisingly rare in the Committee, and when it happens more than occasionally – such as with the Lebanese delegate who repeatedly brought up parallel cases from the preceding days to make his point in the 2015 session – it stands out.

I found the absence of precedents particularly striking in the long discussion about the nomination of 'Darwin's Landscape Laboratory' (United Kingdom) – the house in Kent in which the great naturalist lived his later life and the purposely chosen surrounding landscape in which he made many of his groundbreaking observations – in the 2010 session. Here for once, there had been no sustained lobbying by the State Party so that delegates were free to develop independent positions. The crucial point of contestation was the same as three years earlier when the nomination had first been submitted and then withdrawn: recommending outright rejection, ICOMOS found the house massively restored and therefore inauthentic. The property neither explained Darwin's theories nor carried their physical imprint, and it also had not become a research centre or scientific pilgrimage site where Darwinism had been kept alive. Darwin just lived there once, leaving no traces. However, in over an hour of lively discussion, the French delegate was the only opponent to bring up a parallel case, that of 'Robben Island' (South Africa) – without Nelson Mandela having been imprisoned there for eighteen years, also without leaving physical marks on the property, the island would not have become a symbol of the victory over apartheid, as which it was inscribed in 1999. Yet, nobody thought of referring to 'Wartburg Castle' (Germany) listed in the same year, a nineteenth-century Romantic reconstruction in large part that, without an exiled Martin Luther famously translating the Bible there, would not stand out on a world level and also has few proven marks of Luther's activities. ICOMOS, by contrast, could have pointed to 'Edison National Historic Site', Thomas Edison's residence and laboratory (see Chapter 3), which did not make it onto the List in 1980, even though the physical remains of the inventor's activities are unambiguously present. Ignoring the parallel cases or unaware of them in the first place, the Committee charted its own course.[18] As a countermeasure, the Centre and the Advisory Bodies could think of presenting a couple of reference decisions with parallels to the case at hand. However, this has never been suggested or tried.

It is not that Committee sessions could not use additional structure: time pressure is often deplored and in one session, an average of 14 minutes per decision was reported. If it were not for the approximately one-third of member delegations who, in most of the observed sessions, restricted their input to brief indications of agreement with previous speakers and to interventions they appeared to have been asked for, the agenda would have become even more crammed.[19] Where there was no contention, such as for unchallenged State of Conservation decisions, nominations recommended for inscription, or reports to be noted only, this did not matter much. Proceedings could then appear as well-ordered and reasoned as in the session morning described by Terrill (2014a). For some participants, the lack of drama was almost disappointing: for example, a member of the nomination team for 'Pitons, Cirques and Remparts of Reunion Island' (France) in 2010 told me that, pleased as he was with seeing inscription take place after four years of work, he still found it strange that everything was over in a mere five minutes. However, when members disagreed, the Committee was often bumbling along rather than moving forward on established tracks, and decisions could have an improvised and arbitrary feel even when no member state was strongly invested.

Therefore, nothing could stop the Committee from taking an hour for a confused debate on Schwetzingen with next-to-no practical consequences, which it might have spent on Timbuktu instead. And nothing could prevent the General Assembly of treaty states – which is comparable in its procedures – from spending almost an entire day on the tie-breaking mechanism within the new election procedures adopted in 2009.[20] It is not that participants do not mind messy meetings: a seasoned ICOMOS official found the tie-breaking debate her strangest World Heritage meeting ever, and a North American delegate joked that if he told his superiors how he was spending his time, they would never let him come again. Still, the critics usually single out national self-serving and disregard for expertise as the key problems of the Committee. The conduct of business itself, by contrast, is rarely questioned, and to subject it to the same reflective scrutiny and auditing measures as the secretariat's and the Advisory Bodies' work did not occur to anyone. Beyond the Centre's count of decisions and overrulings presented at the close of the nominations item in some years, no metrics of the Committee's performance are prepared. Focusing on the auxiliary organizations turned attention away from the central body – the more the Centre, ICOMOS and IUCN were pressed to improve their ways, the more liberties the Committee allowed itself. Accustomed to what they witness year after year, delegates have largely lost their capacity to imagine the Committee proceed in a different manner.

However, even the least-invested delegates do not propose dispensing with exchanging arguments entirely: justification for proposed amendments and upgrades is usually given by someone, often Committee members who have been requested to do so. Talking past one another's argument can then become rampant. The 2012 debate about 'Mapungubwe Cultural Landscape' (South Africa) provides a telling case in point: the Committee had inscribed the property in 2003 against ICOMOS advice, with the latter objecting at the time that the submitted map had only shown the boundaries of the property, not those of the buffer zone that was only verbally described in the nomination dossier. Now at long last, the State Party had submitted the map of the buffer zone, but this turned out to be smaller than what had been promised and also conveniently excluded a mining project that reached into the originally described area. To me, ICOMOS's presentation was entirely clear, but South Africa's supporters in the Committee, in no particular order, marshalled the following, partly contradictory objections: the mining project did not reach into the buffer zone (which was true insofar as the latter had ever been formally established), there was a map (which was true for the property, but not for the buffer zone), the State Party curiously knew what ICOMOS did not know, namely the boundaries (which was true if one regarded the buffer zone now submitted by the State Party as already approved, which it was not), the absence of a proper map was a problem (which was true, but this is what ICOMOS had been saying all along), ICOMOS had used the wrong map (which was true if the one just submitted – to which no one had had access – was considered the right map) and the distance between the buffer zone and the project was unclear (which was true, as that zone had never been delineated). Yet ICOMOS's patient clarifications were to no avail. Some of these exchanges looked so staged that no attentive spectator could possibly take them seriously. Still, member states felt compelled to give the appearance of thoughtful consideration and reasoned debate.

A decision-making process with so little that binds the member states does allow for surprising turns, particularly in those cases where nobody has strong stakes in a given issue. But most of all, it is a standing invitation to States Parties to bend it to their own and their friends' advantage. It thus offers little to restrain the turn towards national interests described in Chapter 5. But neither do the basic concepts of the World Heritage arena, as we will see in Chapter 7.

Notes

1. WHC/19/43.COM/8A.Rev, p. 2.
2. http://whc.unesco.org/en/tentativelists (retrieved 27 August 2020). Initially, Tentative Lists were restricted to cultural properties, with natural properties only added from 1996 (Cameron and Rössler 2013: 96).
3. WHC/17/41.COM/18, p. 269. As already mentioned, Israel's contested 'Jerusalem' entry was a major trigger here.
4. WHC-10/34.COM/INF.8B2, p. 9.
5. https://www.icomos.org/en/about-icomos/mission-and-vision/mission-and-vision (retrieved 27 August 2020).
6. WHC-17/41.COM/INF.8B1, p. 5; WHC-18/42.COM/INF.8B1, p. 5.
7. Cf. WHC-2000/CONF.204/4, p. 26.
8. A South Asian Committee delegate complained about mission experts hardly leaving their hotel or being considered as coming from the same world region – a standing selection criterion – despite having lived in Europe for decades. I also heard tales of arrogant and overbearing behaviour, allegations that are all difficult to verify and that I do not assume to be routine occurrences in what will be very diverse encounters.
9. Cf. http://www.icomos.org/images/DOCUMENTS/World_Heritage/ICOMOS_World_Heritage_Panel_2014.pdf; https://www.icomos.org/images/DOCUMENTS/World_Heritage/20160613_ICOMOS_2015_Panel.pdf, https://www.icomos.org/en/home-wh/9074-icomos-world-heritage-panel-2016, https://www.icomos.org/en/home-wh/42724-icomos-world-heritage-panel-2017-2018, https://www.icomos.org/en/home-wh/59209-icomos-world-heritage-panel-2018-2019 for ICOMOS; and WHC/18/42.COM/INF.8B2,pp.3–4andhttps://www.iucn.org/sites/dev/files/iucn_world_heritage_panel_2018-2019.pdf for IUCN (all retrieved 27 August 2020).
10. In the 2006 Committee session, for example, the World Heritage Advisor presenting 'Old Town of Regensburg with Stadtamhof' (Germany) announced that last-minute supplementary material from the State Party had made them drop their reservations about OUV; instead, he claimed that ICOMOS now favoured inscription, rather than deferral as in the written evaluation. The Committee then listed the property. Two informants connected this surprising change of mind with the then ICOMOS President's main employment as head conservator of the federal state of Bavaria in which Regensburg is located and his oft-expressed wish to see the city on the World Heritage List before his retirement (cf. also Schmitt 2011: 203).
11. Cf. WHC-11/35.COM/12C.

12. 'Namib Sand Sea' (Namibia) in 2013, 'Fray Bentos Industrial Landscape' (Uruguay) and 'Rock Art in the Hail Region of Saudi Arabia' in 2015, and a transnational extension to 'Natural and Cultural Heritage of the Ohrid Region' (Macedonia), which now also includes the Albanian part of Lake Ohrid, in 2019.
13. Cf. http://whc.unesco.org/en/upstreamprocess (retrieved 27 August 2020).
14. WHC-14/38.COM/7, p. 1; http://whc.unesco.org/en/reactive-monitoring (retrieved 27 August 2020).
15. Cf. WHC-14/38.COM/7, p. 1
16. https://wikileaks.org/plusd/cables/08PARIS669_a.html (retrieved 27 August 2020).
17. https://whc.unesco.org/en/committee (retrieved 27 August 2020).
18. The nomination was deferred and has not been resubmitted so far.
19. The silent Committee members included Qatar, Iraq, Cambodia, Thailand, Malaysia and Ethiopia, that is, countries with little and/or discontinuous World Heritage engagement who were also emphasizing their learning experience.
20. The already-mentioned revision aimed at including all electoral groups had prompted a general overhaul of the procedure, with the working group proposal requiring an absolute majority for every first voting round only. In second rounds, seats would now go to the candidates in the order of votes even when these failed to reach the previously required absolute majority. None of this was contentious and the only open point was what to do with any candidates tied for the last seat or seats. The proposal suggested to draw lots, but the majority of countries preferred a third voting round. However, would this only include the tied candidates or also the runners-up? One might see this as a marginal question – with more than 190 States Parties each having as many votes as the number of seats to be filled in each round, the probability of such a tie is low in the first place and, short of being either a tied candidate or a runner-up in such a situation, it is impossible to say which of the two options might turn out to one's advantage. It nonetheless took hours to first drop the drawing of lots on Saturday morning and then to realize that no consensus could be reached over the remaining options on Monday morning. Thus, the voting procedure had itself to be subjected to a vote and with that, luck finally left the Assembly completely: the show of hands was close, making a delegate suspect counting mistakes and demand a roll-call vote, and after the laborious exercise of all States Parties present in the room announcing their choice had ended, the count revealed an exact tie, as if to mock the participants. After informal consultations held over the ensuing lunch break, nobody objected to simply adopting the first option (tied candidates only), signalling that most delegations could not care less about the substantive outcome. But they still managed to spend many hours on a matter with next to no practical import.

    In retrospect, two participants suggested to me that this specific confusion was staged, and the irritable Indian ambassador in particular was indeed up in arms early on. However, the objective that one of these participants suspected – preventing the debate on reform measures further down in the agenda – was disproved by the later course of the session, where delegates gave these measures due time and attention, with neither the Indian ambassador nor anyone else trying to suppress anything. I rather think that it all just happened, without a sinister masterplan. A more attentive and proactive chair with more attentive and proactive Centre personnel by their side might have managed to avoid some of the row, simply by reminding delegates of the stakes at issue and the time spent on them. But once such clear-headedness is lost – and it cannot be expected to be present all the time when chairs remain in their seat for dozens of hours – there is little to protect the Committee and the General Assembly from disarray.

Chapter 7

CONCEPTS

Due to the Committee's preferred mode of operation, much of the exuberant textual production described at the outset of Chapter 6 refers to individual cases of World Heritage properties and candidate sites. However, the documents and text also formulate general categories, principles and ideals, and these can, at least to some degree, be considered apart from the procedures just analysed. If we agree to being in the realm of 'heritage belief' (Brumann 2014a) here, this is the theology of the World Heritage arena, consisting of its holy scriptures – the statutory texts – and their exegesis undertaken in manuals or the reports of the expert meetings that the World Heritage bodies and the States Parties keep organizing all year round. This chapter will demonstrate that while this corpus has seen considerable growth and elaboration, it still provides little in terms of certainty: basic categories and concepts have never been clearly and bindingly defined and operationalized. Also, application is inconsistent, not just because of the Committee ignoring the Advisory Bodies, but already at the level of the latter. Yet, most importantly, World Heritage theology evolves in a reactive rather than proactive way, with major clarifications arising only from the dialogue with actual cases of contested candidate sites and properties rather than from general principles. It is therefore not just World Heritage procedures but also the conceptual framework that offers less resistance to vested national interests than might be expected.

I should point out that most of the following discussion refers to the cultural properties. As initially outlined, as a sociocultural anthropologist interested in what my discipline considers an impossible task – establishing universal standards for something as relative as culture – I am particularly drawn to this aspect. Nevertheless, from the first inscription round in 1978, which awarded World Heritage titles to eight cultural but only four natural properties, the large majority of inscriptions have been for cultural sites, and these have been more contested and subjected to reform measures in comparison. The bias is therefore not just mine. However, I do comment on the natural properties too and much of what I describe are general trends. As a second proviso, I focus on the gaps and contradictions of World Heritage theology rather than on admiring its sheer scope and richness. This might appear to be a somewhat malicious exercise to the reader, but my point is that much more energy is invested into expanding the stock of sophisticated views and reflections than bringing consistency into what is already there. I structure my discussion along the basic preconditions for World Heritage inscription, looking at completeness (the aspect referred to as 'integrity'), continuity ('authenticity') and value ('OUV') in turn.

Completeness

As a basic condition required of all World Heritage properties, the Operational Guidelines (§ 88) define integrity as follows:

> Integrity is a measure of the wholeness and intactness of the natural and/or cultural heritage and its attributes. Examining the conditions of integrity, therefore requires assessing the extent to which the property:
>
> a) includes all elements necessary to express its Outstanding Universal Value;
>
> b) is of adequate size to ensure the complete representation of the features and processes which convey the property's significance;
>
> c) suffers from adverse effects of development and/or neglect.

A World Heritage property, in short, should be a unit that makes functional and systemic sense – inscribing just one-half of a cathedral or a volcano is pointless. Also, the property must be intact or 'healthy', as an ICOMOS official once phrased it in an introductory presentation. Enough of it must be left in case its past condition is the main point of interest, and if there are contemporary processes that sustain what is special about it – such as the reproduction of wildlife in a biodiversity hotspot or the agricultural activities behind a landscape of rice terraces – these must be stable and ongoing.[1] As the ICOMOS official also pointed out, integrity and authenticity – the other basic requirement for cultural properties – are not the same thing: reconstructing lost parts of a building will increase its integrity, but, at least from a pre-Nara Document point of view, diminish its authenticity. One could add that the two components of integrity – wholeness and intactness – are not the same thing either: including only the best-preserved buildings of a historic town centre would make the resulting property more intact on average but less whole, a potential trade-off relationship that is rarely systematically explored. So how has the World Heritage arena gone about assessing whether properties are whole and intact?

What has been considered whole enough for the World Heritage List is a very diverse lot. It ranges from the merely 75 square metres of 'Rietveld Schröderhuis (Rietveld Schröder House)' (Netherlands) in Utrecht to 'French Austral Lands and Seas' (France) in the Indian Ocean, which, with 672,969 square km, is nine billion times larger.[2] The 971 properties for which surface areas are given cover a total of 3.7 million square km, more than the territory of India. However, almost one-half of this falls on just the four largest marine properties,[3] and the median is just under 3 square km, a common size for a historical town centre or a free-standing archaeological site, of which dozens feature on the List.

The World Heritage Convention concerns sites, not practices, archival documents or movable artworks, which are the purview of other UNESCO-administered initiatives. The Operational Guidelines also exclude '[n]ominations of immovable heritage which are [*sic*] likely to become movable' (§ 84) and one wonders if anyone thought of the Nubian monuments when drafting this. It is difficult to imagine that a property such as 'Museumsinsel (Museum Island), Berlin' (Germany), a planned architectural ensemble 'inseparable from the important collections that the five museums house',[4] would have made it onto the List if the museums' exhibits were gone, and the fame of many a natural property also depends on migratory species that keep returning to it, but are not present all the time. Yet other than this, immobility applies, sparing the Committee the controversies arising in the 2003 Convention over whether a certain kind of shadow theatre is Turkish or Greek, whether dragon boat festivals are Chinese or South Korean and so on (cf. Aykan 2015; Bortolotto 2016). For all properties but one, the official statistics and the World Heritage Centre website give geographical coordinates,[5] and in all cases except the ones discussed in the previous chapter, their country is uncontested.

As the 150 missing size data indicate, it can be less clear where a World Heritage property ends. '[A]dequately delineated boundaries' (§ 97) are a precondition for listing and these must be demonstrated through detailed maps now, but in the early days, such precision was not strictly enforced. However, in most cases, there is an unambiguous boundary and, interestingly, it is the only thing that is clearly put down, as there are no fixed standards for an inventory of property contents, and the statements of OUV also vary greatly in their specificity.

According to the Operational Guidelines, the boundary of a property should encompass the full area that has OUV. Given the conservation purposes of the Convention, too large dimensions are usually less problematic than too small ones, yet still, 'adequate size' discourages the inclusion of areas lacking OUV. However, separating World Heritage space from ordinary space can be difficult, as an example will illustrate: in 1979 already, the Committee inscribed 'Białowieża Forest' as the largest and best-preserved lowland primary forest of the Central European type. Yet, of the entire forest straddling the border between Poland and the then Soviet Union, the property included only the 50 square km on the Polish side that had national park status.[6] In 1992, an adjoining area of 876 square km across the border in newly independent Belarus was added. However, in 2008, it transpired that the Belarusian authorities had unwittingly limited strict protection of their part to 53 square km fronting the Polish national park, assuming only that bit to be World Heritage. Elsewhere, logging had been permitted, as it also had been outside the Polish national park, so that a first Polish extension proposal in 1999 was rejected by IUCN and the Committee.[7] Yet, in a second attempt in 2014, 545 square km of Polish forest surrounding the original national park were added to the World Heritage property. On that occasion, the boundaries of the Belarusian part were redrawn as well, with one area added but others excluded, yielding a 53 square-km net loss that conversely led to 'much better configured boundaries' according to IUCN's favourable evaluation.[8] There is also a boundary running through the property: a game-proof fence separates the two countries and as this has become a Schengen border in 2004, Committee demands for removing it are unlikely to be fulfilled. Yet, since small and airborne animals as well as plant seeds do not mind the fence, the current generous property dimensions of 1,419 square km, including about 80 per cent of the entire forest and all old-growth sections,[9] help to sustain the untouched 7 per cent (the Polish national park and the Belarusian section that was always strictly protected) and to push back against Polish forestry interests.[10] Still, to argue that value is evenly distributed across this ever-shifting spatial assemblage is difficult. The logged parts of the forest could have been turned into the buffer zone with equal justification, and to claim that there is only one plausible way to draw a boundary here would be far-fetched.

It would also be for other World Heritage sites. Around eighty other properties have seen significant expansions after inscription[11] and many more underwent 'minor boundary modifications' correcting their outlines at some point. Just like Białowieża Forest, many properties also include areas of clearly differential significance. Most people would agree, for example, that the famous Hall of Mirrors contributes more to the OUV of 'Palace and Park of Versailles' than some of the remoter park sections of what are almost 100 square km. However, the World Heritage bodies have not addressed whether OUV can be differentially distributed in space and whether centre-periphery gradients of value might question the idea of a boundary in the first place. It is obvious that for property rights, protective regulations, access and the like, clear delimitations are often a necessity. But reifying them and tacitly presupposing a homogeneous presence of OUV in World Heritage space skirts questions such as those raised over Dresden (see Chapter 3): before deletion, an Australian delegate inquired about the threshold for the irretrievable loss of OUV claimed by ICOMOS. An ICOMOS representative responded that vistas were a key component for a property whose resonance in the arts had been a major reason for inscription. What she did not present is a measure of the visual damage done, either absolutely or in comparative terms, and nor was there an explanation why a bridge not even visible from most of the property could ruin its OUV in its entirety. Of course, the deletion was mainly due to the dynamic of the case and to Germany's lack of responsiveness, yet a clear theory of the distribution of OUV in space to support the step was not put forward.

This lack has become particularly acute in recent debates about 'visual integrity', most of which have concerned modern construction in historical urban environments just like in Dresden. Some of these confrontations led to what were widely perceived as Committee victories, such as the relocation of the Gazprom headquarters skyscraper, Laktha Tower, to a site less visible from 'Historic Centre of Saint Petersburg and Related Groups of Monuments' (Russia) or the cancellation of most of the high-rise developments opposite Cologne Cathedral. However, in cases raised in the sessions I observed, Committee warnings were either ignored or the Committee itself did not take up the Centre's and ICOMOS's proposals for Danger List threats and entries in the first place, such as in the already-mentioned debates over Istanbul, the Tower and Westminster properties in London in 2011 and 'Cathedral, Alcázar and Archivo de Indias in Seville' (Spain) in 2012. Only the Metro Bridge over Istanbul's Golden Horn was somewhat lowered in response. Repeatedly in these debates, Committee delegates denounced 'visual integrity' as a straitjacket, with the Eiffel Tower as the stock example of World Heritage (as part of 'Paris, Banks of the Seine') that, with similar conservatism prevalent at the time, would never have been built. But contested though the concept is, its proponents do little to operationalize it, such as by identifying key vistas and vantage points and assessing how each of these might affected, by constructing proxies,[12] or the like. Instead, Committee debates about visually intrusive construction have veered between blanket assertions of huge damage and equally apodictic ones of an unproblematic or even positive contribution to the historic urban landscape.

However, crucially and unlike Dresden, almost all the controversial developments in the above cases were located outside the property boundaries, and the concerned States Parties and municipalities did not fail to point this out in their defence. In some cases, such as Cologne, a buffer zone that would have covered the development area had been promised earlier but was never installed. However, in Seville, everything was in place. Yet World Heritage does not extend beyond the three famous buildings, from which the Torre Sevilla or Torre Cajasol at a kilometre's distance can only be seen when climbing the rooftops. Had the surrounding old town been included in the property and not only in the buffer zone, there would have been more visual disturbance, but the skyscraper is located, and visually most conspicuous, outside the old town's narrow streets. As also in Cologne, to actually have the new structure spoil World Heritage enjoyment requires the view from afar and from specific angles. However, talk of visual integrity rarely distinguishes these different aspects and sometimes it refers to the view *from* the property rather than on to it, such as when debating the windfarm developments eventually called off in the case of 'Mont-Saint-Michel and Its Bay' (France). Therefore, debates remain case-bound and contingent, with the adversaries seeing what they can get away with in terms of claims rather than referring to agreed standards.[13]

Boundary issues and the distribution of OUV across a property can become even more complicated with serial properties. Despite what 'property' in the singular might suggest, a serial property consists of spatially discrete components, up to as many as 758 in the case of 'Rock Art of the Mediterranean Basin on the Iberian Peninsula' (Spain). Collections of similar sites in greater or lesser proximity under a single World Heritage entry were accepted right from the beginning, with five such properties already inscribed in 1979, so that the real number of discrete World Heritage sites must be closer to 10,000 than to the just over 1,000 official entries.[14] The basic justification for accepting such nominations was that 'the whole series represents more than the sum of its parts', as the 1979 Bureau diagnosed for 'Prehistoric Sites and Decorated Caves of the Vézère Valley' (France) and 'Forts and Castles, Volta, Greater Accra, Central and Western Regions' (Ghana).[15] The Operational Guidelines stipulate that 'it is the series *as a whole* – and not necessarily the individual parts of it – which are of Outstanding Universal Value' (emphasis in original). Included components may thus be unremarkable in their own right as long as they 'contribute to the Outstanding Universal Value of the property as a whole in a … readily defined and discernible way'.[16]

A model case of a sum that is more than its parts is 'Struve Geodetic Arc' (Belarus, Estonia, Finland, Latvia, Lithuania, Moldova, Norway, Russia,

Sweden and Ukraine), a collection of thirty-four out of originally 265 main station points, which stretches over almost 3,000 km from the Black Sea to Hammerfest near the North Cape. Loosely tracking the meridian through Tartu, Estonia, where geographer Friedrich Wilhelm Georg Struve taught, the arc completed in 1855 was crucial for establishing the true size and slightly flattened shape of the earth. While some of the station points are located in buildings or are little monument-like structures in open space, others are not more than simple marks on rocks, and none of them would have a plausible claim to OUV were it not for its role in the entire ensemble. To a degree that is difficult to top, the collection outshines its constituents here and the latter derive their significance from what is elsewhere rather than contained within them. Somewhat in the manner of nations being 'imagined communities' that bond people who never meet (Anderson 1983), series such as this are 'imagined properties', which require compelling narratives to be convincing.[17]

Not all serial World Heritage properties are recognizable as such by their name. 'L'viv – the Ensemble of the Historic Centre' (Ukraine), for example, does not give away its two discrete components, whereas plural forms such as 'Historic Areas of Istanbul' or 'Three Parallel Rivers of Yunnan Protected Areas' are a clearer indicator. In these three cases, the series refers to a lost whole: had the space between the components not been built over or modified greatly in later periods or had it been left as untouched wilderness, a single contiguous property might have been inscribed instead. The serial approach can also be a matter of conservation expediency, such as when choosing only publicly owned buildings or excluding the streets (as in Seville, where the three World Heritage buildings are technically separate components). Still, this leaves a sizeable number of 'truly' serial properties where components never formed a contiguous territory. Their connection can then be functional, such as in the case of 'Mines of Rammelsberg, Historic Town of Goslar and Upper Harz Water Management System' (Germany), where the network of hydraulic installations provided the water power for working the mines and the metals that sustained the town's economy. Interdependence may be that of a central site and its corona, such as 'Sacred Island of Okinoshima and Associated Sites in the Munakata Region' (Japan), where the seven outlying components – adjacent islands, Shinto shrines dedicated to the worship of the island gods and the burial place of the shrine priests – all refer to and depend on the main island in religious terms. But there are also ensembles that were never meant as such, are not particularly interdependent and display no clear hierarchy, such as the mining sites of Wallonia and the decorated farmhouses of Hälsingland on the Saint Petersburg session day (see Chapter 1) or the Meiji industrial sites over which Japan and South Korea fought so intensely (see Chapter 5). Since no kind of link is ruled out a priori, the serial approach is at the forefront of current World Heritage expansion, particularly for countries that have run out of their most obvious single-site candidates by now.

Integrity in a serial property requires the wholeness not just of the components but also of the entire series: the latter must have a plausible link between the components and an adequate set of these that is neither too large nor too small and also consistent, with the included elements being superior to excluded ones (cf. also UNESCO 2011: 67). This raises a second-order boundary problem, in fact the most common source of contention over such nominations, as we already saw: ICOMOS minded the selection of the three Russian kremlins and IUCN that of the thirty-nine components of the Western Ghats in the Saint Petersburg session day, and in subsequent meetings, selection of components was also an issue with 'China Danxia' and 'Camino Real de Tierra Adentro' (see Chapter 4).

However, the Committee inscribed all these series unchanged, as there is usually less contestation over the choice of components than about the suitability of the property as such. This encourages the use of the serial format for bringing World Heritage honours to sites that would not stand on their own, such as in the case of Okinoshima above, where ICOMOS recommended listing only the islands, but the Committee kept the coastal sites in. Particularly suspect in this regard is the post hoc extension of a single site into a series or a small series into a larger one. 'Hadrian's Wall' (United Kingdom), inscribed in 1987 as the most intact part of the border wall of the Roman Empire, for example, was extended in 2005 by adding the Upper German-Raetian Limes in Germany, with the property becoming 'Frontiers of the Roman Empire'. If the Roman border fortifications in their entirety are considered a meaningful whole, the Limes and the further extensions of the property planned by other countries do indeed make the property more complete. However, the Limes features problematic reconstructions and is less intact than Hadrian's Wall, and so is the Antonine Wall in Scotland, the second extension of the property in 2008 and very obviously the United Kingdom's reward for accepting the series approach. Both the Limes and the Antonine Wall were clearly piggybacking on the fame of Hadrian's Wall.

Similarly, 'Darjeeling Himalayan Railway' (India), inscribed in 1999 as the second mountain railway and the first industrial site in Asia, was expanded twice (by adding Nilgiri Mountain Railway in 2005 and Kalka Shimla Railway in 2008) into what thereby became 'Mountain Railways of India'. Only in 2010 did ICOMOS start to object, recommending rejection of a third extension to a series it considered closed, and India withdrew the bid. 'Belfries of Flanders and Wallonia' (Belgium) was already a sizeable collection of thirty such edifices when listed in 1999, but a further twenty-three belfries in northern France and one in Belgium were added in 2005 to what is now 'Belfries of Belgium and France'. ICOMOS approved the extension, but also declared the series closed. However, 'Ancient and Primeval Beech Forests of the Carpathians and Other Regions of Europe' is still open, despite the no less than twelve countries sharing the entry by now. This series was expanded twice, both times against expert advice: IUCN already objected to the first extension of what used to be 'Primeval Beech Forests of the Carpathians' in Slovakia and Ukraine, inscribed in 2007, since it found the five nominated components in Germany too small and too disturbed by human interference in 2011. In 2017, it also rejected the sixty-three components – one as minuscule as 6.5 hectares – submitted by Albania, Austria, Belgium, Bulgaria, Croatia, Italy, Romania, Slovenia and Spain, this time for their lack of selection criteria and of a vision of what else the completed series might contain. Yet, the Committee overruled these concerns both times, contributing to a new record in property multinationalism.

Cases such as these with their obvious motivation to roll out World Heritage fame as broadly as possible were often ridiculed by critical participants, with a single 'Belfries!' saying it all. My interlocutors often suspected inappropriate selection criteria, such as closeness of components to a planned tourist itinerary or the wish to include all provinces or cities in a given region, and in one case, a delegate had heard of illicit payments determining the choice. However, official guidance does not specify what the marginal utility of a component for the wholeness and the intactness of the series has to be and where the cut-off point should be placed.[18] On the contrary, when postponing or rejecting 'Heidelberg Castle and Old Town' (Germany) in 2007, 'Spa of Luhačovice – Area with a Collection of Historic Spa Buildings and Spa-Related Facilities' (Czech Republic) in 2008 and 'Francke Foundations, Halle' in 2016, ICOMOS itself suggested the resubmission of these sites as part of serial nominations for European university towns, thermal baths and progressive educational institutions.[19] Although there is nothing in the statutory texts to support this logic, this advice presupposes that the simple bundling of sites adds value, and States Parties intent on World Heritage fame must take this as an invitation to generate OUV by accumulation.

As to the kind of connections that can justify a series, the Operational Guidelines stipulate that: 'Component parts should reflect cultural, social or functional links over time that provide, where relevant, landscape, ecological, evolutionary or habitat connectivity' (§ 137). In keeping with the reservations about basing OUV on famous individuals (see Chapter 3), ICOMOS has been hesitant to acknowledge a common architect as sufficient for a series of buildings. Accordingly, the first four such cases inscribed between 1984 and 2008 also shared a building type and/or location.[20] However, this reluctance has been dropped, with the 2016 inscription of 'The Architectural Work of Le Corbusier, an Outstanding Contribution to the Modern Movement', a series that spreads over France (ten components),

Switzerland (two), Belgium, Germany, India, Japan and Argentina (each one). With an apartment block, a chapel, a convent, a factory and a museum thrown together with residences, ICOMOS had missed a link beyond the creator in the 2009 and 2011 sessions, recommending deferral. Yet, in 2016, it switched to support, dwelling at length on how the series demonstrates Corbusier's worldwide influence on the Modern Movement.[21] By contrast, in 'Key Works of Modern Architecture by Frank Lloyd Wright' (United States) nominated for the same session, ICOMOS just saw a greatest-works collection and a series 'rather less than the sum of its parts', as selection criteria were unclear and wider influence was insufficiently demonstrated.[22] Yet why exactly the Corbusier series was no best-of list and how, other than by being on a record three continents, it managed to demonstrate influence – something that might instead be found in the works of others – was not stated. The Corbusier series also had not changed a great deal: while six components of the first nomination were dropped, sixteen remained in, including works that ICOMOS had dismissed as secondary or only vaguely related to the architect.[23] And while ICOMOS had suggested the inclusion of works in additional countries, only one of these was added.[24] One therefore wonders about the reason for ICOMOS's turnaround from firm rejection to finding 'a model as to how a series related to the work of an architect might be structured'.[25] ICOMOS may have been bowing to the inevitable here – the lobbying pressure of the French government and the Fondation Le Corbusier had been intense, and without a secret ballot in 2009 and firm Swiss resistance to overriding the ICOMOS recommendation in 2011, the series would long have been listed. But it is also obvious that serial properties move the focus from sites to narratives, as it must have been these, not the hardly modified components, that swayed ICOMOS, as they also did for the Wright series when the latter was re-nominated.[26]

Several of the critics I talked to, including a high-ranking Advisory Body official, argued that such trouble could be avoided by insisting on OUV for each individual component, not just the entire series. ICOMOS itself hinted at particular Le Corbusier and Wright works that might easily make the cut on their own[27] and many of the above-mentioned series do have one or two clear star sites that could also have gone as solo nominations.[28] However, given the affordances of the serial approach for the World Heritage appetites of lesser sites, I am sure that trying to impose limitations here would provoke resistance. Still, I find it striking that no attempt has ever been made to distinguish the functional series such as the Struve Arc and those with spatially close components (the 'lost wholes', so to speak) from the mere collections by association and to then formulate distinct standards for each of these types.

I have already described how cultural routes, canals and railway lines played an important role in realizing the Global Strategy (see Chapter 3).

For such linear properties too, wholeness has been understood in very diverse ways. This already applied for the first two border fortifications in 1987: Hadrian's Wall, while technically a serial property of 193 components, encompassed well over three-quarters of the formerly 118 km of walls so that there was indeed a World Heritage line, with long stretches having only minimal interruptions. Yet, with 'Great Wall' (China), the distance between inscribed sections reaches 1,500 km, and while the two end points of the wall feature among the three components, only around 1 per cent of the 4,000 still-existing kilometres (of originally 6,200) were made World Heritage (cf. Wong 2015). Similarly, while Spain's 'Route of Santiago de Compostela' inscribed in 1993 was not merged with 'Santiago de Compostela (Old Town)' and 'Burgos Cathedral' located on that route, given that the aspects for which these properties had been inscribed went beyond the Camino,[29] it did include the full pilgrimage route with a corridor of a width of 30 metres, thus producing a single contiguous World Heritage space with a length of 738 km.[30] This anticipated the results of the 1994 expert workshop hosted by Spain, where it was agreed that cultural routes require an inclusive approach, with the physical route proper, the infrastructure alongside it and the surrounding landscape all to be considered. However, when France made good on its promise[31] to complement the Spanish part with 'Routes of Santiago de Compostela in France' in 1998, it restricted itself to what was a serial property of sixty-nine individual component sites, most of them churches and only a handful of bridges, lodging places, hospitals and wayside crosses. Not more than 157 km of actual roads – only one-fifth of just one of the four main French routes – were included.[32] Also and despite initial plans,[33] the Spanish and the French parts were never merged.

These divergent views about how dotted a line may become before it ceases to be one persisted over subsequent cultural route inscriptions.[34] They peaked with the colossal cultural routes of the 2014 session: 'Qhapaq Ñan, Andean Road System' included as many as 700 original road kilometres,[35] with the 273 components encompassing all types of road technology (with paved, terraced, walled, raised and sunken segments) and the infrastructure of way stations, inns, bridges, drainages, signposts, barns and storehouses, as well as domestic, ceremonial and administrative buildings.[36] Yet, 'Silk Roads: The Routes Network of Chang'an-Tianshan Corridor' submitted by China, Kyrgyzstan and Kazakhstan contained only fragments. Even though China alone had considered nominating five hundred sites at some point, putting forty-six on its national Tentative List (cf. Williams 2014: 72), only twenty-two of these and a further eleven in the two neighbour countries were eventually submitted to represent a stretch of 8,700 km, with components as far as 800 km apart. Also, the grandiose story of cultural and religious interchange between Zoroastrism, Manichaeism, Nestorian Christianity,

Buddhism and Islam told in the nomination file is strangely out of touch with the actual selection of components, where the Chinese religious sites – temples, caves and pagodas – are exclusively Buddhist.[37] ICOMOS was concerned by the scant representation of road infrastructure where only one posthouse and one watchtower in China had been included.[38] However, what most distinguished this nomination from the Qhapaq Ñan and other cultural routes is the treatment of the actual connection – two mountain passes and a single street segment of a length of 1.3 km in China are all there is.[39] Still, ICOMOS supported inscription, and nobody in the session ventured that the property might not be whole enough for what it was meant to convey. Instead, Committee delegates outdid each other in praise for the historical importance of the Silk Roads and the international cooperation that had led to inscription.[40]

A considerable number of the aforementioned World Heritage serial properties – lines or not – cross national borders. If a joint management framework is in place, there is nothing that prevents such bids; on the contrary, as expressions of transnational friendship, they are encouraged, counted against the nomination quota of only one of the participating countries (which makes them attractive for active nominators such as Germany, France and Italy) and likely to receive special praise. European countries in particular have made an industry of such collective nominations, not just for beech forests, geodetic station points, border walls, belfries and Corbusier works, but also for 'Prehistoric Pile Dwellings around the Alps' (Austria, France, Germany, Italy, Slovenia and Switzerland) or 'Wadden Sea' (Denmark, Germany and the Netherlands). I have been plausibly told that such nominations can lead to rewarding international cooperation. Yet, just as in the single-nation serial properties, giving a boost to the weaker links can be a key motivation. A clear case in point were the 'Viking Age Sites' submitted by Denmark, Germany, Iceland, Latvia and Norway in 2015. In addition to what it found to be a weak overall conception, ICOMOS wondered why two already-inscribed properties in Iceland and Denmark were expected to merge with the series, whereas two other Viking properties in Sweden and Canada were to stay separate and why no further sites in England, Ireland, Russia or Sweden had been considered.[41] The Committee followed the deferral recommendation, with delegates puzzled about a Viking nomination that omitted Sweden. Still, Norway barely managed to restrain its partners, who were all for lobbying Committee members for an inscription upgrade, with Germany – the motor of the nomination – fearing that its own components might not make it otherwise.[42] This confirms that in the absence of a consistent policy how to treat such nominations, going transnational remains, first and foremost, a strategic choice.[43]

There is also no agreed line against using the buffer zone to give a property wholeness. Adopted from nature conservation and generalized to all kinds of World Heritage, establishing such a zone is meant 'to give an added layer of protection to the property'[44] it surrounds, such as by preserving the wider landscape setting, important views or ongoing processes that sustain the property. While initially an alien idea to cultural conservationists and rarely used (Cameron and Rössler 2013: 120), the instrument became an expected feature of World Heritage nominations in the 1990s. Buffer zones have a confusingly hybrid status: not installing one must be justified and regulatory details are to be described in the dossier; a map is to be included; and changing the buffer zone requires a 'minor boundary modification'. Yet still, the buffer zone is distinct from the property and does not have OUV itself.[45] Nevertheless, in the case of serial properties in particular, buffer zones are sometimes used to create the coherence that the property components lack. We already came across an extreme case, 'Rio de Janeiro, Carioca Landscapes between the Mountain and the Sea', on the Saint Petersburg session day. Here, the built city in between the four nominated components of beaches, parks and mountains was turned into the buffer zone, connecting them into a coherent entity. ICOMOS deplored the lack of regulatory detail for the buffer zone, fearing that it might end up 'provid[ing] threats … rather than protection'.[46] However, nowhere did it question the approach as such, although the OUV of the property is supposed to rest on the celebrated interplay between the city and its natural setting, including its echo in the arts. Yet how a property that excludes the actual urban space can express this and be a meaningful 'whole' was never asked, and the brief evaluation section on integrity[47] also ignores this aspect.

Wholeness is a particular challenge for properties that are significant for their associative value, not necessarily for what is present at the sites today. It requires no anthropologist to point out that meanings are not inherent in physical fabric and that a symbolic dimension is common to all heritage, be it mosques, palaces or mining sites. There are nevertheless landscapes and buildings that would be entirely unremarkable were it not for the people, events and ideas associated with them. Therefore, from the outset, the cultural OUV criteria included criterion (vi), the application of which currently requires that a property 'be directly or tangibly associated with events or living traditions, with ideas, or with beliefs, with artistic and literary works of outstanding universal significance'.[48] The sentence has seen numerous revisions, with the initially included 'people' dropped early on (as already mentioned) and with 'living traditions' and 'artistic and literary works' added in the 1990s reform period. However, what is common to all these associated aspects is that they must be known by the visitor in order to be appreciated. What I said about the Struve Arc being an 'imagined' property – given that no single station point captures the significance of the series – applies here too: such sites are significant for what is *absent* from them nowadays, often for good. And when they are 'places of pain and shame' (Logan and Reeves 2008), what once went on there – such as the Holocaust, slavery or nuclear blasts – is meant to never return, neither to these sites of commemoration nor anywhere else. So how can properties defined by absences be complete?

A 2012 expert meeting established and a recent study commissioned by the Committee upheld that in the special case of only this particular criterion, it is the *association* rather than the property itself that must be tested for its 'outstanding universal significance' (note the subtle shift in wording), with a second step then assessing whether the property is linked to that association in a 'direct or tangible' way. Consideration of wholeness and intactness thus shifts to what could be called the physical anchor of the association in the property. This idea remains sufficiently suspicious for application to be restricted: criterion (vi) is the only one to be 'preferably used in conjunction with other criteria'.[49] More than 90 per cent of actual uses of criterion (vi) are in fact in combination with other OUV criteria so that of the integrity of the anchor is not decisive; however, it actually was in the case of the Darwin nomination, where criterion (vi) would have been the only possible route to OUV. Here, ICOMOS recognized the significance of Darwinism (rather than Darwin the person) but found no physical elements that manifested it in the property.[50] Yet, in the case of 'associative cultural landscapes', the aforementioned annex to the Operational Guidelines states that 'material cultural evidence … may be insignificant or even absent',[51] and this reasoning might have been applied to what was nominated as a cultural landscape ('Darwin's Landscape Laboratory'). After all, the scarcity of maroon remains on the Le Morne mountain or of proven Roi Mata traces in his chiefly domain had not inhibited ICOMOS from finding OUV (see Chapter 3).[52]

Absences have reached extremes in the case of 'Hidden Christian Sites in the Nagasaki Region' (Japan) in 2018, a series of ten villages, one castle and one cathedral that, on the recommendation of ICOMOS, the Committee inscribed under criterion (iii), that is, as a 'unique or at least exceptional testimony to a cultural tradition or to a civilization which is living or which has disappeared'. After a first bid for Christian sites in that region had been roundly rejected by ICOMOS in 2015 and then withdrawn by Japan, the new series focused more narrowly on the hidden Christians (*kakure kirishitan*). These were the believers who, after the expulsion of Spanish and Portuguese missionaries and the ban on Christianity in 1614, maintained and transmitted their faith in secret, revealing themselves only in 1865 when the opening of the country brought religious freedom and a new wave of Christian missionaries. However, in the preceding two centuries, leaving *no traces* had been a matter of life and death for their ancestors. Material evidence of hidden Christian faith is therefore restricted to movable artefacts, often things that hide their content below a Buddhist or Shinto surface (such as images of the bodhisattva Kannon with a child, which were read as depictions of the Virgin and Child). Nonetheless, the series is for immobile sites: it includes the villages where the underground Christians lived, the cemeteries where they were buried, the temples and shrines where they were required to demonstrate allegiance to the permitted faiths, just like everyone else, and the sites of the persecution of those found out. With the exception of a few Christian symbols excavated at the castle, none of these sites has revealed any physical traces of the hidden Christians, and the churches built after the ban was lifted, such as the cathedral, are those of other denominations. Most *kakure kirishitan* did join these denominations rather than staying separate, so, if anything, these facilities helped to *end* the cultural tradition. The story of two hundred years of clandestine Christianity is certainly spectacular, but if associations must be carried by tangible material traces, there is little to explain ICOMOS approval of what is largely a blank slate.

For presences too, the threat of their becoming absent always looms and is, of course, the key rationale for heritage conservation. Adequate practical arrangements are a crucial requirement for inscription and I have already reviewed the monitoring mechanisms meant to ascertain that World Heritage properties remain intact. Again, however, how threats are to be measured and exactly which 'serious and specific danger' to a property is grave enough for a Danger List entry has never been clarified. As already described, since the Brasilia session, this list has turned into a voluntary device, with new entries occurring only with tacit or explicit State Party consent. This means that the scope of threats and damages is broad indeed: one finds the 'Ancient City of Aleppo' (Syria) devastated in the Syrian war or the nature reserve 'Manovo-Gounda St. Floris National Park' (Central African Republic), where poaching has annihilated savannah wildlife completely, with no signs of recovery for more than a decade now; yet, one also finds Vienna, where the danger to OUV consists of a couple of high-rise developments outside the property that would be invisible from much of a large and otherwise unthreatened historical town centre. By contrast, one does not find 'Kathmandu Valley', despite the fact that the earthquake of 2015 resulted in unprecedented damage to the historical building stock, with demolitions now rampant and the slow restoration efforts based on methods that ICOMOS considers questionable. The draft decision has foreseen Danger Listing each year, but a Committee majority has kept striking this measure, not wanting to impose anything on a State Party that sees no advantage in Danger Listing, just like it failed to see in the past (see Chapter 3).

A recent audit of World Heritage reactive monitoring, conducted as in almost all cases by arena veterans rather than outsiders, finds that the Danger List has a bad image. It therefore proposes awareness campaigns and the dissemination of success stories in order to bring home that it is a constructive tool for global mobilization rather than a sanction meant to expose States Parties as negligent or incapable (Sheppard and Wijesuriya 2018: 67–72). The fact that properties such as Manovo-Gounda St Floris have lingered on the Danger List for decades without much progress and that the Centre, the Advisory Bodies and the Committee have sometimes moved the goalposts over the course of time – identifying new problems once those that motivated Danger Listing were dealt with – contribute to this sentiment. This has led to countermeasures: a Danger List entry must now be supplemented by a document specifying the 'Desired State of Conservation for the removal of a property from the list of World Heritage in Danger' (DSOCR), which protects the task list from getting revised over time.

It only requires looking at the ease with which session delegates shift between praising Danger Listing as a useful tool (when a State Party asks for it) and condemning it as a demoralizing punishment (when a State Party is opposed to it; cf. also Schmitt (2011: 363–67)) to see that worrying about its image is beside the point. Interpretation of the Danger List as a shameful sanction and 'black mark'[53] was precisely what the domestic opponents of the Cologne and Saint Petersburg high-rises sought in order to discipline the developers, and it clearly contributed to the eventual turnarounds. And while restoring Aleppo might benefit from the 'fully costed action plan' (Sheppard and Wijesuriya 2018: 70) that the audit proposes, abandoning high-rise ventures in Vienna is not a matter of funding. Across all cases, it is obvious that Danger List entries are only taken seriously when a significant proportion of the public and the authorities in charge take them as objective declarations of an emergency situation that puts the title at risk. Yet in such objective terms, arguing that the loss of intactness in Vienna matches that in Aleppo is a challenge.

All this means that both for wholeness (in the case of single-sited, serial, linear and transnational properties and that of buffer zones, visual integrity and tangible traces) and for intactness (in the case of the Danger List threshold), all discussion of integrity has not managed to produce clear standards. If these are expected to arise from the cases, neither the Committee decisions nor the Advisory Body recommendations have ever achieved the level of consistency required for a case law approach. The leeway for States Parties to tailor their properties as it suits them, both for nominations and the conservation regime of already-listed sites, remains large indeed.

Continuity

Indeterminacy also applies with regard to the basic requirement exclusive to cultural properties – that of authenticity. This is the quality of being true, real and genuine, not false or fake, and the word can convey a whole range of additional nuances, such as when applied to people or experiences.[54] In the case of cultural heritage, two different things can compromise authenticity: a Rembrandt painting, for example, can lose authenticity when gradual decay or excessive restorations diminish its original features. And even when in good shape, it can be found to have been inauthentic right from the start because it was done not by the celebrated master, but by someone else in reality, such as the famous 'Man with the Golden Helmet' in Berlin. Since most cultural World Heritage properties are collective or anonymous creations, the latter worry rarely applies. However, the other concern about something remaining what it once was is about continuity rather than authorship. Cultural heritage concerns survivals – things and practices carried forward from pasts whose other constituents have changed or disappeared at some point.

Yet, continuity is not the term used for discussing authenticity in World Heritage statutory texts. Instead, the Operational Guidelines (in § 82) say that:

> properties may be understood to meet the conditions of authenticity if their cultural values (as recognized in the nomination criteria proposed) are truthfully and credibly expressed through a variety of attributes including:
>
> – form and design;
>
> – materials and substance;
>
> – use and function;
>
> – traditions, techniques and management systems;
>
> – location and setting;
>
> – language, and other forms of intangible heritage;
>
> – spirit and feeling; and
>
> – other internal and external factors.

Evidently, the attributes of 'design, material, workmanship, and setting' enumerated in the first Operational Guidelines are considered truthful when these are still close to original conditions rather than having changed a great deal. And with such aspects as spirit and feeling as well, the implicit focus is on what has endured – when 'Stonehenge, Avebury and Associated Sites' (United Kingdom) and other 'power places' attract modern-day New Age adherents, this is unlikely to be given special consideration, but when spiritual practices are assumed to go back centuries, such as for the traditional owners of Tongariro (see Chapter 3), it can justify OUV. In other words, with the more ephemeral authenticity attributes too, continuity counts, and it is primarily this aspect that makes the properties remarkable, not uses, beliefs or feelings of recent vintage (cf. also Khalaf 2019, 2020).[55] Perhaps because all this goes without saying in a heritage context, the World Heritage bodies have stuck to the authenticity term. However, they have never ventured to define it beyond the above passage.

I have already described how the Nara Document on Authenticity of 1994 added to the potential attributes having authenticity, leading to the long list cited above, and how this was meant to expand the possibilities for World Heritage recognition (see Chapter 3). Yet there have been no contrary moves of constraining anything, such as by bringing the attributes into a rank order or specifying separate authenticity conditions for each of the OUV criteria, even though an Australian delegate suggested developing these as early as 1998.[56] Also, while the cited clause asks for 'a variety of' attributes showing OUV, no minimum number is given. Once again, this means that World Heritage policy must be read from actual cases of application. As with integrity, both the ICOMOS evaluations and – since 2007 – the statements of OUV adopted by the Committee contain a section on the authenticity of the property, so there is something to go by.

As already outlined, the Nara Document cleared the path for buildings from nondurable fabric and for vernacular structures (see Chapter 3): relatively recent repairs or the ongoing replacement of materials, often through time-honoured practices, have been characterized as supporting rather than compromising authenticity in subsequent Statements of OUV and while the Nara Document is not explicitly mentioned, its influence is obvious. This holds both for wooden buildings such as East Asian architecture,[57] Eastern European peasant churches[58] or early colonial towns with hybrid architectural features,[59] and for earthen structures in the Middle East and Northern Africa.[60] The new attributes have also facilitated the recognition of cultural landscapes (as cultural properties, these must be authentic too – before integrity was generalized to all properties in 2005, this was all that was asked for). This has been particularly important for the 'organically evolved landscapes' that continue to be re-created today, rather than being mere relicts of past practices, and the 'associative cultural landscapes'. The authenticity of the continuing landscapes is sometimes related to their physical features alone,[61] but in most cases, the Statements of OUV also mention the practices – from agricultural methods to rituals – that sustain these, emphasizing their interplay with the physical landscape. This holds for agricultural landscapes,[62] wine and tequila regions,[63] pastoralist landscapes,[64] sites of coastal resource exploitation[65] and mining areas.[66] In the case of associative cultural landscapes, authenticity is made to rest on the (often indigenous) oral history, ritual practices and traditional knowledge, sometimes in combination with the attributes of built structures,[67] but also independently when tangible traces in the landscape are slight.[68] 'Tombs of Buganda Kings at Kasubi' (Uganda) inscribed in 2001 is a kind of champion here – the nomination manual demonstrates how that property is authentic with respect to *all* the attributes in § 82 above (UNESCO 2011: 61–62).

When browsing the authenticity verdicts, I have trouble detecting a connecting thread: even for properties of the same type such as the European wine landscapes, the Statements of OUV appear to be developed without much reference to one another and the precise line of reasoning tends to be invented anew each time. However, one commonality I do find is the rather benevolent assessment of these categories of heritage: rarely are they rejected for authenticity reasons, and the Statements of OUV tend to content themselves with noticing the presence of traditional cultivation and animal husbandry, a ritual life, oral history and/or practices of incremental restoration, without examining their contemporary scope or whether these features have been continuous over time rather than subject to change or decline. Compared with some of the examinations of stone and steel monuments, there is considerably less critical scrutiny, and pronouncements on authenticity are sometimes as brief and sweeping as for 'Viñales Valley' (Cuba) listed in 1999:

> Viñales Valley is a 'living landscape' with a high degree of authenticity in terms of location and setting, forms and designs, materials and substances, uses and functions, traditions and techniques, and spirit and feeling. It has been able to preserve its specific character, while adapting to modern conditions of life and receiving flows of visitors. The property's attributes thus express its Outstanding Universal Value truthfully and credibly.[69]

The very existence of this traditional agricultural landscape with its scenic karst formations appears to prove its authenticity, regardless of the scope of the mentioned adaptations, and one wonders what would have to happen for this property to ever become inauthentic.

However, the treatment of authenticity is most uneven where one would least expect it, namely for stone structures of the conventional kind, often in Europe. Curiously too, the Nara Document has played a larger discursive role here than for the new kinds of structures and building materials that it was meant to authorize. As already outlined, the initial World Heritage formulations on authenticity derive from the 1964 Venice Charter and its emphasis on original materials and the retention of all historical layers. These principles mark the belated triumph of figures such as William Morris

(1834–96) and John Ruskin (1819–1900), who in their time fought the ideal of 'restoration' that dominated the early stages of public heritage conservation. For the mainly medieval European buildings that were to be saved from long periods of neglect and half-ruination, the restoration approach emphasized the realization of the original builders' presumed intent, both by completing unfinished parts and by removing the additions of later periods. As its champion, Eugène Viollet-le-Duc (1814–79), put it in a widely cited phrase, 'to restore an edifice is not to maintain it, repair or rebuild it, but to re-establish it in a complete state that may never have existed at a particular moment'.[70] Yet, against this Romantic and empathetic stance and no less widely cited, Ruskin claimed that 'it is … no question of expediency or feeling whether we shall preserve the buildings of past times or not. We have no right whatever to touch them. They are not ours. They belong partly to those who built them, and partly to all the generations of mankind who are to follow us' (1920 [1849]: 206).[71] This ideal of minimum intervention ultimately prevailed in European conservation, finding its way into the World Heritage Operational Guidelines, which continue to stipulate today that: 'Reconstruction is acceptable only on the basis of complete and detailed documentation and to no extent on conjecture' (§ 86).

Nevertheless, when Poland nominated 'Historic Centre of Warsaw', this principle was violated already right after opening the World Heritage List. To more than 85 per cent, this is a postwar reconstruction of what the Nazis had deliberately laid to ashes,[72] but after initial hesitation, an ICOMOS expert argued that 'a systematic 20th Century reconstruction [can] be justified for inclusion on grounds, not of Art but of History'.[73] Swayed by the symbolic weight of the rebuild it recognized under criterion (vi),[74] the Committee inscribed Warsaw on the World Heritage List in 1980.[75] However, the Bureau session of the same year hastened to emphasize that 'there can be no question of inscribing in the future other cultural properties that have been reconstructed'[76] (cf. also Cameron 2008: 20–21). And, indeed, ICOMOS rejected a number of reconstructions in the following decade, including a first bid for Dresden.[77] Yet, the Committee ignored ICOMOS concerns and did list 'Rila Monastery' (Bulgaria) and 'Rhodes' (Greece),[78] and ICOMOS itself did not object to inscribing 'Speyer Cathedral' (Germany) in 1981, despite extensive neo-Romanesque restorations, and even saw their influence on the evolution of the principles of restoration as a justification for OUV.[79] The ban on reconstruction was thus imperfectly enforced and with the records being scarce for this period, one can only speculate about the reasons.

Yet, the Nara Document has not boosted consistency. On numerous occasions in the two decades that followed, ICOMOS or the Committee have admonished nonstandard reconstructions and 'over-restoration' of historic town centres,[80] religious sanctuaries,[81] imperial palaces and forts,[82] and archaeological sites,[83] sticking to the Venice Charter strictness embodied in § 86 above without wondering whether the Nara Document might provide a different take on the matter. The most far-reaching case, which led to de facto deletion, is that of the reconstructed Bagrati Cathedral in Kutaisi, part of 'Bagrati Cathedral and Gelati Monastery'. As the Georgian authorities ignored ICOMOS advice on reusing remaining original stone blocks and not implanting new building parts of reinforced concrete in an irreversible manner into the historic fabric, the Committee put the property on the Danger List in 2010 when the plans were announced.[84] However, the rebuild was completed in 2012 and following the insistence of ICOMOS for something it saw as 'altered to such an extent that its authenticity has been irreversibly compromised',[85] Georgia had the cathedral removed through a 'significant boundary modification' three years later, with the formerly serial property reduced to a single-sited one – just 'Gelati Monastery' – now.

In comparison to these almost exclusively non-European properties, ICOMOS and the World Heritage Committee have been significantly more tolerant of reconstructions of European properties. Sometimes, the ICOMOS evaluation simply talks past the fact, in the manner of Speyer Cathedral mentioned above.[86] Nineteenth-century Romantic reconstructions of medieval structures nominated in around 2000 were also downplayed as 'typical of the time'[87] or not problematized at all,[88] and nor did ICOMOS prevent attempts to get around questionable authenticity through a cultural landscape approach in the cases of 'Upper Middle Rhine Valley' (Germany) with its many reconstructed castles and of Dresden Elbe Valley.[89] Even currently ongoing construction is no automatic threat to authenticity: in 2005, the famous La Sagrada Familia Cathedral in Barcelona was added to the already-inscribed serial property, which was renamed as 'Works of Antoni Gaudí' (Spain). The ICOMOS evaluation simply notes the authenticity of the parts of the church completed by Gaudí, but is unconcerned with the contemporary building process, as this follows the architect's plans on the basis of 'scientifically elaborated guidelines'.[90]

For still other European properties inscribed between 1996 and 2008, the Romantic restorations were explicitly valued as testimonies of an important stage in the history of heritage conservation rather than being seen as a challenge to authenticity.[91] And when 'Historic Fortified City of Carcassonne' came back in 1997, Viollet-le-Duc's restorations were qualified as a landmark in conservation history, leading to inscription.[92] It is in support of these properties that ICOMOS evaluations actually brought up the Nara Document[93] and that a Polish delegate, in the acceptance speech for 'Castle of the Teutonic Order in Malbork', welcomed the demise of 'our West European fundamentalism … that has brought us to the definition of authenticity limited exclusively to the material substance; an idea that we have wanted to impose on the other cultural regions of the world'.[94]

In 2005 and as the most contested reconstruction so far, the 'Old Bridge Area of the Old City of Mostar' (Bosnia and Herzegovina) was inscribed on the World Heritage List. The iconic bridge collapsed in 1993, after being deliberately shelled in the Balkan War, and was reconstructed with the support of UNESCO and the World Bank between 2001 and 2004. While ICOMOS had recommended inscription from the outset, the lack of a management plan, uncontrolled building activity in and near the old town, unclear nomination criteria and the fact that bridge completion was still pending led to the postponement of a decision in 1999, 2000 and 2003.[95] In 2005, Committee debated listing what had become a reduced section of the Old City for several hours, with some delegates complaining that the ICOMOS evaluation had overly focused on the bridge and its reconstruction – 'it was not only the Bridge and certainly not the International Community that were being inscribed, but the old town of Mostar', as one of them stressed.[96] Yet, here again, other delegates cited the Nara Document for support, and they also played with metaphorical uses, such as the bridge between religious communities that had now been reinstated.[97] In the end, the Committee struck a compromise between Venice and Nara: the current buildings that were are all of 2000s vintage, the special features of the original architecture and the quality of the reconstruction – which some delegates doubted – were not included in the Statement of OUV, against the advice of ICOMOS, and were only mentioned elsewhere in the decision text. Instead, inscription came to rest solely on criterion (vi) – that is, associative value – despite the official preference for that criterion's combined use.[98] The justification transforms the reconstruction into the very basis of OUV, arguing that it strengthened the Old City's value as a symbol of peaceful multiethnic and multireligious coexistence over centuries, and linking it vaguely with international peace and cooperation[99] (cf. also Cameron 2008: 22–23). The Committee and ICOMOS thus approved another political reconstruction, coming full circle from where it started with Warsaw. Participants' recollections make me assume that for the Committee majority that backed inscription of Mostar, the step of inscribing a symbol of peace and reconciliation *felt* authentic, but that does not make the current bridge any more continuous with the past.

Overall, I do not find more consistency here than in the way in which integrity and completeness have been treated in the World Heritage arena. Following the Nara Document, the list of attributes that can demonstrate the required continuity with the past was extended, and this has had a liberating effect on properties from nondurable building materials and those cultural landscapes where processes sustaining OUV are ongoing, producing incremental change.

Yet conventional stone buildings were treated in contradictory ways even before Nara, with reconstruction being outlawed in principle, but repeatedly tolerated in practice. Also, the decades since Nara have seen a rehabilitation of nineteenth-century Romantic restorations, complete with gestures to the Nara Document that was not meant for such use. In extreme cases, reconstruction itself appears to generate OUV, such as for Warsaw and Mostar and their symbolic weight for healing the wounds of war.

However, all of this remains case-bound and has an improvised feel, with little reference to general principles. Committee decisions and admonitions and ICOMOS evaluations run the gamut from unflinching Venice Charter purism to considerable laissez-faire. ICOMOS and/or the Committee could declare once and for all that Romantic restorations are no less valuable than medieval fabric or could permit reconstruction that enhances authenticity attributes such as use, spirit and feeling and/or has a supreme 'We Shall Overcome' kind of symbolic significance. But they have not done so, nor has it occurred to the States Parties to defend modifications in this way, not even when these are meant to strengthen the very function for which the property is on the List.[100] Integrity and authenticity continue to be mashed up, even in recent ICOMOS evaluations and to the dismay of arena veterans, such as the Japanese who deplored that the difference is still not clear in a working group session in 2012.[101] The potential trade-off relationship between the two – such as a restored Carcassonne losing authenticity as a medieval monument, but gaining integrity as a complete fortress – remains unaddressed. I also see contradictions between the strictness against the Bagrati Cathedral reconstruction or 'over-restoration' outside Europe and the more lenient treatment of European properties, as I cannot explain why the aforementioned first nomination of Dresden's Baroque monuments was rejected as consisting of postwar reconstructions when 'Speicherstadt and Kontorhaus District with Chilehaus' in Hamburg (Germany) passed as authentic in 2015, despite half of the Speicherstadt buildings being postwar reconstructions as well. None of these many positions is intrinsically unreasonable, but to derive a consistent line from how the World Heritage arena has applied them in practice is a challenge.[102]

Value

Certainty is also not easily attained for the third requirement, 'outstanding universal value', or, as it is most often referred to, 'OUV', deploying the technical ring that an acronym conveys. Value was not part of initial World Heritage plans and nor was a celebratory list: a mere register of sites in urgent need of international safeguarding efforts appeared sufficient to many (Titchen 1995: 147–51), also given that consulted experts predicted difficulties in trying to decide which sites might have supreme importance on a global scale (ibid.: 101–2).[103] Despite these reservations, the drafting committee for the Convention settled on keeping a list, and the text mentions 'outstanding universal value' as the precondition for inscription no less than thirteen times.[104] However, the Convention does not define OUV, passing on the task of establishing criteria to the World Heritage Committee (§ 11.2). This body, in the first revision of the Operational Guidelines made in 1977,[105] clarified that the Convention 'is only for a select list of the most outstanding [properties] from an international viewpoint',[106] even though in the Committee session that year, 'the representative of ICOMOS … recognized the difficulty of drafting criteria to be applied to cultural property throughout the world and of translating concepts into words that were meaningful on a universal scale'.[107] Subsequent versions have insisted that 'Outstanding Universal Value means cultural and/or natural significance which is so exceptional as to transcend national boundaries and to be of common importance for present and future generations of all humanity'.[108]

As a consequence and already by 1979, the Committee deferred 'Eidsvoll Building' – the manor where the first Norwegian Constitution was adopted – since it was 'undoubtedly of national importance but the justification of the universal value … should be further developed'.[109] A supreme role for the respective nation has made it into the OUV justifications of World Heritage properties a number of times, such as for Wartburg Castle or 'Royal Hill of Ambohimanga' (Madagascar),[110] but usually in addition to other, more broadly conceived reasons for inscription. 'Universal' may thus be seen as operationalized by both the Operational Guidelines and actual listing practice to some extent.

Turning to what is 'outstanding', this quality must be gleaned from the six criteria for cultural sites and four criteria for natural sites formulated in the Operational Guidelines in 1977.[111] Slightly revised and reordered over the years (see Chapter 3 and above) they currently read as follows:

> The Committee considers a property as having Outstanding Universal Value … if the property meets one or more of the following criteria. Nominated properties shall therefore:
>
> (i) represent a masterpiece of human creative genius;
>
> (ii) exhibit an important interchange of human values, over a span of time or within a cultural area of the world, on developments in architecture or technology, monumental arts, town-planning or landscape design;
>
> (iii) bear a unique or at least exceptional testimony to a cultural tradition or to a civilization which is living or which has disappeared;

> (iv) be an outstanding example of a type of building, architectural or technological ensemble or landscape which illustrates (a) significant stage(s) in human history;
>
> (v) be an outstanding example of a traditional human settlement, land-use, or sea-use which is representative of a culture (or cultures), or human interaction with the environment especially when it has become vulnerable under the impact of irreversible change;
>
> (vi) be directly or tangibly associated with events or living traditions, with ideas, or with beliefs, with artistic and literary works of outstanding universal significance. (The Committee considers that this criterion should preferably be used in conjunction with other criteria);
>
> (vii) contain superlative natural phenomena or areas of exceptional natural beauty and aesthetic importance;
>
> (viii) be outstanding examples representing major stages of earth's history, including the record of life, significant on-going geological processes in the development of landforms, or significant geomorphic or physiographic features;
>
> (ix) be outstanding examples representing significant on-going ecological and biological processes in the evolution and development of terrestrial, fresh water, coastal and marine ecosystems and communities of plants and animals;
>
> (x) contain the most important and significant natural habitats for in-situ conservation of biological diversity, including those containing threatened species of Outstanding Universal Value from the point of view of science or conservation.[112]

It will be immediately apparent that these items, rather than being criteria in the strict sense, refer to substantive types of heritage sites – in particular (iv), (v), (viii), (ix) and (x) – or features of sites that crosscut these types – (i), (ii), (iii), (vi) and (vii). Instead of elaborating the 'O' of OUV, they all use 'outstanding' or other, similar qualifiers themselves – 'masterpiece', 'important', 'unique', 'exceptional', 'superlative' and 'significant' – without operationalizing any of them. Whoever finds 'outstanding universal value' vague will be none the wiser after reading the criteria.

This also dawned on the Committee in the 2000s, making it commission guidelines on OUV definitions and standards from the Advisory Bodies. For the cultural sites, one of the ICOMOS World Heritage Advisors reviewed the past application of the respective criteria and the treatment of specific candidate properties (Jokilehto 2006). However, something resembling an operational definition of OUV is nowhere in evidence in his text. For the natural sites, a team around the Director of IUCN's World Heritage

Programme also took stock of the past application of the natural criteria and the treatment of landmark candidates. Yet while the IUCN authors claim a more systematic approach of determining OUV than for the cultural sites and increasing strictness over time (Badman et al. 2008: 4, 6), I cannot find a precise delimitation of the OUV threshold either. The nomination manual confirms the centrality of OUV and lists common mistakes in demonstrating it in candidate files (UNESCO 2011: 56–58), but how to recognize OUV when one sees it is not explained.

In comparison, the 'thematic studies', which the Advisory Bodies have been conducting for the World Heritage Committee, do more in terms of clarifying OUV and where to find it. IUCN has identified 'gaps' in World Heritage coverage, that is, too few and/or too small-sized properties, for both nearshore and offshore (Abdulla et al. 2013: 29–44) and for terrestrial biogeographic provinces (Bertzky et al. 2013: 19-24). It has also done so for areas of high biodiversity and endemism; the most irreplaceable protected areas for amphibian, bird and mammal species – both for all species and for the threatened ones (Bertzky et al. 2013: 24–52); and, somewhat less elaborately, for terrestrial and marine wilderness areas (Kormos et al. 2017: 27–48). This continues an earlier proactive approach: starting with a list of 219 sites to be considered for World Heritage nomination prepared in 1982 (Cameron and Rössler 2013: 51), IUCN has never hesitated to make suggestions. However, to a considerable extent, this can build on registers, lists, databases and indexes developed outside the immediate World Heritage context, such as the World Database on Protected Areas[113] or the IUCN Red List of threatened species. IUCN staff have also contributed to developing numerical indicators, such as the irreplaceability indexes by which areas protected for their biodiversity can be ranked (Le Saout et al. 2013). Given that in 2013, as many as 156 out of 217 natural properties were on the List in part or exclusively because of their exceptional biodiversity (Bertzky et al. 2013: xi), this goes a long way. For geological properties such as volcanos too, a certain rank order exists and I was repeatedly told that there is generally much more agreement about what is outstanding among nature experts than among culture experts. Accordingly, IUCN even goes as far as giving a final count of natural World Heritage properties. That figure of 300 may not be the last word, one official told me, but it was evident to him that the current List is much closer to containing two-thirds of all conceivable natural World Heritage properties than, say, merely one-fifth.

The thematic studies by ICOMOS,[114] by contrast, appear more specialized, often both thematically and regionally, such as when one of them zooms in on water-related cultural heritage in North Africa and the Middle East. Compared to the 2000s, production has also slowed down in the 2010s, with several years passing without any such publication. Rock art has been taken up most comprehensively, with studies of five different world regions, but there is no systematic identification of gaps. Moreover, even when single chapters are structured like nomination files for the considered site, such as in a study of astronomical and archaeoastronomical heritage, the authors hasten to emphasize that this is in no way meant to prejudice Committee deliberations (Ruggles 2017: iv). A more comprehensive analysis from 2004 does identify a number of topics as missing on the List almost completely, such as pastoralism, or being represented only in a narrow way, such as when almost all modern heritage is located in Europe and the Americas or when religious sites are overwhelmingly Christian (ICOMOS 2004: 41–43). Yet in contrast to IUCN's efforts, something like a bucket list of the most suitable candidate sites for rounding out the cultural part of the List is absent. Also, and in defiance of the general 'avalanche of indicators in global governance' (June and Wedel 2010), the construction of numerical measures or indexes of OUV has never been attempted, and nor has anyone tried to devise proxies, such as the prominence of sites in the literature or on the internet.[115]

On what else, then, have the Advisory Bodies and the World Heritage Committee based their judgements of OUV? Since an intensional or connotative definition that characterizes OUV by its features has never been provided, this leaves an extensional or ostensive definition, that is, by reference to properties on the World Heritage List. Evidently, these could not serve as a yardstick for the very first list inscriptions, but many of the early candidates were the famous and iconic sites that many people would expect to feature on a world list in any event, such as the Galápagos Islands, the Grand Canyon, the historic centre of Rome, Machu Picchu or the Taj Mahal. Soon enough, a solid body of World Heritage properties to relate to had assembled. Thus, in the absence of solid general criteria, proceeding by precedent is conceivable: given that all sites on the World Heritage List are assumed to have OUV, it logically follows that a given candidate property shown to be at least as deserving as one of them must have it too.

The Committee has been encouraging this approach by putting increasing weight on the 'comparative analysis', which must now be part of each nomination file[116] and is addressed in dedicated sections in the ICOMOS and IUCN evaluations. The Operational Guidelines stipulate relatively briefly that this analysis 'shall explain the importance of the nominated property in its national and international context'.[117] The nominations manual further specifies that:

> The purpose of the comparative analysis is to ascertain, first, whether there is scope in the World Heritage List for the inclusion of the nominated property, and second, to demonstrate that there are no comparable properties in the same geo-cultural area (cultural properties) or globally (natural properties)

> with similar values that might be nominated in the future … a conclusion should be drawn as to how it compares with other properties and why, if there are other similar properties, the nominated property should be seen as the best exemplar or representative. (UNESCO 2011: 67–68)

The wording is less than crystal clear and 'scope in the World Heritage List' is undefined, but I interpret this as saying that a candidate property must be shown both to belong on the List, relative to what is already there, and to be superior to other possible candidates of a similar nature.

However, it is surprising how often this kind of reasoning is bypassed or forgotten in the Advisory Body evaluations, particularly those of ICOMOS. If we turn to the candidate properties of the Saint Petersburg session day for a demonstration, IUCN clearly applies the comparative logic in its evaluations: it finds the lakes of Ounianga unequalled as a lake system in a hyper-arid environment and the Western Ghats beating all other continental tropical areas in terms of biodiversity and endemicity. It also argues that while some other nature reserves have higher biodiversity than Sangha Trinational, the size, intactness and remoteness of that property make it special. The Chenjiang fossil site is not just shown to stand out among kindred fossil sites, but also to compare favourably with a particularly similar listed property, which it beats in terms of age.

Comparative reasoning that leads to a pronouncement on OUV is also apparent in several ICOMOS evaluations. For Hälsingland and for Rio, ICOMOS concludes that their special combination of features cannot be found elsewhere in Northern Europe and Switzerland (the scope of comparison for the vernacular houses) or in the entire world (Rio). For Çatalhöyük, ICOMOS notes Bronze Age sites that are left unmentioned in the nomination file, but finds the site to be unique on the basis of ICOMOS's own comparisons, even when these are only briefly outlined. For the Russian kremlins, ICOMOS misses proof of what the three nominated sites add to other already-listed kremlins and how they stand out from the unlisted ones, also applying a clear comparative logic. However, for Schwetzingen, ICOMOS only notes that 'properties with similar values … are already inscribed on the World Heritage List',[118] perhaps implying but not clearly saying that the candidate site is not their match and therefore lacks OUV. For 'Vineyard Landscape of Piedmont', a clear argument how the candidate property is to be ranked in relation to other wine regions is entirely missing, and for 'Mining Sites of Wallonia', it is claimed that while the analysis in the nomination file is incomplete, ICOMOS's own thematic studies *allow* for such a comparison, without its conclusion actually being spelt out. Even the most well-meaning reader will have trouble explaining how exactly ICOMOS establishes OUV by comparison in these two cases. And when browsing other

ICOMOS evaluations over the observed sessions, one cannot help noting a trend: there certainly are those cases where ICOMOS applies the expected comparative logic in a consistent way, but there are also a large number of cases where the 'Comparative analysis' section in the evaluation appears to take the respective section in the nomination file as its subject rather than the candidate site proper. Then, the sites compared and the similarities and differences found in the analysis are summarized and the overlooked sites are mentioned, often leading to a statement about the completeness, or lack thereof, of the presented *analysis*. Yet what all this means for the OUV of the *property* and what precisely grounds the recommended decision is often implied at best, if not left entirely unstated. Also, in a number of cases, the comparative exercise is undermined in the first place: properties are declared entirely incomparable, clearly not just as a compliment about how much they stand out,[119] but as an assertion that nothing similar exists, in clear defiance of the nominations manual, which requires a defined context.[120] It is surprising how often comparison is neglected or even entirely suspended and how rarely this is criticized in the sessions, in a framework where – in the absence of independent indicators – comparison with other sites is the only thing on which OUV can be based.

This leaves the decontextualized assertion of OUV for the nominated site. And, indeed, the Statement of OUV to be formulated for each property has been much emphasized in recent years, given that conservation is difficult to accomplish without an agreed baseline, and has also been demanded retrospectively for those early inscriptions the value of which was never officially verbalized. However, most of these statements are extensive descriptions of the property's history and features rather than systematic argumentations of its OUV. Beyond the use of similar adjectives as in the ten criteria above, such as 'outstanding' or 'unique', it is the sites that are being presented here, not what gives them value.

Perusing subsequent evaluations of candidate properties that were resubmitted, one also notes startling reversals of judgement, particularly on the part of ICOMOS. Some of these suggest lobbying pressure, perhaps at the level of the ICOMOS World Heritage Panel (see Chapter 6), or maybe also the resigned acceptance that the nominator's thirst for inscription cannot be reined in anyhow, whatever ICOMOS proposes. This is suggested by the treatment of the 'Hill Forts of Rajasthan' nominated by India: during the first time round in 2012, ICOMOS was so sure of its judgment that it advised noninscription[121] – terminal rejection if adopted and therefore a declaration of there being not even the slightest doubt about OUV being absent. Yet, after a Committee upgrade to referral in the session, the second evaluation in the following year swung round to advising inscription, despite only minor changes of the resubmitted bid and a similar list of deficiencies in the main text.[122] For a further case, after the Committee upgraded an ICOMOS noninscription recommendation for 'Bolgar Historical and Archaeological Complex' (Russia) to a referral in 2013, unusually confirming its OUV in the decision text, ICOMOS also backed inscription of the resubmitted property the following year.[123]

Eventual inscription is fairly common to begin with: Victoria Reyes counts 1,353 nominated sites between 1978 and 2010 (2014: 52), when the cumulative number of inscribed properties stood at 913.[124] The eventual success rate at that time – before overrulings became rampant – stood at only 55 per cent for natural properties, but at 67 per cent for cultural properties (Reyes 2014: 53–55),[125] despite the fact that IUCN's more proactive guidance regarding suitable candidates should reduce the proportion of hopeless bids. ICOMOS is thus less strict overall. It certainly does not budge in all cases, as we already saw for the Darwin nominations, and I have the impression that the firmest rejections are often reserved for what the organization knows best, namely the European sites such as the already-discussed cases of Schwetzingen and the Viking site series or 'Naumburg Cathedral' (also Germany), which was inscribed at the third attempt in 2018 when ICOMOS still stuck to noninscription, despite a Committee upgrade to referral in 2017 that had confirmed OUV in the decision text.[126] By contrast, non-European submissions for the newer, non-elite categories such as cultural landscapes are more likely to receive deferrals when their OUV appears in question, allowing for a second attempt. But other than this, I am at a loss to explain why some assessments of re-nominations diverge so much from the initial ones when others do not, also given that justifications for the turnarounds are not provided.

ICOMOS could invest more into pre-emptive clarification. As already discussed (see Chapter 3), expert meetings were summoned before deciding over the first railway line or canal property, and convening one was a reason to postpone a decision on the Lake District until such time as the cultural landscape category had been properly conceptualized. However, in other cases such as the Route of Santiago or Brasilia, expert meetings on cultural routes and modern heritage only followed the inscription of the first case, and none of the texts produced in connection with these activities provided very detailed instructions how to deal with future cases, let alone pronouncements on the OUV potential of specific sites.

This trend continues even after all the debates and overrulings of the 2010s. Of the four publications currently hyperlinked on the nominations page of the World Heritage Centre website,[127] two concern sites of memory directly (ICOMOS 2018; International Coalition of Sites of Conscience 2018) and one deals with the associative criterion (vi), which is often used for capturing commemorative aspects (Cameron and Herrmann 2018). The reason for this abundance is that almost a dozen nominations for properties associated with recent violent conflicts are in the pipeline, and since these wars and genocides turned countries and ethnic groups into mortal enemies, such submissions have the potential of breeding division and confrontation rather than peace. These materials provide many useful clarifications; they also duly support an approach that is inclusive of all relevant views and voices. Whether sites of the Rwandan genocide are too recent for the World Heritage List or too much like Auschwitz-Birkenau, which was inscribed in representation of all similar sites (see Chapter 3); whether the conflicts must have a strong global dimension – such as the world wars – to qualify; whether in such cases, the war cemeteries and memorials of just two countries (France and Belgium) can be allowed to go forward as a candidate property when other protagonists and victims of the conflict have not yet joined; or whether properties chiefly presented and interpreted in the mode of human tragedy (such as in that binational nomination project) are preferable to those submitted as symbols of patriotic heroism (such as in the short descriptions for the Tentative List entries of the Stalingrad memorial in Volgograd, Russia, and an Angolan liberation war memorial; cf. ICOMOS (2018: 16, 20–21)) is not stated. The ICOMOS text on evaluations is expressly called a 'discussion paper' only and suggests holding an expert meeting or a whole series of these right on the first pages (ICOMOS 2018: 4). But since some of the candidate sites are already being evaluated, the results of such efforts will arrive too late. Yet again, this means that principal points will only be clarified indirectly, in dialogue with the nominations and under the contingencies of their order of arrival.

In the same vein, the report on criterion (vi) (Cameron and Herrmann 2018) provides the most comprehensive overview of its uses in World Heritage inscriptions I have seen. However, the 'guidance' promised in the title – clear pronouncements about proper/desirable and improper/undesirable uses – is almost entirely absent, except in the final recommendation that 'further guidance should be developed'. Also, while revisions to the nominations manual and other texts are suggested (Cameron and Herrmann 2018: 21), no wording for these is provided, and studying nominations where criterion (vi) was struck by the Committee or that were unsuccessful in the first place did not occur to anyone. The working group (cf. Cameron and Herrmann 2018: 88–89) whose efforts fed into the report consisted almost entirely of World Heritage veterans and included many Advisory Body, Centre and State Party representatives whom I know to be critical of the post-Brasilia turn of the Committee. Yet, seeing their de facto refusal to make World Heritage policy even when expressly invited to do so, one wonders what familiarity with the arena does to one's capacity to imagine it differently.

It is as if the properties are expected to magically reveal their OUV, or lack thereof, if only considered in sufficient depth, and I indeed noticed a belief in inherent OUV even among the more critical minds. When conversations and interviews turned to questionable Committee decisions, such informants invariably began to point to examples such as Chichén Itzá, Yellowstone, the Mezquita of Cordoba or the early List inscriptions in general to show that some sites are indeed outstanding beyond any doubt, whatever else might be washed onto the List by a questionable process. The intensional or ostentative definition is there again, but only for a select part of the World Heritage properties rather than their entirety. And if only the Committee would return to focus on OUV, these critics often continue, all the current problems would simply fall away (rather than start, from having to work with an undefined quality).

The assumption that the 'wow sites', as they are often referred to, will simply blow away everyone who sees them transforms OUV into something that, while resistant to objectification, can nevertheless be intuitively sensed. This was also implied in what an ICOMOS office-holder told me about his visit to Darwin's house. As the second deferral decision of 2010 still allows for a possible return of 'Darwin's Landscape Laboratory' in the future, he decided to put the site on his itinerary during a family holiday in England. And when visiting the house with the little museum, he was relieved to feel that there was no OUV, something he texted immediately to other ICOMOS and IUCN officials who had been involved in processing the nominations. Likewise, a Western European State Party delegate, after a visit to one of the German beech forests already mentioned, would proclaim to me that: 'This is not OUV!'

I have some affection for the fact that people as fully aware of all the contingencies of the World Heritage process as these two informants still believe in a quality that is accessible to direct, unmediated experience. And this is, after all, the promise that the World Heritage title also holds for visitors who likewise hope to be overwhelmed when stepping into something that is on the list of glory. If the myth of immediately perceptible site magic were gone, this might harm the World Heritage endeavour more than the current inconsistencies and suspicions of arbitrariness. After almost forty years of trying to grasp what OUV judgements are based on, the ultimate reliance on sensing rather than proving it may appear regressive, but a value that is so intrinsic and complex as to defy objectification and measurement is perhaps the most outstanding of all.

The difficulties and inconsistencies in devising general quality measures for culture will hardly surprise most anthropologists. How could any claim of universal value be more than a subjective reflection of positioned interests and of particular professional and disciplinary biases, they would say. But even when working from questionable premises, many would – in the age of indicators – probably still expect the World Heritage institutions to describe and then find this elusive 'OUV' in a more systematic, better theorized and less improvised way than they actually do. Without the widespread conviction that at least some World Heritage sites are great beyond question, transcending all cultural bias, and that such greatness is amenable to everyone's experience – even when defying verbal expression or quantification – I think the system could not be sustained. To a certain degree, this may be true for all frameworks of cultural heritage designation, but when comparing the World Heritage system to what Natalie Heinich describes for the French national 'Inventaire' of historical architecture (2009), the mystical element is significantly stronger at the global level. Maybe a UN body is more in need of myths, having so few tools of a more substantive nature to do its work, and maybe the world must be enchanted and hold the promise of greatness to remain bearable?

Going Case by Case

To summarize, it is not just the procedures of World Heritage evaluation, monitoring and decision-making that can be bent to vested interests; standards and concepts also offer little resistance. Authenticity, integrity and, most of all, OUV as the key ingredients of World Heritage have been defined and applied in a less than clear and consistent way, especially for the cultural properties, which make up most of the List. The degree to which ICOMOS in particular improvises in dealing with candidate properties, rather than constructing clear indicators or applying a systematic logic of precedent, is surprising. As a consequence, short of taking the costly step of submitting a full-fledged nomination, reasonable certainty over the prospects of a candidate site is difficult to obtain. This has repeatedly been criticized, and not just by Terrill, who argues that 'surprise is not good system design' (2014b). Leading IUCN and Centre officials too have proposed separating the examination of OUV from that of other requirements, since these – in particular, integrity and the conservation and management status – can be more easily fixed, as they say.[128]

But once a nomination is submitted, there is the weight of the invested time and resources, which makes it difficult for the often large number of people involved to accept a disappointing outcome. Also, while the Advisory Bodies strongly assert the impartiality and scientific validity of their advice and while most participants have only a fragmented sense of past decisions, a dim awareness that ICOMOS and IUCN have reversed their judgements before is fairly widespread among participants. That these then feel encouraged to do whatever it takes to realize desired outcomes, even brushing aside the experts, is not a huge surprise. It is also no surprise that States Parties do not cry out for restrictions, given that this extends their room for lobbying and negotiation. Instead, it should be in the Advisory Bodies' own interest to protect themselves by establishing more specific standards and guidelines. Whether such efforts could be successful in the current mood is a different question, but the thirty years preceding the Brasilia session have also seen little systematization beyond the specifics of the case and what its isolated consideration inspires among its evaluators. To a considerable degree, this means that World Heritage continues to be redefined with every nominated property.

However, there is a final factor to be considered and without which, again, the already considered factors would have had a lesser impact. This is the way in which an uneven world has been represented on an uneven List.

Notes

1. '89. For properties nominated under criteria (i) to (vi), the physical fabric of the property and/or its significant features should be in good condition, and the impact of deterioration processes controlled. A significant proportion of the elements necessary to convey the totality of the value conveyed by the property should be included. Relationships and dynamic functions present in cultural landscapes, historic towns or other living properties essential to their distinctive character should also be maintained. 90. For all properties nominated under criteria (vii)–(x), bio-physical processes and landform features should be relatively intact' (Operational Guidelines, 2019 version, http://whc.unesco.org/en/guidelines (retrieved 3 September 2020)). Paragraphs 92–95 then outline in more detail what integrity should look like for these four natural OUV criteria; for the cultural criteria, examples are said to be 'under development' (marginal comment to § 89).
2. For property statistics, cf. http://whc.unesco.org/en/list/xls/?2018 (retrieved 3 September 2020).
3. All 'marine' properties on the World Heritage List contain islands under national jurisdiction – otherwise, they would not fulfill the nomination conditions.
4. http://whc.unesco.org/en/list/896 (retrieved 3 September 2020).
5. Cf. http://whc.unesco.org/en/list/xls/?2018 (retrieved 3 September 2020). The one exception is 'Western Tien Shan' (Kazakhstan, Kyrgyzstan and Uzbekistan).
6. https://whc.unesco.org/document/155139, p. 7; http://whc.unesco.org/en/soc/1412 (retrieved 3 September 2020).
7. Cf. WHC-99/CONF.209/22, p. 10, 32 COM 7B.20, http://whc.unesco.org/en/soc/642, the State of Conservation reports from the 2000s at http://whc.unesco.org/en/list/33/documents and 29 COM 7B.15 (retrieved 3 September 2020).
8. WHC-14/38.COM/INF.8B2, p. 77.
9. Ibid., p. 79.
10. 'Sanitary cuttings' in the property – purportedly against bark beetle outbreaks and for maintaining roads and fire breaks (cf. the State of Conservation reports from the 2000s to the present at http://whc.unesco.org/en/list/33/documents (retrieved 3 September 2020)) – led to a Danger Listing threat in 2016 (WHC/17/41.COM/18, pp. 76–77).

Sitting on the Committee itself, Poland did not block the decision, suggesting domestic turmoil in which conservation authorities were looking for Committee support (cf. also WHC-14/38.COM/INF.8B2, p. 81).

11. There are eighty-seven properties with an entry for 'secondary date' in the table in http://whc.unesco.org/en/list/xls/?2018 (retrieved 3 September 2020), meaning that they were subjected to evaluation procedures equivalent to those for new candidate sites a second time. In almost all cases, this concerns extensions treated as 'significant boundary modifications', but there are also a small number where only new OUV criteria were added, such as for 'Tongariro National Park' (New Zealand) and 'Uluṟu-Kata Tjuṯa National Park' (Australia).
12. If internet image search hits are taken to reflect the typical visitors' visual experience, for example, one might determine the proportion of images in which the planned construction would actually be visible.
13. Typically, talk of Danger Listing disappeared from the draft decisions of the Centre and ICOMOS once the controversial buildings were completed, despite the visual disturbance persisting.
14. I am not aware of any official count. Statistics for the 2006–17 period mention 467 World Heritage nominations, of which 203 were for serial properties, with a total of 2,832 components. This makes for a grand total of 3,096 discrete sites, that is, an average of 6.6 per nominated property, motivating my estimate (cf. WHC-06/30.COM/8B, p. 22; WHC-07/31.COM/8B, p. 30; WHC-08/32.COM/8B, p. 31. WHC-09/33.COM/8B, p. 30; WHC-10/34.COM/8B, p. 40; WHC-11/35.COM/8B, p. 32; WHC-12/36.COM/8B, p. 36; WHC-13/37.COM/8B, p. 43; WHC-14/38.COM/8B, p. 50; WHC-15/39.COM/8B, p. 46; WHC/16/40.COM/8B, p. 42; WHC/17/41.COM/8B, p. 41).
15. CC-79/CONF.005/COL.I, p. 3.
16. 2019 version, § 137 (https://whc.unesco.org/en/guidelines (retrieved 3 September 2020)). When serial properties are nominated in successive instalments, OUV must be already present from the start (§ 139).
17. To a degree, the narrative trumped the property here: while all ten original countries are represented, they nominated only the 'most … prominent sites' (https://whc.unesco.org/uploads/nominations/1187.pdf, p. 8 of the PDF file (retrieved 3 September 2020)), including both terminals. Even when accepting that some station points have not been found, others are known to be lost and conserving them all may be 'impractical' (at p. 32), a more inclusive approach might have offered itself. Yet in its evaluation, ICOMOS found integrity 'almost not applicable because of the special characteristics and value of the nominated property' (WHC-05/29.COM/INF.8B.1, p. 97) and the Committee did not mind this strangely evasive take when adopting the recommended inscription in 2005.
18. This includes the report of an expert meeting on natural serial properties (cf. https://portals.iucn.org/library/sites/library/files/documents/Rep-2009-005.pdf, pp. 7–9 (retrieved 3 September 2020)). One expert meeting on serial properties managed to distinguish an 'all-inclusive approach' from 'representativeness' – somewhat like wholeness versus intactness – and applied each of these to the natural OUV criteria (i.e. types of natural properties), finding that a series approach can bring greater integrity and sustainability to ecosystems and biodiversity hotspots (criteria (ix) and (x) – obviously, components close to each other are assumed here), whereas it is less suited to superlative natural phenomena or geological sites (criteria (vii) and (viii)); see https://whc.unesco.org/document/124860, p. 31 (retrieved 3 September 2020).

19. WHC-07/31.COM/24, p. 173; WHC-08/32.COM/INF.8B1, p. 124.
20. 'Works of Antoni Gaudí' in Barcelona (Spain), 'City of Vicenza and the Palladian Villas of the Veneto' (Italy), 'Major Town Houses of the Architect Victor Horta (Brussels)' (Belgium) and 'Fortifications of Vauban' (France), all with between four and twenty-five components.
21. WHC/16/40.COM/INF.8B1, p. 232.
22. Ibid., p. 200.
23. Maison Guiette in Antwerp, Belgium and Firminy-Vert, France (WHC-09/33.COM/INF.8B1, pp. 137, 138), respectively.
24. This was the Capital Complex in Chandigarh, India, whereas Corbusier works in the United States, Tunisia and in Russia were left out. A reason for the omissions was only given for Centrosoyuz Building in Moscow, as this was found to lack proper conservation and protection (WHC/16/40.COM/INF.8B1, pp. 136, 220).
25. Ibid., p. 232.
26. Upon resubmission in 2019, inscription of what now were eight of the original ten components was decided (WHC-19/43.COM/INF.8B1.Add, p. 35).
27. For Le Corbusier, it named the apartment block Unité d'habitation in Marseille, Villa Savoye and the Chapelle Notre-Dame-du-Haut in Ronchamp (WHC-11/35.COM/INF.8B1.Add, p. 55); for Wright, it suggested Robie House, Fallingwater House and the Guggenheim Museum in New York (WHC/16/40.COM/INF.8B1, p. 203).
28. Such as Lascaux Cave among the Vézère Valley sites or Elmina Castle among the Ghanaian forts.
29. WHC-93/CONF.002/14, p. 41.
30. This was clarified in 2001, when Spain argued that transplanting wayside structures would mitigate the submergence of several kilometres of the route by the planned Yesa Dam reservoir (WHC-2001/CONF.205/10, p. 44; WHC-01/CONF.208/4, p. 29; WHC-01/CONF.208/24, p. 131; WHC-04/28.COM/26, pp. 122, 269). With the 2015 extension into 'Routes of Santiago de Compostela: *Camino Francés* and Routes of Northern Spain', the total length of the property now exceeds 2,000 km.
31. WHC-93/CONF.002/14, p. 42.
32. The inferior state of conservation when compared to the Spanish route was given as the reason (http://whc.unesco.org/archive/advisory_body_evaluation/868.pdf, pp. 22–24, 25 (retrieved 3 September 2020)).
33. Cf. http://whc.unesco.org/archive/advisory_body_evaluation/868.pdf, p. 25 (retrieved 3 September 2020); WHC-98/CONF.203/18, pp. 28–29.
34. 'Sacred Sites and Pilgrimage Routes in the Kii Mountain Range' (Japan) included more than 300 km of pilgrimage trails, while Israel's contested Incense Route nomination (see Chapter 5, note 57) incorporated only a 50 km stretch of actual road, combined with three separate archaeological sites. Mexico's Camino Real de Tierra Adentro bid (see Chapter 4) featured just 70 out of originally 1,400 road kilometres, provoking ICOMOS complaints about a mere 'chain of buildings' in the session. Five earlier World Heritage inscriptions are given as component parts of the Camino Real, but curiously retain their original entries (cf. WHC-10/34.COM/INF.8B1.Add, pp. 107–8, property website, maps). Two of them, 'Historic Centre of Mexico City and Xochimilco' and 'Protective Town of San Miguel and the Sanctuary of Jesús Nazareno de Atotonilco', are themselves serial properties.
35. WHC-14/38.COM/INF.8B1, p. 320.
36. Once again, the relationship of the route to already-listed World Heritage properties was neglected. The fact that 'Quebrada de Humahuaca' (Argentina), a cultural landscape

that the ICOMOS evaluation had characterized as 'possibly a cultural route' in its own right before inscription in 2003, was part of the Qhapaq Ñan is simply passed over in the nomination file and evaluation, even when the dossier describes the relationship of Qhapaq Ñan segments with that property. However, neither ICOMOS nor the Committee commented on the fact that the new property included part of an old one: the Plaza Inca Hanan Hauk'aypata – the point of origin of the Qhapaq Ñan – is none other than the Plaza de Armas, the central square of 'City of Cuzco', Peru's first World Heritage inscription of 1983. The nomination dossier betrays conscious obfuscation: nothing in the maps of the included parts of the plaza (http://whc.unesco.org/uploads/nominations/1459.pdf, pp. 1754–59 (retrieved 3 September 2020)) gives away that this is in the middle of Cuzco, and they show only Quechua names, not a single Spanish one. The Qhapaq Ñan series thus quietly accomplished what had been a problem in 2009 when San Luis Potosí, Mexico, featuring in two serial properties at the same time, ultimately had to be dropped from one of them (see Chapter 4).

37. The Kyrgyz and Kazakh components do feature Muslim sites and a Manichean church, but China did not include the Daqin Pagoda – testimony to a Nestorian monastery in Chang'an (today's Xi'an), the capital of the Tang Empire – and even the Great Mosque of that city, although this sanctuary has been operational since the eighth century and its completely Sinicized architecture, dispensing even with minarets, embodies cultural fusion like little else on the route. Both these sites had been on China's Tentative List, and a comparative table included in the nomination file (http://whc.unesco.org/uploads/nominations/1442.pdf, pp. 535–36 (retrieved 3 September 2020)) shows the Great Mosque to meet as many selection criteria as other Xi'an sites such as the Great White Goose Pagoda that did make the final cut (Cheung 2019: 119–21). However, a flurry of last-moment revisions produced what, in times of separatist fears over the Uighurs in Xinjiang province, was clearly a political selection: Islam was ousted and Tentative List sites in Ningxia Hui Autonomous Region – named after the Han Chinese Muslims, the Hui – were excluded too (cf. http://whc.unesco.org/en/tentativelists/5335 (retrieved 3 September 2020)). By contrast, Buddhism, as the one Silk Roads religion most strongly associated with China, enjoys pride of place (Cheung 2019: 119–21). This aligns with a Sinocentric narrative in the nomination file and the inclusion of several very China-centred sites, including the tomb of Zhang Qian (died 113 BCE) – celebrated as the imperial envoy whose travels 'opened' the Silk Roads, which is anachronistic and converts a gradual development into a discrete and unilateral event (ibid.: 122–26) – and the Western Han and Tang palaces in Xi'an – nodes of power that presided over empires unfolding towards the east rather than primarily known for their connection with the Silk Roads in the west.
38. WHC-14/38.COM/INF.8B1, pp. 156, 165.
39. http://whc.unesco.org/uploads/nominations/1442.pdf, p. 294 (retrieved 3 September 2020).
40. China, the five Central Asian states and half a dozen of their neighbours – Iran, Afghanistan, India, Nepal, South Korea and Japan, with Turkey listing tentative properties – had been involved in joint preparations for more than a decade (cf. Williams 2014: 71–74). During these, a study commissioned by ICOMOS found almost seven hundred extant caravanserais, way stations and other stopping places (Williams 2014: 36–37), and what else it lists in terms of Silk Roads remains must run in the thousands (cf. Williams 2014: 34–55). Therefore, the participating countries agreed to break up the twenty so-called 'corridors' recommended by the study into smaller yet still transnational packages. Another one of these, 'Silk Roads: Penjikent-Samarkand-Poykent Corridor' nominated by Uzbekistan and Tajikistan, was also nominated in 2014, but only received

a referral, up from ICOMOS's deferral recommendation. The evaluation missed basic information, a selection rationale for components and a clarification of how the ten serial components were related to the already-inscribed properties 'Historic Centre of Bukhara' and 'Samarkand – Crossroad of Cultures' (both Uzbekistan), of which the latter even featured in the name of the candidate property.

ICOMOS may have had no choice to accept at least one of the two Silk Roads bids because of its own involvement and the logistical challenge of evaluation: instead of the usual single inspection trip to the candidate property, there were two each for the Silk Roads components and no fewer than eight for the Qhapaq Ñan all coming in the same year, some of them to such remote areas that one mission expert got lost in the desert for one day and night, as the ICOMOS representative reported in the session. The nomination file for the Chang'an-Tianshan Corridor had 800 pages and almost 5,000 pages of maps, supporting documents, etc.; the file for the Qhapaq Ñan had 1,400 pages and as many pages of supplementary material. China and its partners announced the future submission of extensions, a common concession when a series appears to be incomplete. However, they have not yet done so, and the revised Uzbek-Tajik nomination as well as bids for Silk Roads segments in other countries are also still pending.

Interestingly, it was not about these routes but 'The Grand Canal' (also China), long anticipated and then celebrated in the same session as 'the world's largest and most extensive civil engineering project ensemble prior to the Industrial Revolution' (http://whc.unesco.org/en/list/1443 (retrieved 3 September 2020)), that ICOMOS came close to insisting on wholeness. The earlier canals 'Canal du Midi' (France) and 'Rideau Canal' (Canada) had been inscribed in their totality, including the pioneering works of engineering such as bridges, tunnels, locks, aqueducts and fortifications connected with them. However, the nominated sections of the Grand Canal included only half of that waterway, even when this still amounted to more than 1,000 km (WHC-14/38.COM/INF.8B1, p. 111). Despite what would thereby become the longest World Heritage property at the time, ICOMOS therefore saw 'a paradoxical situation ... on the one hand, the repetitive succession of long sections of canal does not seem to make a decisive contribution to the Outstanding Universal Value; on the other hand, the continuity of the course of the canal across China, and the continuity of its hydraulic systems, is not well highlighted by a discontinuous series' (at p. 116). Given that much of the excluded sections have been modernized for what still is very active use, the selective approach traded wholeness for authenticity. Yet, the general question implicit here was not clarified, since the Committee overruled the referral recommendation. The latter had been mainly due to a different lack of wholeness, namely the omission of the footpaths, trees and buildings next to the canal and its banks over much of its course (at pp. 121, 124).

41. WHC-15/39.COM/INF.8B1, pp. 215–26.
42. Ironically, this concern was disproved in the aftermath: with the series initiative falling apart, Germany nominated 'Archaeological Border Complex of Hedeby and the Danevirke' by itself, being rewarded with inscription in 2018. While the ICOMOS evaluation noted that the property had been part of the earlier serial bid, one searches in vain for comments on the comparative merits of the two approaches and what they might do in terms of creating plausible wholes (cf. WHC-18/42.COM/INF.8B1, pp. 212–21).
43. There is also no rule if contiguous properties straddling borders should be one entry or several: Sangha Trinational from the Saint Petersburg session day and Białowieża Forest

are a single transboundary property, but 'Sundarban National Park' (India) and 'The Sundarbans' (Bangladesh) are two separate ones. Of two famous waterfalls sitting on borders, 'Mosi-oa-Tunya/Victoria Falls' (Zambia and Zimbabwe) is a single property, but 'Iguazú National Park' (Argentina) and 'Iguaçu National Park' (Brazil) are separate.

44. Operational Guidelines, 2019 version, § 104 (http://whc.unesco.org/en/guidelines (retrieved 3 September 2020)).
45. Ibid., §§ 104–7. The use of 'core zone' for 'property', while discouraged, continues to be widespread in World Heritage materials, suggesting to the uninitiated that it is the totality of both 'zones' that constitutes a World Heritage property.
46. WHC-12/36.COM/INF.8B1, p. 386.
47. Ibid., pp. 382–83.
48. Operational Guidelines, 2019 version, § 77 (http://whc.unesco.org/en/guidelines (retrieved 3 September 2020)).
49. Ibid., § 77. Joint use even became mandatory in 1997 and criterion (vi) was restricted to 'exceptional cases only' in 1980. However, this made it too much of a poor cousin to the other criteria, impeding the freshly introduced associative cultural landscapes and the Global Strategy goal of opening up towards indigenous cultural heritage. It also motivated what even the above-mentioned study characterizes as a 'cosmetic' search for a second criterion when inscribing Robben Island in 1999 – the statement of OUV (http://whc.unesco.org/en/list/916 (retrieved 3 September 2020)) must be the longest existing description of Robben Island that manages *not* to mention Nelson Mandela. 'We kind of manufactured something', as a session participant told me. Therefore, the current version talking about preferability was adopted in 2005.
50. For Wartburg Castle, the 'rich cultural associations, most notably its role as the place of exile of Martin Luther, who composed his German translation of the New Testament there' were recognized under criterion (vi), despite equally shaky physical traces of that sojourn. Here, criterion (iii) was added for 'an outstanding monument of the feudal period in central Europe', but as the official description itself says, there are only 'some original sections' from that time in what is largely a nineteenth-century reconstruction.
51. Operational Guidelines, 2019 version, Annex III, § 10 (http://whc.unesco.org/en/guidelines (retrieved 3 September 2020)).
52. Also, when asserting the 'indissoluble link' with Maori religious beliefs and oral traditions upon inscription of Tongariro National Park as the first cultural landscape in 1993 (by adding criterion (vi) to the natural criteria), no mention was made of any physical marks that such spirituality might have left on the landscape. With 'Val d'Orcia' in Tuscany, Italy (inscribed in 2004) or Rio de Janeiro as well, it was the imprint of the landscape in art that justified use of criterion (vi), meaning that these properties themselves produced traces outside their boundaries rather than the other way round.
53. https://wikileaks.org/plusd/cables/05QUITO1799_a.html (retrieved 3 September 2020).
54. To cite from the first Google phrase search hit for 'authentic person' (28 September 2019): 'Authenticity is about presence, living in the moment with conviction and confidence and staying true to yourself. An authentic person puts the people around them at ease, like a comforting, old friend who welcomes us in and makes us feel at home. There's never any doubt or questioning the integrity of an authentic individual. Their behavior, in terms of ethics and morals, is as predictable as snow during wintertime in Minnesota. You know what you're going to get' (https://medium.com/personal-growth/the-5-key-ingredients-of-an-authentic-person-259914abf6d5 (retrieved 3 September 2020)).

    For a long time now, anthropologists in particular have enthusiastically engaged in deconstructing popular notions of authenticity, usually pointing out the political, commercial or other vested interests behind what they see, in essence, as an arbitrary claim, often connecting it with elite interests and an expert point of view (Bruner 1994; Gable and Handler 1996; Handler and Linnekin 1984). However, as I have argued elsewhere (Brumann 2014a: 80–81), authenticity is also very much a popular and lay concern: Bortolotto vividly describes how it creeps in through the back door, even in an environment from which it has been officially ousted, such as into the nominations to the lists of the 2003 Convention for the Safeguarding of the Intangible Cultural Heritage (Bortolotto 2011a, 2020).

55. Khalaf (2017: 270) and I (Brumann 2017b: 285) arrived at this insight independently.
56. WHC-98/CONF.201/9, pp. 20–21. There is only one not very clear Operational Guidelines caveat that draws distinctions: 'Attributes such as spirit and feeling do not lend themselves easily to practical applications of the conditions of authenticity, but nevertheless are important indicators of character and sense of place, for example, in communities maintaining tradition and cultural continuity' (2019 version, § 83; http://whc.unesco.org/en/guidelines (retrieved 3 September 2020)).
57. 'Seokguram Grotto and Bulguksa Temple' (South Korea), 'Historic Villages of Shirakawa-go and Gokayama' (Japan), 'Changdeokgung Palace Complex' (South Korea), 'Itsukushima Shinto Shrine', 'Historic Monuments of Ancient Nara' and 'Shrines and Temples of Nikko' (all Japan), all inscribed between 1995 and 1999.
58. 'Wooden Churches of Maramureş' (Romania; cf. the ICOMOS evaluation), 'Wooden Churches of Southern Małopolska' (Poland), 'Wooden Churches of the Slovak Part of the Carpathian Mountain Area' (Slovakia) and 'Wooden *Tserkvas* of the Carpathian Region in Poland and Ukraine' (Poland and Ukraine), all inscribed between 1999 and 2013.
59. 'Luang Prabang' (Laos), 'Hoi An Ancient Town' (Vietnam), 'Lamu Old Town' (Kenya) and 'Historic Inner City of Paramaribo' (Suriname), all inscribed between 1995 and 2002.
60. 'Ksar of Ait-Ben-Haddou' (Morocco), 'Ancient *Ksour* of Ouadane, Chinguetti, Tichitt and Oualata' (Mauritania), 'Bam and Its Cultural Landscape' (Iran), 'Tomb of Askia' (Mali) and 'Historic City of Yazd' (Iran), all listed between 1996 and 2017 or, in the case of the Moroccan property listed earlier, with a section on authenticity dating from that period.
61. Such as for 'Agricultural Landscape of Southern Öland' (Sweden), 'Fertö/Neusiedlersee Cultural Landscape' (Austria and Hungary), 'Tokaj Wine Region Historic Cultural Landscape' (Hungary), 'Þingvellir National Park' (Iceland), 'Iwami Ginzan Silver Mine and Its Cultural Landscape' (Japan) and 'Le Colline del Prosecco di Conegliano e Valdobbiadene' (Italy), all inscribed since 2000.
62. Such as 'Rice Terraces of the Philippine Cordilleras' (the Philippines), 'Portovenere, Cinque Terre, and the Islands (Palmaria, Tino and Tinetto)' (Italy) and 'Sukur Cultural Landscape' (Nigeria), all inscribed in the late 1990s, and 'Coffee Cultural Landscape' (Colombia), 'Konso Cultural Landscape' (Ethiopia), 'Cultural Landscape of the Serra de Tramuntana' (Spain), 'Cultural Landscape of Bali Province: The Subak System as a Manifestation of the *Tri Hita Karana* Philosophy' (Indonesia), 'Landscape of Grand-Pré' (Canada), 'Cultural Landscape of Honghe Hani Rice Terraces' (China), 'Palestine: Land of Olives and Vines – Cultural Landscape of Southern Jerusalem, Battir' (Palestine), 'Kujataa Greenland: Norse and Inuit Farming at the Edge of the Ice Cap' (Denmark) and 'The English Lake District' (United Kingdom), all inscribed since 2011.
63. More or less directly for 'Costiera Amalfitana' (Italy), 'Wachau Cultural Landscape' (Austria), 'Alto Douro Wine Region' (Portugal), 'Landscape of the Pico Island Vineyard

Culture' (Portugal), 'Agave Landscape and Ancient Industrial Facilities of Tequila' (Mexico), 'Lavaux, Vineyard Terraces' (Switzerland; mentioned in the OUV criteria in this case), 'Vineyard Landscape of Piedmont: Langhe-Roero and Monferrato' (Italy) and 'The Climats, Terroirs of Burgundy' (France), all inscribed between 1997 and 2015.

64. Such as 'Pyrénées – Mont Perdu' (France and Spain), 'Orkhon Valley Cultural Landscape' (Mongolia), 'Koutammakou, the Land of the Batammariba' (Togo), 'Richtersveld Cultural and Botanical Landscape' (South Africa), 'The Causses and the Cévennes, Mediterranean Agro-pastoral Cultural Landscape' (France), 'Bassari Country: Bassari, Fula and Bedik Cultural Landscapes' (Senegal) and 'Cultural Landscape of Maymand' (Iran), all inscribed between 1997 and 2015.
65. Such as 'Curonian Spit' (Lithuania and Russia), 'Vegaøyan – The Vega Archipelago' (Norway), 'Saloum Delta' (Senegal), 'Aasivissuit – Nipisat. Inuit Hunting Ground between Ice and Sea' (Denmark) and 'Budj Bim Cultural Landscape' (Australia), all inscribed between 2000 and 2019.
66. Such as 'Erzgebirge/Krušnohoří Mining Region' (Germany and the Czech Republic) listed in 2019, where mining has ended but the associated cultural practices continue.
67. Such as 'Royal Hill of Ambohimanga' (Madagascar), '*Sacri Monti* of Piedmont and Lombardy' (Italy), 'Sacred Sites and Pilgrimage Routes in the Kii Mountain Range' (Japan) or 'Osun-Osogbo Sacred Grove' (Nigeria), all inscribed between 2001 and 2005.
68. Such as 'Le Morne Cultural Landscape' (Mauritius), 'Chief Roi Mata's Domain' (Vanuatu), 'Sacred Mijikenda Kaya Forests' (Kenya), 'Sulaiman-Too Sacred Mountain' (Kyrgyzstan), 'Papahānaumokuākea' (United States), 'Taputapuātea' (France), 'ǂKhomani Cultural Landscape' (South Africa), 'Pimachiowin Aki' and 'Writing-on-Stone/Áísínai'pi' (both Canada), all inscribed between 2008 and 2019.
69. http://whc.unesco.org/en/list/840 (retrieved 3 September 2020).
70. '… restaurer un édifice, ce n'est pas l'entretenir, le réparer ou le refaire, c'est le rétablir dans un état complet qui peut n'avoir jamais existé à un moment donné' (Viollet-le-Duc 1875: 14).
71. Interestingly, this posits 'mankind' as a rights-holder long before the idea emerged in international law.
72. CC-79/CONF.005/6, Annex II, p. 5.
73. CC-79/CONF.003/11, Annex.
74. 'The Historic Centre of Warsaw is an exceptional example of the comprehensive reconstruction of a city that had been deliberately and totally destroyed. The foundation of the material reconstruction was the inner strength and determination of the nation, which brought about the reconstruction of the heritage on a unique scale in the history of the world' (http://whc.unesco.org/en/list/30 (retrieved 3 September 2020)).
75. CC-80/CONF.016/10, p. 4.
76. CC-80/CONF.017/4, p. 4.
77. In addition to the below properties, these included Viollet-le-Duc's pet restoration project of 'The Town of Carcassonne' (France; SC-85/CONF.007/9, p. 11) and 'Dresden (Baroque Ensemble)' (German Democratic Republic; CC-90/CONF.003/12, p. 13).
78. SC/83/CONF.009/8, p. 6; SC-88/CONF.001/13, p. 17. ICOMOS had dismissed Rhodes as 'pseudo-medieval' and 'grandiose pastiches … devoid of archaeological rigor' (http://whc.unesco.org/archive/advisory_body_evaluation/493.pdf, p. 3 (retrieved 3 September 2020).
79. CC-81/CONF/003/6, p. 4, http://whc.unesco.org/archive/advisory_body_evaluation/168.pdf (retrieved 3 September 2020. For more German churches, ICOMOS also

changed its mind (with no details given) when 'The Church of St Michael, Hildesheim' it had rejected in 1982 (CLT-82/CONF.014/6, p. 6) returned as 'St Mary's Cathedral and St Michael's Church at Hildesheim' in 1985, to now support the inscription (http://whc.unesco.org/archive/advisory_body_evaluation/187bis.pdf, p. 2, SC-85/CONF.008/9, p. 8 (retrieved 3 September 2020).

80. Such as for 'City of Quito' (Ecuador; WHC-08/32.COM/24, pp. 138–39), Istanbul (WHC-94/CONF.003/16, p. 40; WHC-06/30.COM/INF.19, p. 131), 'Historic Town of Zabid' (Yemen; WHC-96/CONF.201/4, p. 20; WHC-97/CONF.208/4B, p. 27), 'Historic Centre of Bukhara' and 'Samarkand – Crossroad of Cultures' (both Uzbekistan; WHC-97/CONF.208/4B, p. 26), and 'Tabriz Historic Bazaar Complex' (Iran; http://whc.unesco.org/archive/advisory_body_evaluation/1346.pdf, p. 5 (retrieved 3 September 2020).
81. Such as for 'Historical Monuments of Mtskheta' (Georgia; WHC-07/31.COM/24, p. 112; WHC-09/33.COM/20, p. 139) and 'Baroque Churches of the Philippines' (WHC-98/CONF.203/5, pp. 61–62), mosques in 'Historic Cairo' (Egypt; WHC-95/CONF.203/16, p. 25) and 'Ancient City of Damascus' (Syria; WHC-97/CONF.204/11, p. 36) and temples in 'Lumbini, the Birthplace of the Lord Buddha' (Nepal; WHC-04/28.COM/26, p. 264; WHC-05/29.COM/22, p. 79).
82. Such as for 'Imperial Palaces of the Ming and Qing Dynasties in Beijing and Shenyang' (WHC-06/30.COM/19, p. 101), 'Historic Ensemble of the Potala Palace, Lhasa' (both China; WHC-96/CONF.201/21, p. 39; WHC-03/27.COM/INF.24, p. 80) and 'Bahla Fort' (Oman; WHC-95/CONF.203/16, pp. 23–24; WHC-96/CONF.201/21, pp. 23–24; WHC-2000/CONF.202/17, p. 14).
83. Such as for Abu Mena (Egypt; WHC-13/37.COM/20, p. 33), the Negev desert cities on the Incense Route (Israel; WHC-04/28.COM/26, p. 219), 'At-Turaif District in ad-Dir'iyah' (Saudi Arabia; WHC-10/34.COM/INF.20, p. 585), 'Samarra Archaeological City' (Iraq; WHC-11/35.COM.INF.20, p. 60), 'Al Zubarah Archaeological Site' (Qatar; WHC-13/37.COM/20, p. 181) and 'Group of Monuments at Hampi' (India; WHC-99/CONF.208/8, p. 36).
84. WHC-10/34.COM/20, pp. 143–46.
85. WHC-13/37.COM/20, p. 44.
86. For example, after both world wars, significant parts of 'Mehmed Paša Sokolović Bridge in Višegrad' (Bosnia and Herzegovina) listed in 2007 were reconstructed with non-authentic materials and in nonauthentic forms, but ICOMOS found that 'the visible alterations to form and material are secondary, and can be put right by appropriate restorations' so that, in a confusing turn, 'authenticity seems excellent' (WHC-07/31.COM/INF.8B1, p. 115). When the German Limes was added to Hadrian's Wall, ICOMOS bemoaned that 'in many cases … the authenticity has been compromised by unacceptable reconstructions'. Yet in an inventive move, these reconstructions were declared the buffer zone of the World Heritage property (http://whc.unesco.org/archive/advisory_body_evaluation/430ter.pdf, pp. 165, 167 (retrieved 3 September 2020)), a vertical one in fact that sits on top of the original wall sections under the ground.
87. http://whc.unesco.org/archive/advisory_body_evaluation/708.pdf, p. 16 (retrieved 3 September 2020), concerning 'Historical Monuments of Mtskheta' (Georgia).
88. Such as Wartburg Castle (WHC-99/CONF.204/INF.7, p. 68), 'Three Castles, Defensive Wall and Ramparts of the Market-Town of Bellinzona' (Switzerland; WHC-2000/CONF.204/INF.6, pp. 150–51) and 'Kuressaare Fortress' (Estonia, deferred for other reasons and then never re-nominated; WHC-04/28.COM/INF.26, p. 217).

89. For the Upper Middle Rhine Valley, it found that: 'In the case of a cultural landscape of this type the quality of integrity is as relevant as that of authenticity' (WHC-02/CONF.201/INF.2, p. 45), implying that the former makes up for deficiencies of the latter, and for Dresden, it argued that 'while recognizing the unfortunate losses in the historic city centre during the Second World War, the Dresden Elbe Valley, defined as a continuing cultural landscape, has retained the overall historical authenticity and integrity in its distinctive character and components' (WHC-04/28COM/INF.14A, p. 88). Similar excuses of reconstructions because of their role in the religious landscape were also brought forward for the Chinese properties 'Mount Wuyi' and 'Mount Wutai', listed in 1999 and 2009, respectively (http://whc.unesco.org/en/list/911 (retrieved 3 September 2020; WHC-09/33.COM/INF.8B1, p. 4).
90. WHC-05/29.COM/INF.8B.1, p. 172.
91. Such as for 'Luther Memorials in Eisleben and Wittenberg' (Germany; WHC-96/CONF.201/21, p. 66), 'Castle of the Teutonic Order in Malbork' (Poland; WHC-97/CONF.208/17, p. 47) and 'San Marino Historic Centre and Mount Titano' (San Marino; WHC-08/32.COM/24, p. 178). In the latter case (an upgrade inscription), ICOMOS had recommended referral, but found OUV and authenticity present.
92. WHC-97/CONF.208/17, p. 42.
93. Such as with regard to Carcassonne (http://whc.unesco.org/archive/advisory_body_evaluation/345.pdf, p. 28 (retrieved 3 September 2020)) and San Marino (WHC-08/32.COM/INF.8B1, p. 187).
94. In the French original, 'notre fondamentalisme ouest-européen … qui nous a amenés à la définition de la notion d'authenticité, limitée exclusivement à la substance matérielle; une idée que nous avons voulu octroyer aux autres régions culturelles du monde' (WHC-96/CONF.201/21, Annex VI.2).
95. WHC-99/CONF.208/8, pp. 70–71, WHC-2000/CONF.202/17, p. 47, WHC-2000/CONF.204/21, p. 51, WHC-03/27.COM/24, p. 116; http://whc.unesco.org/archive/1999/whc-99-conf209-inf7e.pdf, p. 141 (retrieved 3 September 2020).
96. WHC-05/29.COM/INF.22, p. 185.
97. Ibid., pp. 183–87.
98. WHC-05/29.COM/22, p. 140–41.
99. 'With the "renaissance" of the Old Bridge and its surroundings, the symbolic power and meaning of the City of Mostar – as an exceptional and universal symbol of coexistence of communities from diverse cultural, ethnic and religious backgrounds – has been reinforced and strengthened, underlining the unlimited efforts of human solidarity for peace and powerful co-operation in the face of overwhelming catastrophes' (WHC-05/29.COM/22, p. 141).
100. Such as for the new pilgrimage facilities at 'Lumbini, the Birthplace of the Lord Buddha' (Nepal), inscribed as a pilgrimage centre (http://whc.unesco.org/en/list/666 (retrieved 3 September 2020)).
101. For 'Pampulha Modern Ensemble' (Brazil), inscribed in 2016, for example, the positive ICOMOS evaluation discusses the continued presence of original building features under 'integrity' (cf. WHC/16/40.COM/INF.8B1, p. 256) and this has also been retained in the Statement of OUV, where the section on 'authenticity' curiously fails to mention a single feature that is authentic, instead speaking only of threats (cf. https://whc.unesco.org/en/list/1493 (retrieved 3 September 2020)).
102. Labadi (2010) has also taken up authenticity and what she calls 'post-authenticity', but mainly in the nomination dossiers for candidate properties submitted by States Parties

rather than the ICOMOS site evaluations and Committee decisions, that is, the texts authorized by the World Heritage bodies.

103. The very first Committee session of 1977 also records that: 'The representative of ICOMOS ... recognized the difficulty of drafting criteria to be applied to cultural property throughout the world and of translating concepts into words that were meaningful on a universal scale' (CC-77/CONF.001/9, p. 4).
104. Cf. http://whc.unesco.org/en/conventiontext (retrieved 3 September 2020).
105. http://whc.unesco.org/archive/opguide77b.pdf (retrieved 3 September 2020).
106. Operational Guidelines, October 1977 version, § 5.ii; now § 52, 2019 version (http://whc.unesco.org/en/guidelines (retrieved 3 September 2020)).
107. CC-77/CONF.001/9, p. 4.
108. Operational Guidelines, 2019 version, § 49 (http://whc.unesco.org/en/guidelines (retrieved 3 September 2020)).
109. CC-79/CONF.005/6, Annex II, p. 7.
110. 'Criterion (vi): The Wartburg Castle ... is also a powerful symbol of German integration and unity' (http://whc.unesco.org/en/list/897 (retrieved 3 September 2020)); 'Criterion (iii): The Royal Hill of Ambohimanga is the most significant symbol of the cultural identity of the people of Madagascar' (http://whc.unesco.org/en/list/950 (retrieved 3 September 2020)).
111. Operational Guidelines, October 1977 version, §§ 7, 10 (http://whc.unesco.org/en/guidelines (retrieved 3 September 2020)).
112. Operational Guidelines, 2019 version, § 77 (http://whc.unesco.org/en/guidelines (retrieved 3 September 2020)).
113. https://www.protectedplanet.net/c/world-database-on-protected-areas (retrieved 3 September 2020).
114. https://www.icomos.org/en/about-the-centre/publicationsdoc/monographic-series-3/198-thematic-studies-for-the-world-heritage-convention (retrieved 3 September 2020).
115. Schmutz and Elliott (2017) detect an increasing use of scientific-sounding vocabulary in ICOMOS evaluations, but this is not the same as a scientific argumentation.
116. IUCN provided such analyses right from the start (Cameron and Rössler 2013: 181).
117. Operational Guidelines, 2017 version, § 132 (http://whc.unesco.org/en/guidelines (retrieved 3 September 2020)).
118. WHC-12/36.COM/INF.8B1, p. 228.
119. Such as in the following examples: 'ICOMOS considers that the comparative analysis suggests that there are three buildings within the nominated series that stand out as being incomparable: the Unité d'habitation, Marseille, the Villa Savoye, and the Ronchamp chapel' (WHC-09/33.COM/INF.8B1, p. 139); 'ICOMOS considers that the comparative analysis adequately demonstrates that the nominated series in relation to its proposed Outstanding Universal Value does not have comparators either on or outside the World Heritage List' (WHC/16/40.COM/INF.8B1, p. 220).
120. '... the aim of the comparative analysis is not to demonstrate that the property is unique, but that it has an exceptionally strong claim to be of Outstanding Universal Value in a defined context' (UNESCO 2011: 59).
121. Cf. WHC-12/36.COM/INF.8B1, pp. 121–35.
122. Cf. WHC-13/37.COM/INF.8B1.Add, pp. 22–37.
123. Cf. WHC-13/37.COM/INF.8B1, pp. 251–58; WHC-14/38.COM/INF.8B1.Add, pp. 20–30. In the latter evaluation, the formulations very much suggest acceptance of the inevitable in a case where lobbying pressure reached rare heights, as ICOMOS personnel confirmed to me. See also the account by Meskell et al. (2014: 11–13). However, contrary

to what is claimed there, Bolgar was first nominated for the 2013 session, not for 2012. Since OUV had been confirmed in the decision text of the 2013 referral, the ICOMOS mission to follow took the novel format of the 'advisory mission' assisting the State Party, not that of an evaluation mission judging the site (which does not take place in the case of referrals). The presence of World Heritage Centre personnel that astonishes Meskell – as they are not part of evaluation missions – might relate to this aspect.

For further turnarounds, ICOMOS also ended up proposing inscription for the already-discussed 'Hidden Christian Sites in the Nagasaki Region' (Japan) in 2018 and for 'Erzgebirge/Krušnohoří Mining Region' (Czech Republic and Germany) in 2019, despite advising noninscription before for these candidate sites, which led to their withdrawal a mere two (Nagasaki) or three (Erzgebirge/Krušnohoří) years earlier. I have not heard of similarly strong lobbying pressure as for Bolgar being applied here.

124. A total of 911 World Heritage properties (http://whc.unesco.org/en/list/stat#s6 (re-trieved 3 September 2020)) and two deleted properties.
125. Reyes considers 870 properties, a confusing number since the total for World Heritage properties was already larger in 2008 and was never exactly that figure in any given year.
126. Cf. WHC-18/42.COM/INF.8B1.Add, pp. 34–41.
127. http://whc.unesco.org/en/nominations (retrieved 3 September 2020).
128. As the IUCN official memorably phrased it, 'the reality is if you don't have the values, you can't create them. Integrity you can kind of fix, protection and management you can certainly put in place. The values bit, you either got it or you don't, and if you haven't got it, you can't invent it'.

Chapter 8

Global North and South

We are back at the Saint Petersburg Committee session of 2012, three days before the session day described in Chapter 1. As announced when the plenary session broke for lunch, buses are waiting outside for the participants wishing to attend the reception of the Tanzanian delegation, and I board one of them for what is a short ride of not more than a block's distance. Inside the posh rooms of one of the aristocratic residences that past imperial glory has bequeathed to the city, Tanzanian representatives welcome everyone with a handshake and there is an abundance of canapés, salmon skewers and other snacks that, together with the drinks served, make up for the missed lunch at the main meeting venue. Only a few Europeans and Americans have bothered to come and most of the other sixty or so guests have African and Asian countries on their name tags. I chat with a Thai delegate and a Cuban delegate before a Tanzanian lady calls our attention. She introduces a government minister and we are treated to a 10-minute film that introduces the delegation's main concern. This is receiving permission from the World Heritage Committee for cutting out a small portion of a World Heritage property, the Selous Game Reserve – a nature reserve larger than the Netherlands – so that this area can be mined for uranium. The very professionally made footage emphasizes how much Tanzania needs development, and testimonials of a carefully picked array of locals – a village headman, a park ranger, a 'mother', even a bishop in full vestments – express high hopes for the jobs and revenue from the future mine. A schematic map shows that only 0.7 per cent on the edge of the reserve is slated for mining.

When the film ends, the minister – cutting a very elegant figure – takes over, introducing several other dignitaries lining up next to him, even including one from the political opposition, as he points out. He emphasizes Tanzania's commitment to the World Heritage Convention and to conservation, but it must also pursue sustainable development, particularly when expected profits of US$3 billion will help to manage the remaining reserve – no less than one-third of the revenues, a Swiss delegate tells me later, is said to be foreseen for managing the park. The delegation is coming with an open mind, the minister stresses, but at the same time – and here he reaches the dramatic climax of his carefully crafted sentences – he hears 'the cry of the people', and Tanzania as a sovereign nation will not have any condescending advice. Still, he assures us, nothing is decided yet, and by no means should we assume that corporate interests are dictating the course of government action (a somewhat strange remark that comes out of the blue). We are then left to finish our snacks and drinks, and when boarding the buses back to the afternoon plenary session, a South African delegate cannot help shaking his head and laughing in recognition of what, he confirms to me, was a masterly PR stunt – one in which we heard next to nothing about the site proper, as a Malian delegate notices. It is also the final event of a whole series during recent Committee meetings (paid for by the foreign mining firms behind the project, as session rumour has it) since four days later, the World Heritage Committee approves the 'minor boundary modification' proposed by Tanzania. The German ambassador to UNESCO and other supporters have managed to convince the doubters that Tanzania will have its way in any event, so that linking the excision of the mining area with protective demands for the remaining property is the most realistic option for conservation. And although this is not mentioned, Tanzania did make a major concession a year earlier when calling off plans to build a highway through the Serengeti National Park, which is also a World Heritage property (see Chapter 2).

On a session day between the reception and the decision, the Committee – to the loud jubilation of French delegates in the hall – approves the inscription of the 'Nord-Pas de Calais Mining Basin' on the World Heritage List. As the ICOMOS representative outlines in her introductory presentation, this is a very large collection of buildings and landscape features left by two hundred years of coal mining. The included fifty-one slag heaps, with the tallest rising more than 140 metres, 'symbolize the landscape identity of the Mining Basin' according to the evaluation.[1] The decision is taken hurriedly, as everyone is eager to depart for the host nation's evening reception in the

Peter and Paul Fortress; however, the French delegation delivers its short acceptance speech, with a local representative marvelling that their mountains are made of slag.

The inscription means that past destruction of nature – the slag heaps that now dominate an otherwise flat landscape – qualifies as a cultural achievement having OUV so that twenty-two years after the closure of the last pit, the northern French mining area acquires a second life as heritage. By contrast, had the Committee upheld the decision recommended by IUCN, it would have denied Tanzania a mining future in Selous. The country could then not pile up its own slag heaps and future generations could not, at some point, re-appreciate them as valuable testimonies of a bygone age rather than as mere eyesores. This means that converting ravaged nature into cultural heritage is the prerogative of the Global North, whereas if the Committee had stuck with the draft decision for Selous, it would have denied the Global South the experience.

The French landscape was agricultural before the advent of mining, of course, not the presumably pristine haven of wildlife and biodiversity that Tanzania nominated for World Heritage inscription in 1982, fully aware of the protective obligations. Also, IUCN is concerned that Tanzania's proposal is another attempt of subverting the mining industry's pledge not to touch World Heritage properties, following on the Three Parallel Rivers of Yunnan in the Brasilia session (see Chapter 4). In the eyes of IUCN, the consequences of uranium mining – massive landscape upheavals, radioactive emissions, and the use and disposal of large quantities of sulphuric acid – on several hundred square kilometres of a nature reserve are anything but 'minor'. Yet still, the contradiction remains: the destructive exploitation of nature in the contemporary Global South is suspect of putting World Heritage at risk, whereas when having occurred in the right location and age – in the Global North and well before the advent of institutionalized conservation – it may have World Heritage value in its own right.

The issue comes up again towards the end of the meeting when the Committee discusses the follow-up to the periodic reporting exercise that has just been completed for the African countries. A Senegalese delegate cites Émile Zola's novel *Germinal* before reminding everyone of the inscription of a mining site that demonstrates that Europe has had its development. Africa, however, still needs it, so this must be taken into account in the context of a future-oriented convention, he says, with the French UNESCO ambassador rushing to his support. Other African delegates I talk to during these days also keep returning to Selous and argue that without compensating Tanzania for the foregone profits, the international community cannot insist on self-restraint. Nobody strongly objects, let alone offers such funds.

The Slanted World of World Heritage

These incidents from 2012 illustrate a key tension that riddles World Heritage: that between Global North and South, and the time lags that determine how it plays out. Not only are African and other Southern countries late in having what they see as their fair share of economic development, they are also late in bringing their World Heritage candidates to the Committee. And then Southern countries are also historically late in producing the things that, after requiring the necessary patina, can win World Heritage titles (which includes uranium mines, as in the case of 'Erzgebirge/Krušnohoří Mining Region' (the Czech Republic and Germany), listed in 2019). Many observers have noted the significance of North–South divisions in the World Heritage arena (e.g. Bertacchini, Liuzza and Meskell 2017; Hølleland and Wood 2019: 6–7; Meskell 2013a: 488–89; Meskell, Liuzza and Brown 2015; Schmitt 2009: 112–14), but none of them, in my view, has fully grasped how much they have fuelled the recent transformation. I see the post-Brasilia mode of operations as the direct consequence of the Committee being late, not addressing global inequalities and converting a slanted into a level world in due course. A widespread sense that the benefits arising from the World Heritage List have not been fairly distributed over the globe is crucial for what has transpired.

I already described that the large-scale sharing of conservation responsibilities and expenses prefigured by the Nubian campaign never arose in the World Heritage context, so that the World Heritage title is the major perk the Committee has to offer. It is therefore the allocation of inscriptions that must be looked at, and here multiple objections could be raised to what I just said. Three of the first twelve listings of 1978 were in Africa, no fewer than in Europe, and when it became difficult to ignore the preponderance of European and other Western countries' listings, the World Heritage arena embarked on a reform course, adopting the Global Strategy in 1994 and 'anthropologizing' cultural heritage. As I demonstrated with the diversity of inscriptions in the 2008 session, ranging from chiefly domains over sacred groves to maroon hideouts (see Chapter 3), this removed obstacles for non-elite heritage not conforming to a European-derived sense of monumentality. Going by these precedents, almost any country in the world can now find plausible candidates within its borders and, sure enough, the number of countries with World Heritage properties rose from 102 in 1994 to 149 in 2009.

However, this is only one way of looking at post-Global Strategy developments. The newly approved cultural heritage categories were simply added to the canon, but did not replace anything in it, and a lack of fit with the new policy was never a sufficient reason to reject a candidate site. The very reform years of the 1990s saw record numbers of European inscriptions of often rather conventional fare – when Italy nominated twelve candidates and had a record ten of them listed in 1997, for example, six of these properties contained palaces, royal residences, a cathedral, and the archaeological remains of cities, temples and elite country estates, that is, precisely what the Global Strategy had just called to transcend. Both State Party delegates and Centre officials deplored the disconnect between declared goals and the reality of European dominance in the 1998 and 1999 sessions when two other UNESCO regions – Sub-Saharan Africa in 1998 and the Arab countries in 1999 – presented no candidate sites at all.[2] Since calls for self-restraint proved to be ineffective, the Committee imposed a quota of one candidate per state and year in 2000. Yet while this ended single-country listing sprees, it did nothing to affect the numerical lead of a continent whose maze of national borders had expanded even further in the 1990s.

Also, the basic premise of treaty states using their nomination slot year after year for candidates of their own sovereign choice remained untouched, with calls for voluntary restraint being regularly ignored. I have been repeatedly told that Global Strategy pronouncements on what is overrepresented and what is missing on the List do indeed guide some States Parties' nomination policies. Yet, aside from a second national nomination slot reserved for either natural heritage or cultural landscapes from 2005 on, nothing in the way of restrictions on well-represented heritage or, conversely, preferential treatment of missing regions and kinds was ever installed. The implementation of the Global Strategy thus depended almost entirely on the uncoordinated action of individual States Parties.

However, since these would often prioritize their own national interests, no turnaround resulted. The European wine region, not the Oceanian sacred mountain, became the most common World Heritage cultural landscape. The Nara Document, as already mentioned, was brandished in favour of nineteenth-century re-enactments of the European Middle Ages, not adobe structures. Industrial and mining sites and icons of architectural modernity cluster no less on the continent than classical monuments, so Europe received the most listings for these categories too. Therefore, in the 2008 session I singled out for its broad and 'anthropological' take (see Chapter 3), eight out of nineteen inscriptions would still go to Europe, more than to any other UNESCO region. And in the 2009 session that saw Dresden leave, no less than two-thirds of nominations – twenty out of twenty-nine – were from Europe. All this meant that between 1994 and 2009, the European share of the World Heritage List did not shrink, but actually increased, from 49.4 to 52.5 per cent for the cultural properties chiefly targeted by the Global Strategy and from 40.8 to 45.7 per cent overall.[3] Twelve of the top twenty countries in terms of inscriptions were European (see Table 8.1).

Table 8.1. States Parties with most World Heritage List inscriptions in 2009.

Italy	44
Spain	41
China	38
France	33
Germany	33
Mexico	29
United Kingdom	28
India	27
Russia	23
United States	20
Australia	17
Brazil	17
Greece	17
Canada	15
Japan	14
Sweden	14
Poland	13
Portugal	13
Czech Republic	12
Peru	11
Total for all countries	892

Note: Derived from http://whc.unesco.org/en/list/xls/?2019 (retrieved 5 September 2020).

A regular objection raised by arena participants and in the literature (Gfeller 2015: 379) is that the European count includes a number of overseas properties of the former colonial powers, Israel, Russian properties east of the Ural (around one-third) and Turkish properties in Anatolia (all except two). However, taken together, these account for a mere tenth of European listings. Another objection is that Western Europe accounts for most listings, whereas the former Soviet Bloc is less prominent (cf. also Meskell, Liuzza and Brown 2015: 454–56). More commonly, the numbers exercise as such is questioned, also by critics of the post-Brasilia turn, given that many of the non-European properties are large in size. Seven World Heritage properties might not be much for the heir of an ancient civilization, an Egyptian delegate told me, but 'Memphis and Its Necropolis – the Pyramid Fields from Giza to Dahshur' alone measures more than 160 square km and includes lots of outstanding monuments. One might indeed wonder whether Switzerland – with its currently twelve inscriptions – is 'better represented' than Tanzania when one of the latter's seven properties, the Selous Game Reserve, is larger than Switzerland. Yet still and unless one subscribes to the most Eurocentric optic, it is difficult to justify why between 1996 and 2009, the whole of Sub-Saharan Africa had fewer cultural World Heritage properties than the single country of Italy.[4] Likewise, although the category of cultural landscapes was conceived with the non-Western world in mind, with thirty-four out of sixty-five listings in 2009, Europe ended up just as dominant as with other cultural heritage (cf. Mitchell, Rössler and Tricaud 2009: 122–23).[5] I think it is naïve to deny the normalizing power of the actual contents of the List: these shape the global public's sense of what proper World Heritage is supposed to be. Committee press releases also give equal attention to every new property, be it the size of Switzerland or of a suburban garden. Counting inscriptions is therefore not as irrelevant as some critics would have it.

Although outshone by others on the List, Africa did stand in the limelight when it came to conservation problems of already-inscribed properties: as Table 8.2 demonstrates for 2009, African – and, somewhat less so, Arab – countries had a larger share of State of Conservation Reports and particularly of Danger-Listed properties than of inscriptions as such. By contrast, the only clearly underrepresented region in terms of conservation worries was Europe and North America, with properties being more than twenty times less likely to appear on the Danger List than African ones.[6]

As a result, the dominant discourse across the observed sessions was that the World Heritage List did not capture the world's heritage in the representative and balanced way that the Global Strategy had envisioned. In every new meeting, delegates from the Global South would deplore the situation and call for remedial action already on the first or second day, with some Northern delegates and leading UNESCO officials such as the Director-General or the Director of the World Heritage Centre chiming in. By contrast, I cannot recall open challenges to this mainstream view, such as by claiming the superiority of classical antiquity or European architectural achievements. When a German delegate came close, arguing that the concentration of durable monumental heritage in Europe was an indisputable fact and suggesting that his country's rich experience could help others less blessed in this regard, audible grumbling to what some delegates perceived as a condescending tone rose in the hall. 'When he messes up, he messes up big time', an African delegate would later tell me about the delegate, whom he found to be usually well-intentioned.

However, the consequences drawn from this state of affairs differed greatly: limited visibility of the South or some of its regions was a recurring argument in favour of overrulings when these spread from the Brasilia session onwards.

Table 8.2. Properties with State of Conservation Reports and Danger-Listed properties discussed and/or decided by the Committee in 2009.

UNESCO region	Properties	In %	Properties with State of Conservation Reports	In %	Danger-Listed properties	In %
Africa	78	*8.7*	19	*12.9*	12	*40.0*
Asia and the Pacific	186	*20.9*	31	*21.1*	6	*20.0*
Latin America and the Caribbean	121	*13.6*	22	*15.0*	4	*13.3*
Arab countries	66	*7.4*	14	*9.5*	5	*16.7*
Europe and North America	441	*49.4*	61	*41.5*	3	*10.0*
Total	892	*100.0*	147	*100.0*	30	*100.0*

Note: Cf. WHC-09/33.COM/7A, WHC-09/33.COM/7B and http://whc.unesco.org/en/list/stat#d6 (retrieved 5 September 2020). Statistics prepared by the World Heritage Centre are by the original five regions, lumping together the two electoral groups 'Western Europe and North America' and 'Eastern Europe'.

The Advisory Bodies and the countries supporting them, by contrast, while not denying List bias, would still insist on proper procedure and quality standards, which they saw as essential for the credibility of the venture. They pointed to the African World Heritage Fund, the Upstream Process and other support mechanisms as the way to address imbalances, despite the fact that what the richer States Parties invested there was just a fraction of what they spent on their own bids. Importantly, however, this never translated into any kind of prohibitions: voicing understanding for the less visible countries and world regions would not prevent the very same delegates from pushing their own nominations or those of others that were already well endowed. A coalition to stop the likes of Italy or Spain from bringing in yet more candidates and thereby bending everyone's sense of the List to a European optic never arose. Counteraction was restricted to the official PR materials, where invariably properties from the Global South feature more prominently than they do on the List (as also observed by Schmitt (2011: 113–14); cf. e.g. UNESCO (2011: 20–30) for an example).

Arriving in 2009, I found widespread frustration with this state of affairs among participants from the Global South. It expressed itself, for example, in the Brazilian chairperson going into the details of UNESCO regional percentages when deploring List bias in his opening remarks in 2010 or an Indian delegate speaking of a spreading sense that the Convention had not really been made for the developing countries and that these were not as welcome as others. 'We're not wanted', an East African participant told me, with the Europeans thinking that they own the venture and do better in fulfilling its requirements. A Brazilian delegate found an informal conspiracy at work: rich Northern countries best equipped to go through the materials before the session would open State of Conservation reports of Southern countries for discussion, thus deflecting attention from problematic Northern cases such as the Lascaux caves – part of 'Prehistoric Sites and Decorated Caves of the Vézère Valley' (France) – that were also in serious trouble. An Egyptian heritage conservationist was convinced that when Northern delegates cited the credibility of World Heritage to slow down what they perceived as premature inscriptions, they were merely thinking of protecting the value of the World Heritage 'brand' for their own earlier listings. The Advisory Bodies in particular were regularly singled out for their lack of understanding of Southern plights, tacit Eurocentric perspectives and club-like closedness. As an Arab conservation architect described the situation right before Brasilia to me: 'I noticed at that time that there was a real position in the members of the Committee etc., from China, from India, from Arab states, saying that ICOMOS is a European white occidental thing. They don't understand us, they don't understand our heritage … it is colonialism.' A Brazilian diplomat compared the Advisory Bodies to the Free Masons when I interviewed him:

> Since I've started to follow the Committee, I have been shocked by the black box … created in the World Heritage system in which the vocabulary, the concepts and the procedures are understood only by a few. There are some experts who have been in this for many years, twenty years, twenty-five years and who find themselves to be the owners of the World Heritage system … only initiated people can follow.

Also in an interview, an Indian civil servant dismissed criticism of his and other state delegates' assertive behaviour in the most recent Committee sessions: 'They use a word that is amazingly retrograde. "Oh, it's becoming politicized!" All these years when you sat around with a bottle of wine and decided nominations, it was *not* political? Really! Don't hide behind this, this sounds like some Soviet era relic.' It was clear that the targeted organizations were perceived as Northern, with representatives whom the Brazilian saw as falling back on restrictive Northern viewpoints time and again, despite their best intentions.

I think that in all this, the Advisory Bodies, the Centre and the most often Northern States Parties siding with them were not aware of how much the World Heritage Committee sessions are a global stage that puts both triumph and embarrassment on stark display, rather than being only a bureaucratic arena, and that this is more keenly felt by those with a lesser record of success in the game. The above-mentioned East African heritage manager, for example, told me that no matter how justified the individual case might be, it was still shameful for African participants to see their conservation failures being picked on year after year, under conditions where state control of large territories is fragile to begin with. Speaking in the session, one of the above-mentioned Indians also claimed that Southern countries are prone to perceive negative decisions as a loss of face.[7] Despite some of these comments being made for instrumental purposes, they reflect a genuine and widespread feeling that Southern countries have been withheld their due in the World Heritage arena.[8]

Juxtaposing 'Global North' and 'Global South' may feel antiquated in the post-Cold War world order, and Kalb and Steur (2015) have usefully summarised the empirical and political deficiencies of this conceptual pair. However, the division of the world into the two halves of North and South (or Western/Euroamerican vs. non-Western, formerly colonizing vs. colonized, rich vs. poor, etc.) is very much taken for granted by most World Heritage Committee participants, both in formal session interventions and informal conversations and quite irrespective of their own provenance. Where exactly Japan, the post-Socialist EU countries, Turkey, the oil-rich Gulf states or the BRICS powers are to be placed is often left open. Yet, where the Netherlands, Burkina Faso and most others must be put is straightforward. Dividing the world in this way is expected practice in intergovernmental settings and also as a structuring principle, such as when the World Bank formally distinguishes 'Part I' and 'Part II' (i.e. borrowing) countries (cf. Wade 2011). While certainly making for a broad brush, the divide reflects the vastly uneven distribution of political and economic resources in the contemporary world and the fact that a number of countries are quite successful in defending their long-standing political and economic privileges at the cost of others. I therefore think that 'Global North' and 'Global South' retain discursive mileage not just in practical geopolitics but also analytically, as also do other anthropologists (Comaroff and Comaroff 2012; Greenough and Tsing 2003; Hart and Sharp 2015). In any event, within the World Heritage arena, the binary counts as an established fact.

North and South in Brasilia and after

Deploring Northern bias has been a recurrent trope of session discourse for a long time and some past confrontations have been perceived as juxtaposing Northern and Southern countries, such as the aforementioned removal of

Tipasa, Algeria, from the Danger List in 2006 (see Chapter 3) where, despite Schmitt's reservations (2009: 113–19), the US diplomatic cable clearly confirms the salience of the divide in the confrontation.[9] But for the 'Brasilia rebellion', the North–South tension was not just one influence among others, but the dominant structuring factor in my eyes. As described earlier, Sweden, Estonia and Switzerland were the only ones to hold up the flag of proper procedure and expert advice consistently. All three are Northern countries that belong to the EU or its single market, but have only small populations, and none of them is in the top ranks of international relations, unlike the United States and Canada, which had just ended their Committee terms. Their opponents, by contrast, were a rare collection of non- and peri-European heavyweights, including the BRICS states Brazil, Russia, China and South Africa; Mexico as a further G20 member; and Egypt, Nigeria, Ethiopia and Thailand as further populous countries dominant in their world regions. Together, the Southern Committee members represented one-third of humanity, a hundred times more than the three guardians of the Convention.[10] If France and Australia had joined the three, this would have made for a weightier set, but, as already outlined, these two held back much of the time, a fact that some participants ascribed to their wish to protect their own nominations.[11]

Reading this constellation from the point of view of past World Heritage benefits (see Table 8.3) is instructive, as every arena participant with whom I shared this observation agreed: the three European resisters were leading all others on the Committee in terms of World Heritage density by population and mostly also by size – on average, they had fared more than a dozen times better than the others, and an equally successful China would have been close to five hundred (going by size) or even two thousand properties (going by population), not just thirty-eight. 'They're protecting their investment', said a former Committee chairperson from the South when I told her. The three European resisters' personnel were certainly not aware of this fact and appeared sincere to me in their support for ICOMOS and IUCN. Southern participants too, while pointing out that these countries can more easily afford large teams of experts, careful preparation and (in the case of Sweden and Switzerland) political neutrality, valued their competency and relied on them for special tasks, such as by proposing them for ad hoc drafting groups. Certainly, none of the three delegations was intent on constraining List access for its own sake. Yet very evidently, these Northern countries have drunk their wine in the past – with almost empty Tentative Lists and the domestic pressure for more inscriptions nowhere near that of some of their Southern peers such as China (Yan 2018), preaching water to everyone else becomes easier. Repeatedly, delegates from Southern countries pointed out how much less demanding List inscription used to be in the past. In their eyes, Northerners insisting on proper standards and procedures were

Table 8.3. Populations, surface areas and World Heritage property numbers of 2010 Committee members.

	Population in 1,000s	Surface area in square km	World Heritage properties (in 2009)	Per million inhabitants	Per million square km
Estonia	1,340	45,100	2	1.49	44.3
Sweden	9,394	449,964	14	1.49	31.1
Switzerland	7,790	41,284	10	1.28	242.2
Australia	22,327	7,741,220	17	0.76	2.2
Bahrain	807	694	1	1.24	1,440.9
Barbados	257	430	0	0.00	0.0
Brazil	194,946	8,514,877	17	0.09	2.0
Cambodia	14,138	181,035	2	0.14	11.0
China	1,338,300	9,596,961	38	0.03	4.0
Egypt	84,474	1,001,449	7	0.08	7.0
Ethiopia	84,976	1,104,300	8	0.09	7.2
France	64,877	551,500	33	0.51	59.8
Iraq	32,297	438,317	3	0.09	6.8
Jordan	6,093	89,342	3	0.49	33.6
Mali	15,370	1,240,192	5	0.33	4.0
Mexico	108,523	1,958,201	29	0.27	14.8
Nigeria	158,259	923,768	2	0.01	2.2
Russia	141,750	17,098,242	23	0.16	1.3
South Africa	49,962	1,221,037	8	0.16	6.6
Thailand	68,139	513,115	5	0.07	9.7
United Arab Emirates	4,707	83,600	0	0.00	0.0
Top 3 countries	18,524	536,348	26	1.40	48.5
Bottom 18 countries	2,390,202	52,258,280	201	0.08	3.8
World	68,100,000	148,900,000	890	0.13	6.0

Note: Cf. https://siteresources.worldbank.org/DATASTATISTICS/Resources/POP.pdf, pp. 6–9; https://unstats.un.org/unsd/environment/totalarea.htm; http://whc.unesco.org/en/list/xls/?2019 (retrieved 5 September 2020).

therefore 'kicking away the ladder' they themselves had already climbed, much as in other North–South contexts to which Chang (2002) has applied the phrase.

Not just did the North–South divide structure the confrontation in Brasilia, it also shaped the results: of the nine inscriptions supported by the Advisory Bodies, only four were in Southern countries, while four were in Europe and one was in the United States. However, of the twelve upgrade inscriptions, eleven were in Southern countries and only one was in the North (in its outlier Australia). Most regular session participants were certainly aware that outspoken non-European countries were among the driving forces behind the unprecedented conduct of the session, but the surprise that my figures encountered almost every time when I shared them reveal how much the fragmented nature of the Committee sessions (see Chapter 6) occludes the bigger picture. In this case, it is a fairly straightforward one: countries from the Global South finally joined forces to challenge expert rule and its Northern supporters, with countries from the Global South being the main beneficiaries. With the single exception of 1982, the European piece of the inscription pie had never been smaller.

With the three resisters rotating out of the Committee and the more lenient Germany and Japan coming in over the next few years, the line increasingly ran between *all* Committee states and the Advisory Bodies rather than between North and South. However, outcomes returned to the usual Committee pattern: while the share of European and North American inscriptions remained low in 2011, it rose over the following years to reach one-half of the listings in the peaceful 2015 session, making for the highest proportion in fourteen years. In the generally lenient climate, this largely reflected who had bothered to submit candidates. However, generous treatment of Northern nominations was an additional factor. Submitted by Denmark, 'Parforce Hunting Landscape in North Zealand' was recommended for inscription in the ICOMOS evaluation, hailed as the first park on the List specifically designed for (royal) hunting. The main text mentions a whole litany of deficiencies – an arbitrary selection of nominated components, insufficient comparative analysis, the lack of detailed surveys, problems of authenticity, inadequate management and insufficient supporting research.[12] These are very similar to the problems identified for two African candidates, 'Thimlich Ohinga Cultural Landscape' (Kenya) and 'Nyero and Other Hunter Gatherer Geometric Rock Art Sites in Eastern Uganda' (Uganda).[13] Therefore, an ICOMOS insider told me, all three candidates were foreseen for deferral in the first evaluation draft discussed by the internal ICOMOS World Heritage Panel. Yet the Danish authorities made use of the newly introduced feedback process, receiving extensive hints on how to improve their bid, and the final ICOMOS evaluation not only changed the initial judgement but also suggested the use of an additional OUV criterion,[14] thus providing a very exceptional service. For the two African candidates, by contrast, no feedback support is mentioned in the evaluations. I am sure that spatial closeness to ICOMOS headquarters and network ties to European-dominated ICOMOS circles played a role for the Danes making use of this option when the Kenyans and Ugandans did not. When I pointed out the discrepancy between the evaluations, a Southern participant to the ICOMOS process told me in Bonn that Denmark, as a rich country with a long history of heritage conservation, could be relied on to address remaining issues after inscription in a way that the African countries could not. Yet, this amounts to justifying double standards, which were not even concealed: as mentioned above, the critical analysis of deficiencies was not removed from the evaluation text for the Danish site.

In the session debate, a Lebanese Committee delegate argued that the African nominations would have been much more in need of the feedback process than Denmark, resulting in an unusual burst of spontaneous applause in the hall. His Senegalese colleague from the only Sub-Saharan African country on the Committee pressed for inscribing the Kenyan property, given that it could be seen as an archaeological site rather than a cultural landscape and that ICOMOS saw some potential for OUV in this regard. He implored his Committee peers that postponing the nomination would amount to 'sectarianism and exclusion', leaving one of the UNESCO regions entirely without new titles, and together with Algeria, whose representative likewise declared herself perplexed by the divergent treatment, he submitted an amendment for inscription. Yet with Kenya not having done much lobbying, the other Committee members, including the Southern ones, relished the rare opportunity to show commitment to expert advice. Uganda had not even bothered to send a delegation to Bonn (a factor that, again, may have been due to a lack of resources). The Committee did upgrade both nominations in the end, but only to referral, not to outright inscription. The friendly free-for-all Bonn session was therefore the first since Brasilia to pass over a UNESCO world region entirely, despite regular assertions in the sessions that the List was short of African properties, particularly cultural ones. Sure enough, 'Thimlich Ohinga Archaeological Site' was eventually listed in 2018. However, by that time, Denmark had already obtained its next two inscriptions, whereas the Ugandan rock art property is still to return.

The composition of the Bonn Committee would have lent itself to a different outcome: no permanent Security Council member was on board, Germany and Japan were the only G7 members, and neither of these two economic rather than military giants nor the other Northern members (Finland, Portugal, Croatia and Serbia) are known for a very assertive stance in contemporary geopolitics. The fifteen Southern members of the Committee – including oil state Qatar and G20 states Turkey, India and South Korea – might thus have found little resistance had they united to push for a more principled anti-Eurocentric agenda. But they did not, focusing instead on their national interests. Thus, for the first time in a decade, not a single World Heritage title was awarded to the forty-eight Least Developed Countries (LDCs).[15] The poorest World Heritage title recipient of the session, Committee member Jamaica, was still ahead of almost eighty other countries in GDP per capita (no. 113 in the world in 2015), and while Jordan (no. 95), Mongolia (no. 94) and China (no. 87) each had a site listed too, the other non-Western inscriptions went to the significantly richer Uruguay, Mexico, Saudi Arabia, South Korea, Japan, Singapore and (two each) to Iran and Turkey.[16] Delegates continued to talk North versus South and Europe/the West versus the rest when it suited their purposes, but solidarity along this fault line and rooting for the likes of Kenya (no. 153) or Uganda (no. 167) was a different matter.

Therefore, in the post-Brasilia decade of the World Heritage Committee, while the Europe and North American proportion of new listings has decreased, this is still the UNESCO region with the most titles (see Table 8.4). Of the other four regions, Asia and the Pacific is the only one to have significantly expanded its share. However, those of Latin American and the Caribbean and Africa have actually shrunk. Comparing the most successful countries of this decade with those until 2009 (see Table 8.5) shows that Italy and Spain have dropped somewhat, while China (now tied with Italy for the overall pole position), Iran and Turkey have surged. Yet, the earlier top ten are still all in the top nineteen and there are only five new names among the latter. The post-Brasilia decade has thus continued the Bonn trend, having only limited effect on North–South imbalances in the World Heritage List. At the same time, Africa and the Arab countries dominate State of Conservation Reports and the Danger List just as they did before.[17]

Why, one wonders, did the Brasilia rebellion not achieve more for the Global South it started out benefiting so much in terms of World Heritage titles? I think that there are two reasons, one being the resilience of Northern bias passing under the radar, especially among the Advisory Bodies, and the other, more fundamental one being the general lack of Southern solidarity in the years when after the economic crisis, windows of Southern opportunity suddenly opened across global governance. Let me demonstrate each of these points in turn.

Culture Chaos and the Resilience of Northern Bias

So what, as the first point, do the Advisory Bodies – in particular ICOMOS, as the main target of the complaints – do to invite perception as Northern rather than as the international bodies their names lay claim to? After all,

Table 8.4. World Heritage inscriptions per UNESCO region before and since 2010.

UNESCO world region	1978–2009	In %	2010–19	In %	Total	In %
Europe and North America	441	*49.4*	88	*38.1*	529	*47.1*
– Europe	**408**	***45.7***	**79**	***34.1***	**487**	***43.4***
Asia and the Pacific	186	*20.9*	82	*35.5*	268	*23.9*
Latin America and the Caribbean	121	*13.6*	22	*9.5*	143	*12.7*
Arab States	66	*7.4*	21	*9.1*	87	*7.7*
Africa	78	*8.7*	18	*7.8*	96	*8.5*
Total	892	*100.0*	231	*100.0*	1,123	*100.0*

Note: Derived from http://whc.unesco.org/en/list/stat#d6 (retrieved 5 September 2020). For the European figures, I subtracted those for Canada (cf. http://whc.unesco.org/en/statesparties/ca (retrieved 5 September 2020)) and the United States (cf. http://whc.unesco.org/en/statesparties/us (retrieved 5 September 2020)), taking into account their two joint inscriptions ('Kluane/Wrangell-St Elias/Glacier Bay/Tatshenshini-Alsek' (joint since 1992) and 'Waterton Glacier International Peace Park').

ICOMOS played a leading role in the reforms in the 1990s that broadened the scope of cultural heritage. Among the World Heritage advisors drafting and presenting the evaluations in the observed sessions and representing ICOMOS in expert meetings, advisory missions and the publication of manuals and guidelines, one would find a cultural geographer known for a sensitive ethnographic study of the Umayyad Mosque of Damascus (Rudolff 2010) or a historian of civil engineering, not art historians specializing in European churches. The most senior Advisor, a historian and archaeologist who started out with studying the English Lake District, went on to publish on African traditional architecture (Denyer 1978) and work in India and China. In our conversations, the Advisors struck me as cosmopolitan in orientation, certainly more intent on supporting the global mission of World Heritage than on reproducing Western supremacy, and, as already explained for the evaluation process, they were not hindered from choosing desk reviewers with all kinds of qualifications from all over the world.

Yet, it is a different matter whether they were *perceived* as cosmopolitan by session participants, starting with appearances. All three Advisory Bodies have their headquarters in Western European countries, as most participants know. In most of the observed sessions, ICOMOS personnel presenting the evaluations was exclusively white and I assume that most participants – to whom, as already mentioned, their details were not introduced – took them to be Europeans, even when they actually included a Mexican national as one

Table 8.5. States Parties with the most World Heritage inscriptions before and since 2010.

1978–2009		2010–18	
Italy	44	China	17
Spain	41	Iran	14
China	38	Germany	13
France	33	France	12
Germany	33	India	11
Mexico	29	Italy	11
United Kingdom	28	Japan	9
India	27	Turkey	9
Russia	23	Spain	7
United States	20	Denmark	6
Australia	17	Mexico	6
Brazil	17	Russia	6
Greece	17	Brazil	5
Canada	15	Canada	5
Japan	14	South Korea	5
Sweden	14	Portugal	4
Poland	13	Saudi Arabia	4
Portugal	13	United Kingdom	4
Czech Republic	12	United States	4
Peru	11	14 other countries	3
Total for all countries	892	Total for all countries	231

Note: Derived from http://whc.unesco.org/en/list/xls/?2019 (retrieved 5 September 2020).

among four or five main faces. Since 2015, when names have been published, the ICOMOS World Heritage Advisor team has always included between three and five Europeans, which alone or together with the Australians and Canadians would constitute a majority. By contrast, individuals from African and Arab countries were only employed in some years, and Asian and Latin American nationals still wait to be called. ICOMOS officials often cite the difficulty of finding qualified people with good English and French, the two working languages, but one wonders whether this is sufficient justification. IUCN introduced a greater variety of nationalities and appearances somewhat earlier, but the main speaker in the sessions was a majority Englishman, with other Europeans and North Americans also presenting. The orientation meeting on the afternoon preceding the Paris session in 2011 was typical in this regard: all three Advisory Bodies had picked a British or American national as their main speaker. None of the three was a surprise, as they were the most experienced representatives of their respective organizations, being familiar faces to and respected by many in the World Heritage arena. Yet still, this meant that first-timers would not just learn the ropes of Committee business, but also that it is normal for prototypical Northerners – white native speakers of English – to explain World Heritage matters to the rest of the world. Even if they were aware of this aspect, the impression this might make did not bring any of the Advisory Bodies to reconsider their choice.

Northern dominance continues in the ICOMOS World Heritage Panel, the internal body finalizing the recommendations. Reacting to Committee criticism, ICOMOS has published the composition of the Panel since the evaluation cycle leading up to the 2015 session, complete with statistics giving the number of panel members per world region. These show that in all five years, the eleven to seventeen Europeans and North Americans were an absolute majority of the nineteen to thirty-one voting panel members. No other UNESCO region ever sent more than seven members and there could be only a single person from a region (Latin American and the Caribbean in 2018), less than the two or more voting members that some – invariably Northern – single countries sent (Germany in 2015, the United States in 2016 and France in 2016 and 2019). As if to excuse the lopsided picture, the statistics juxtapose the percentages of panel members by world region with those of candidate sites by world region, demonstrating rough proportionality.[18] But the implicit claim in this arrangement – that experts from the same world region are best prepared to judge the OUV of a candidate property – undermines the purported universality of both OUV and the evaluation process. The figures for World Heritage Advisors also must mean (and my other information confirms this) that while quite a few evaluations of Southern candidate sites are drafted by individuals from the Global North, the reverse rarely applies. And since Panel members are not expected to hold back on candidates outside their world region, recommendations for both Northern and Southern candidates are decided by a body with a stable Northern majority. As already mentioned, IUCN usually does not give the nationalities of its Panel members. However, going again by what is available about these individuals on the public internet and in the LinkedIn network, six out of ten voting members in the 2016–17 cycle were European or North American, with an Australian in addition, and in the 2018–19 cycle, four out of eight were.[19] The World Heritage Centre is even more extreme in this respect: in 2010, a Mexican delegate deplored in the session the fact that 62 per cent of staff were European or North American, without being corrected about this claim.

I do think that distribution matters not just because it normalizes Western dominance but also because of the role of intuition and sensing OUV when assessing candidates. This holds particularly for the cultural properties, where, as outlined in Chapter 7, there is so little else in terms of standards or a solid comparative methodology to go by. Yet, ICOMOS's insistence on the scientific nature of its recommendations is not only questionable in this regard; its World Heritage Panel is also not a particularly scientific body to begin with. In the Panel for 2015, for which I checked members' credentials, no more than six out of nineteen Panel members and only one out of six (nonvoting) Advisors held a doctoral degree, much as this is taken to certify the competent scientific practitioner the world over. The ICOMOS President himself had to correct participants in the 2012 Committee session after these had repeatedly referred to him as 'Dr Araoz', saying that he was 'just a normal person'.

Instead, the ICOMOS World Heritage Panel is a political and practitioners' body. In 2015, it included no less than half of the twenty elected members of the current ICOMOS Board, among them the president, four of the five vice presidents, the secretary-general and the treasurer. The membership's trust in these individuals is therefore proven, but may stem from a lot of other considerations than merely their capacity to judge OUV. Also, as many as thirteen members were conservation architects or conservation planners, whereas there was only one trained historian, one geographer and no anthropologist or sociologist on the Panel. Despite fervent calls in the 2011 General Assembly to diversify Advisory Body expertise, not much had changed compared to the composition in the 2006–8 cycles, the regional and disciplinary uniformity of which had been strongly criticized in an external audit commissioned by the Committee (Tabet 2010: 29). The leading discipline is still one that has formed in Europe and in dialogue with European built monuments.

All this is not to deny the impressive practical experience, publication records (largely in applied fields) and university teaching (usually not in the most famous ones) of many of these individuals. Yet still, this is not a global panel of academically acclaimed cultural heritage authorities, but a body where the average member is a practising conservationist from the Global North who has managed to get elected into ICOMOS office. That member also might wish to stay in office. In an organization with many more voting members from the Global North than from the South (where quite a few Convention treaty states do not have a national ICOMOS committee to begin with, so that their ICOMOS members cannot vote, even if they managed to pay their way to the ICOMOS General Assemblies with their considerable participation fees), this might create the temptation to cater to what the electorate might wish to see. One insider saw a clear conflict of interests of which, he said, even Panel members were critical.

I should hasten to say that I was able to convince myself of the integrity and cosmopolitan commitment of many a recent ICOMOS World Heritage Panel member during my fieldwork. The Panel is also not automatically partial to European and North American nominations – for the 2015–19 period, the rate of recommended inscriptions was the second-lowest for all regions.[20] Still, I think it is inevitable that what is more familiar to the majority receives a more nuanced and welcoming treatment, particularly when a body is so uniform. Apart from the effects of the personal encounters during the feedback process, this is how I explain to myself the much more lenient treatment of a Danish cultural landscape featuring European Baroque than of the two African ones. Two years earlier in 2013 as well, another such Baroque park, 'Bergpark Wilhelmshöhe' (Germany), had met with ICOMOS approval, largely because of the sophisticated waterworks. By contrast, ICOMOS recommended noninscription – implying certainty about the absence of OUV – for a cultural landscape composed of a series of Buddhist temples, Shinto shrines and other sites in Kamakura, Japan's third most important historical city. In the eyes of most Japan specialists, the Kamakura sites are no less important than those of Nikkō or Hiraizumi, which do feature on the World Heritage List. The evaluation objected that the city's ancient sites were not on a par with those of Kyoto or Nara,[21] but analogous reasoning would require the Kassel palace or, for that matter, the Danish hunting park to be the match of Versailles, Schönbrunn or the Escorial, something that few specialists would venture to claim. Tellingly too, the eighteenth-century monumental sculpture of Hercules in the Wilhelmshöhe gardens – not much of a landmark in art history – received special praise in the evaluation including for its technical features.[22] By contrast, the thirteenth-century bronze image of the buddha Amida in the Kōtokuin temple in Kamakura – one of the most famous monumental sculptures of Buddhism and a technical challenge as well (cf. the historical documentation displayed on site) – was not even mentioned. Clearly, there is more readiness to appreciate the subtle differences between European Baroque parks than those between Japanese historical cities, a tendency to which a more regionally and professionally diverse ICOMOS World Heritage Panel with no ICOMOS office-holders might be less susceptible.

In the latter case, there might also be less 'culture chaos', my term for the unconscious retention of a conventional culture concept even when a broader one is officially expected to be applied. As already described, the Global Strategy followed the historical course of anthropology when moving beyond a monumental and elitist concept to include everyday and vernacular expressions and the culture of the common person (see Chapter 3). However, it never outlawed established categories. This must have motivated leaving in place the established Advisory Body that had coalesced around European architecture and (to a lesser degree) archaeology rather than searching for fresh expertise. And this means that in the ICOMOS World Heritage Panel, a majority of practising conservationists from Northern, most often European countries deliberate the OUV of African sacred groves or Oceanian chiefly domains, a task to which they bring no more expertise than many others. Within their specializations and countries, the individuals running for ICOMOS office and volunteering for the Panel are certainly at the cosmopolitan end, a far cry from those ICOMOS experts who in a 1986 expert meeting 'made disparaging comments about the heritage of disadvantaged countries such as Guinea Bissau and the Central African Republic, countries they saw as lacking in heritage of universal value' (Gfeller 2015: 370, 383); the blatant Eurocentrism that suffused a number of presentations during the conference preceding the 2008 ICOMOS General Assembly in Quebec, which I attended; and the conservatism of a leading ICOMOS office-holder who as late as the 2000s would tell an African interlocutor that the World Heritage List is for European cathedrals and parks, not for African sacred forests. But they are not immune from slipping back into the elite mode when assessing the properties, as not just the Baroque parks but also the following case demonstrates.

For the 2011 session, Palau and the Federated States of Micronesia jointly nominated the 'Yapese Stone Money Sites', taking up one of the best-known cases of prestige money (Furness 1910) that is also an anthropological textbook staple. The property included not just the dancing ground on Yap where most of the huge stone discs are on display, but also two quarries on Palau where the Yapese cut the discs in the past to sail them home over hundreds of dangerous kilometres, with several half-finished exemplars still sticking from the rock today. Also included was the former residence of a US American sea captain on Yap whose steamship had encouraged an age of decadence with ever-larger discs but decreasing prestige value. One would be hard-pressed for better testimony of long-distance connections in unlikely places. Being even more anthropological than the nominators, ICOMOS missed details of the related traditions and rituals and of the association of the dancing ground with the discs[23] and proposed the inclusion of nearby settlements in the proposed property, as these were 'intangibly functionally connected'.[24] And here indeed, the nomination file – largely drafted by archaeologist Scott M. Fitzpatrick (cf. Fitzpatrick 2010: 144) – gives little information and fails to evoke comparable examples of daring Oceanian seafaring in pursuit of prestige such as the famous *kula* ring (Malinowski 1953 [1922]): given the general romanticism for things indigenous in this arena (see Chapter 5), I think that this was a crucial omission.

However, what most motivated ICOMOS to recommend deferral was the lack of a 'carefully documented inventory of the stone money discs … recording location, history, ownership, and category in a typology of production from the use of shell and bone tools to the use of metal tools, based on observation of size, shape, tool markings and shape of the central hole' since this 'has the potential to assist with dating of the pieces' (ICOMOS 2011:196).[25] I see this as a – perhaps unconscious – slippage from anthropological into elite culture, as the discs are now treated like the artworks in a museum or the finds in an archaeological dig (where dating could indeed be important). The inventory becomes an end in itself, not a mere means for conservation and against threats such as theft or physical damage: given the material and weight of the discs, the remoteness of Yap, the almost complete lack of tourism and the visibility of any intruders to locals, a collection of photos documenting the current location and condition of the discs would fully suffice as a practical baseline for conservation. Yet, in the session debate, nobody thought of questioning the demand, irrespective of the burden it would place on two small island states lacking in specialized personnel and resources, and the exasperation of Palau's representative. The latter was new to the forum and his last-minute lobbying attempts among Committee members – once he saw what was common practice – had failed to elicit support, as he told me. In my assessment, contacting Committee members earlier could indeed have led to an inscription upgrade, also given the fact that several declared themselves impressed by the sites in the debate. Therefore, it was indeed the inventory question that kept the stone money sites off the List – at present, the nomination is still to return.

Interestingly, Committee delegates from the very Global South that suffers most from 'culture chaos' were sometimes complicit. Looking back on the Saint Petersburg session day, an Indian official wondered how the Swedish farm houses could possibly have OUV when other sites that embodied nation-building (Russian kremlins) or were the oldest urban settlement (Çatalhöyük) did not. 'These must be wonderful farm houses, but if we look at civilizations, come on, let's be clear about what we do', he told me, with more than a nod to monumentalism.

As I have shown in more detail elsewhere (Brumann 2018: 1221–23, 1226), 'culture chaos' is by no means restricted to the World Heritage arena, but is also prevalent in many other UNESCO-related activities concerning culture, despite a declared will to open it up, and in general contemporary public discourse. We find it, for example, when ethnographic museums exhibit their collections in the decontextualized manner of works of art or when in the German political controversy about a majority *Leitkultur* ('lead culture'), which immigrants should adopt, both the protagonists and the opponents shift between the elevated (such as symphony orchestras or the 'Judeo-Christian tradition') and the trivial (such as sauerkraut and garden gnomes) all the time when giving concrete examples for culture, with no obvious awareness of a shift. I am therefore not surprised to see such slippages in World Heritage Committee sessions. Yet if superseding them is a key objective, relying on Northern conservation architects for assessing the world's cultural heritage is not well suited to the task.[26]

The Limits of Southern Solidarity and of UNESCO

Let us return to the peaceful Bonn session and the one controversy that divided not just Japan and South Korea or Israel, Palestine and Jordan, but the entire Committee. In the midst of the UNESCO budget crisis and in an effort to adapt work time and costs to limited means, two working groups meeting in the session breaks came up with the proposal to limit World Heritage nominations to one candidate per state and twenty-five overall for each annual cycle, down from the prior two and forty-five, respectively. After all, since the option to present two candidates within the same year was introduced in 2005, only eleven out of the 193 treaty states had used it more than once – the Northerners France (six 'double nominations'), Germany (three), Italy (three), Israel (six) and Japan (two); Russia (two) and Turkey (three) too; and the Southerners Iran (five), India (four), China (eight) and Mexico (three).[27] Except for Israel, all these countries already had at least seventeen World Heritage sites before Bonn, placing them in the top 10 per cent of treaty states. '25/1', as the proposal was referred to, would thus have hurt only a well-endowed minority, while it might have helped to shift the attention and resources bound by nominations to the less weighty states of the Global South, for whom a single nomination every couple of years is already a challenge. In the working group sessions, the proposal enjoyed broad support. But when it was finally brought to the plenary, a non-Western trio – Committee members India, Turkey and Japan, all of them with long Tentative Lists of candidate sites avidly awaiting nomination – forcefully argued for more deliberation rather than a hasty decision, with Germany offering only lukewarm opposition. Other delegates grumbled about this turnaround, but in the generally nonconflictive mood, '25/1' was postponed.

Table 8.5 demonstrates the consequences of this: while for the period until 2009, it requires the top nineteen countries to reach one-half of all inscriptions, for the 'post-Brasilia' decade of 2010–19, it requires only twelve countries, with four of them each accounting for more than 5 per cent of listings, a mark that not even Italy reached previously. As a consequence, treaty states' inequality in terms of World Heritage properties, measured by the Gini coefficient, has hardly changed (0.642 in 2009 and 0.636 in 2019)[28] and continues to be higher than their inequality in terms of GDP per capita (down from 0.549 to 0.527 in the same period).[29] Some Southern countries have increased their pieces of the pie, but the distribution of World Heritage is as skewed as ever.

So what is the reason for the rebellion coming in 2010 rather than earlier, and for so little change in the aftermath? In order to provide an answer to this question, it is instructive not to isolate the World Heritage arena, but to look at the context of what happened in other global bodies at very much the same time. In the autumn of 2008, the US subprime mortgage crisis and the collapse of the Lehman Brothers investment bank triggered the worst financial meltdown since the Great Depression in many North American and European countries. However, many of the leading Southern economies ended up less affected, so that countermeasures included attempts to integrate them more strongly into global financial governance. In a world where by 2030, the US and European share of global GNP is expected to have dropped to one-half of what it was in 2010 (Laïdi 2014: 356–57), it appeared to be high time to adapt global financial governance to the changing circumstances. Therefore, the G20, the World Bank and the International Monetary Fund (IMF) initiated reform processes that were meant to give the Global South more of a 'voice'.

The G20, which began as a consultative forum of finance ministers in 1999, was upgraded to heads-of-government level meetings in 2008 ('G20L'), on a par with the G7 summits. In comparison with the G7 or (when including Russia) G8, the G20 indeed provide for a broader representation of global economies, as the included countries – when counting in the joint EU seat – cover almost 90 per cent of global GDP and two-thirds of the world's population. Yet the composition continues to draw criticism and divides the Global South – going by the official yardsticks of GDP and population, the excluded Nigeria, Egypt and Colombia had a no less plausible claim than Argentina, Saudi Arabia, South Korea and Australia, which were included. Also and even after the G20L upgrade, the agenda continued to be dominated by the United States and the other G7 members. Southern members became more assertive, but were still mainly reacting to Northern initiatives rather than taking it themselves (Wade 2011: 354–59).

Between 2008 and 2010, the World Bank redistributed its voting shares that couple a country's financial investment with its number of votes in bank decisions. This was officially determined to give the poorer countries the parity they demanded instead of their prior 42.8 per cent. Yet the final agreement went only halfway, raising their share to 47.2 per cent, and this value was artificially boosted by including states that do not borrow from the World Bank, such as Poland, Hungary, Saudi Arabia, Kuwait, South Korea and Singapore, among the 'development and transition countries' (DTC). Some Northern countries, including most G7 members, did indeed lose shares, yet the revision gave little to the poorest states, benefited China, South Korea, Turkey and Mexico but less so India, and did not abolish the up to eightfold overrepresentation of some Northern and particularly obstinate countries. Belgium and the Netherlands combined still have almost as many shares as China and India combined, even though they only have one-sixth of their GDP, and Russia's and Saudi Arabia's pieces of the pie remain large, since these countries threatened to block the entire reform otherwise. Note that 'overrepresentation' refers here to the respective country's share of global GDP, which is supposed to determine 75 per cent of the distribution, so that decision-making power in the World Bank continues to be even more skewed than the world economy as such (Vestergaard and Wade 2013: 153–59; Wade 2011: 359–63; 2013: 20–23). Yet while China passed on half of its entitled shares to other DTCs (Vestergaard and Wade 2013: 159, Wade 2011: 362), 'months of nearly 24/7 negotiations' seeking 'compromises at the third decimal point' in 'a messy and opaque process … that caused a lot of bad blood' were dominated by national self-serving (Wade 2011: 361–62). Further Southern solidarity did not arise, much as the whole reform was meant for the good of the Global South. Given that subsequently, five of the G7 members ended their former practice of voluntarily forgoing part of their voting rights and that some Southern countries did not subscribe to their allotted increase (not wanting to make the corresponding investment), the actual shift of shares from high- to low- and middle-income countries dwindled to a mere 0.46 per cent (Vestergaard and Wade 2013: 159; Wade 2013: 23).

The IMF governing body, urged by G20 demands, likewise decided on a new distribution of quota shares in 2010. This indeed moved 6 per cent of shares, but here too, the category of 'advanced countries' lost less than had been initially expected in the negotiations, and the winners were China, Russia, India, Brazil, Turkey and Mexico rather than the poorer countries, of which the African group even lost shares (Wade 2011: 363–65). And when a new World Bank president had to be appointed after Robert Zoellick's resignation in 2012, the final choice fell on American Jim Yong Kim. G20 leaders had agreed that the customary arrangement of filling the World Bank presidency with an American and the IMF presidency with a European should end, and the nationality of the candidate should not matter any longer. Robert Wade – apparently well informed because of his own former employment in the World Bank – recounts in detail how taking this directive to heart, the eleven (out of a total of twenty-four) voting executive directors hailing from the Global South managed to agree on a joint candidate at one point. As this Nigerian was a former managing director of the bank, more experienced in finance and economics than her contender and a woman in addition, it would have provided an 'opportunity for Obama to enter the history books' in Wade's assessment (2013: 27). Yet in an election year, the US President was under pressure to keep a symbol of American pre-eminence, and the Europeans sided with him because of American return support on other issues. Crucially, however, Southern backing for the Nigerian candidate also crumbled one by one, in Wade's assessment through horse-trading with the Americans so that in the months following Kim's election, a string of senior World Bank appointments went to nationals of big Southern countries such as China and India (Wade 2013: 24–29). For Wade, the whole episode 'shows how the developing countries' distrust of one another makes it easy for the Americans to split them with bilateral deals' (2013: 29).

On the basis of these and similar observations, such as a study of how Southern countries refrain from taking the lead in transnational institutional innovations, even those specifically designed for the benefit of the Global South (Hale and Held 2011: 25–26), Wade argues that many of the emerging states 'wish to be *at* the top table, *in* global governance, but are wary of *more* global governance (which would curb their sovereignty) and wary of proposing initiatives that would put more responsibilities on their shoulders' (2011: 366, emphasis in original). In the terminology he develops, they thus subject themselves to some degree of 'hegemonic incorporation' under the continued leadership of the North and fall back on 'Westphalian assertion' (that is, the pursuit of national interests and sovereignty) most of the time, but do not arrive at genuine 'multilateral cooperation'.

I find the parallels with how the Brasilia rebellion played out in the World Heritage arena striking. This includes not only the timing but also the list of Southern winners. As shown above, the World Bank and IMF voice reforms saw seven G20 countries not part of the original G7/G8 setup succeed in increasing their shares or preserving an already high level in either or both institutions – China, India, Brazil, Turkey, Mexico, South Korea and Saudi Arabia. Five of these countries also rose in terms of World Heritage inscription rankings in the 2010s, often significantly and with Iran being the only other Southern country to do so. Mexico and Brazil dropped somewhat in rank, but in terms of their proportional share of listings, Mexico lost only slightly while Brazil actually won (see Table 8.5). However, African countries lost out, both in terms of IMF shares and World Heritage properties (see Table 8.4).

My reading of the parallel is therefore as follows: emerging countries such as the seven just highlighted ones had already been outperforming the West in terms of economic growth for a while, making demands for better representation in global bodies that were increasingly difficult to ignore. Then, the unexpected financial crisis threw the established geopolitical and economic world order even further into doubt. The idea of broadening global governance now also won support from Northern countries intent on using the South's economic momentum to overcome the crisis. The moment of transition gave leading Southern countries a new sense of importance and potential, and when the realization that experts could simply be ignored spread in the Brasilia Committee session, they grasped the opportunity, just as they did in the voice reforms. Southern exclusion had led to this point and continued to be targeted as a problem in Committee session rhetoric, but when the window of opportunity opened, perhaps too suddenly, the leading Southern countries thought of themselves, not of the Global South (which, of course, merely mimicked what Northern countries had been doing all along).

In Wade's terms then, Westphalian assertion has become the agreed mode of operation of the Committee ever since, and the Northern-dominated expert bodies were also finally brought in line. Yet in many ways, this is only hegemonic incorporation, given that both the Eurocentric residues of the Advisory Bodies and the Northern states' habit of flooding the World Heritage List with their sites remain unchallenged, provided that the specially interested Southern countries are not hindered from doing so too. However, Southern solidarity in the absence of preparatory lobbying, such as for a Ugandan cultural landscape, is not forthcoming. What arises instead is an informal 'G11' of Northern and Southern List leaders that team up to cement their List dominance with double nominations, irrespective of the effect on the system's capacities and the complaints about geographical imbalance. All other things being equal, states from the Global South should prefer eradicating Northern bias and increasing Southern influence in the World Heritage system, just like they should prefer a World Bank presidency open to Southern nationals. Yet, this potential shared return is delayed and requires special efforts, so that it loses out against the immediate benefits to be reaped from the status quo: when doing a modicum of lobbying or when sitting themselves on the World Heritage Committee, interested Southern countries can realize their national interests even without reform, just as those represented on the World Bank executive board, by agreeing to yet another US presidency, could corner second-tier positions for their own nationals.

In all this, we have to factor in the time lags initially mentioned: had the World Heritage arena not gone on for so long to preach universalism and a Global Strategy while actually practising Northern supremacy, particularly in the aggregate outcomes, redistributing only very little in terms of financial resources at the same time, there might have been less of the frustration among Southern representatives I have described. This might then have made for more patience to renegotiate general terms rather than scrambling for national inscriptions as soon as the opportunity arose. Yet as the World Bank and IMF parallel demonstrates, it was very much the sign of the times for the stronger Southern countries to think of their own benefit. Since these countries basically got what they wanted in the World Heritage arena by overruling the experts, rather than rethinking expertise and revising World Heritage fundamentals, Northern bias has remained in place. As a result, the Committee has ended up practising business as usual, with a couple of Southern winners, but Northern hegemony less affected than might be expected. Tim Winter (2015) has recently postulated 'heritage diplomacy' as a separate field with its own trajectories. Yet, what we see in the recent transformation of the World Heritage arena is not distinct from what went on in other global bodies at very much the same time. It has little to do with World Heritage proper, beyond the title being a coveted resource.

It is here where I most disagree with Lynn Meskell's book-length account (2018). Meskell places the World Heritage Convention and its development within the history and mission of UNESCO, and, in doing so, is admirably broad in her coverage, unearthing a wealth of connections between facts and incidents unfolding over many decades. However, I have trouble finding a clear social scientific message in her account beyond the general, somewhat tragic sense – 'a future in ruins' – that things have gone the wrong way: UNESCO and World Heritage, often blended into a single entity, have failed to exhaust their potential and live up to the UN agency's mission of global peace. This diagnosis has merits as an exercise in intellectual history. Yet, in terms of explaining what has been going on in the World Heritage Committee, evidence is needed of how larger UNESCO discourses and policies manage to reach the World Heritage arena and influence the individuals and organizations playing a role there.

I do not want to deny that such discourses and policies are present to a certain extent. The Director-General of UNESCO is the depositary of the Convention, acts in its name in emergency situations and appears at the annual sessions, usually to appeal to Committee members and other States Parties to honour the spirit of the Convention and the Global Strategy. The World Heritage Centre as the Convention secretariat is part of the UNESCO headquarter bureaucracy and while there are some officials who stay in the Centre for a long time, others rotate in and out. UNESCO member state diplomatic representatives – the Permanent Delegates and their personnel – usually cover all UNESCO activities, meaning that in crash-course fashion, they become acquainted with the general habits and discourses of this UN agency. UNESCO specificities certainly exist: I have already pointed out how the involvement of topics such as culture and education leads to a greater diversity of home ministries and educational backgrounds among the diplomats dispatched to UNESCO than in other UN agencies (see Chapter

5). The tone of discussions is often idealistic, such as when in the General Conference of 2011, one member state representative after another waxed lyrical about UNESCO's peace mission and cited famous writers in order to underline its importance, only to then regret their inability to help out with the huge deficit that the US suspension of dues had caused. UNESCO has also been perceived as an ally of the Global South in the past, lastingly concerned with redress for past and present Northern wrongs (Hauser-Schäublin 2013), committed in its fight against the residues of scientific racism (Maurel 2007) and an active promoter of non-Western perspectives, such as in projects for a universal history (Brisson 2015: 545). Adoption of the MacBride Commission report 'Many Voices, One World' that proposed a 'New World Information and Communication Order' against US and UK protests made these pull out of UNESCO in 1985 and 1986, respectively (Wells 1987), not to rejoin before 2003. Much of the practical activities of UNESCO such as the dissemination of educational standards are also more relevant for Southern states, whereas schools and universities in rich countries have less need for guidance, making the organization a larger presence in the Global South than the North, as I was told. All these past developments have certainly helped to set a general tone and rhetoric where Northern countries have learnt to throw their weight around in a less blunt fashion than, say, in the Security Council, the WTO, the IMF or other UN bodies where the political and economic stakes are higher; tellingly too, globalization critics target these latter bodies rather than UNESCO (Schmitt 2011: 144).

As a further specific, UNESCO has been described as particularly far-flung and fuzzy in its portfolio of activities, with a tendency to adopt every initiative that the member states bring to its door, to the point of becoming a 'palace of the invisibles' in a widely read polemic (Fichtner 2008). Both State Party delegates familiar with other UN organizations and UNESCO personnel tend to depict it as poorly organized, with a crippling amount of bureaucratic red tape, lots of advance communication required to get anything done and extensive self-referential reporting for internal monitoring. The UNESCO staff I met, not just inside but also outside the World Heritage Centre, appeared highly motivated to me, calling the almost absurd heights of indolence, inertia and nepotism reported in the (admittedly satirical) account of a retired official (Courrier 2005) into question. Still, what I saw myself suggested a less effective apparatus than, say, the average national ministry bureaucracy of the more resourceful countries. Given the breadth of backgrounds and nationalities, this is hardly a surprise.

All this influences the World Heritage arena and, as already pointed out, happenstance – such as the fact that Palestine went for full state membership at UNESCO first, so that the ensuing US withdrawal and disastrous budget loss struck this particular UN agency rather than others – can deeply affect it. But it needs to be stressed yet again that both the World Heritage General Assembly and the World Heritage Committee are independent of UNESCO, its officials and its governing bodies – currently, even their composition is different, with the United States and Israel remaining Convention States Parties while having left UNESCO. Had the Director-General any authority over the Committee, she would not need to repeat her exhortations to observe proper procedure and to practise restraint year after year, and had the general UNESCO climate been decisive for the Committee at an earlier stage, the European dominance of the World Heritage List would never have developed. In order for longer-term UNESCO trends and developments – such as the turn from archaeological research towards heritage conservation, a move that particularly irks Meskell (2018: 37–44, 57–58) – to exert an influence inside the World Heritage arena, someone must carry them there.

If World Heritage decisions were taken by individuals sworn to honour general UNESCO policies, this would indeed be conceivable. But as I have emphasized, decision-making authority rests with nation states and the diplomatic personnel they dispatch to UNESCO and the sessions. While these may end up adopting a loose UNESCO style and perspective to some degree, they are neither specialists of the agency's history nor particularly knowledgeable about the evolution of the World Heritage arena – often, they seem to be the least informed of all participants in this regard. I find it unlikely that this suffices to commit them to a UNESCO and/or World Heritage agenda in any depth, unless there is exceptional individual interest and commitment. By contrast, the close parallels of World Heritage developments with those at the World Bank or the IMF confirm to me that across global bodies, diplomats are very much aware of larger geopolitical trends, the place of their country in it and the orders that their national governments issue to them in what are coordinated foreign policies.

So if it comes to explaining the transformation of the World Heritage area and its timing – from a site of Southern rebellion in 2010 to one of global peace in 2015 – I think that delving into the UNESCO deep past is unnecessary.[30] Brasilia has much more to do with the financial crisis and the sudden possibilities for the Global South it precipitated across global governance; Bonn and the aftermath have much more to do with the time it took to beat recalcitrant Northern expert bodies into submission, rotate Northern countries supporting them out of the Committee, and satisfy leading Southern countries, which, when getting what they most wanted, saw no need to push for a more fundamental overhaul. The intellectual historian may long for a more complex and sophisticated account. Yet parsimony is a scholarly virtue, and explaining the transformation of the World Heritage arena, once its North–South tensions and time lags could no longer be suppressed, does not require more than I presented.

Notes

1. WHC-12/36.COM/INF.8B1, p. 210.
2. Cf. WHC-98/CONF.203/18, pp. 2, 33; WHC-99/CONF.204/15, p. 1; http://whc.unesco.org/en/list/stat#d6 (retrieved 5 September 2020); see also Cameron and Rössler (2013: 58, 98–101).
3. From 161 out of 326 cultural properties to 362 out of 689 and from 179 out of 439 properties of all kinds to 408 out of 892, figures I derive from the total numbers for Europe and North America (http://whc.unesco.org/en/list/stat#d6 (retrieved 5 September 2020)), subtracting those for Canada (cf. http://whc.unesco.org/en/statesparties/ca (retrieved 5 September 2020)) and the United States (cf. http://whc.unesco.org/en/statesparties/us (retrieved 5 September 2020)), but taking into account their two joint inscriptions ('Kluane/Wrangell-St Elias/Glacier Bay/Tatshenshini-Alsek' and 'Waterton Glacier International Peace Park').
4. Cf. http://whc.unesco.org/en/statesparties/it and http://whc.unesco.org/en/list/stat#d6 (retrieved 5 September 2020). See Elliot and Schmutz (2016: 169–70) for an analysis of how African candidates were also particularly likely to face ICOMOS resistance regarding the OUV criteria proposed in the nomination files.
5. Interestingly, this confirmed the prediction of an IUCN representative in the Bureau session of 1990: no non-European country had asked for the introduction of cultural landscapes, he said, a step that would only increase European dominance (SC-91/CONF.002/2, p. 27; SC-91/CONF.002/15, p. 25).
6. Since most properties with issues keep returning (see Chapter 7), regional distributions tend to change only slowly over the years.
7. I think this also motivates Committee delegates' proposals to open for discussion a regionally representative selection of State of Conservation Reports so as not to clamp down on Africa too much (Sheppard and Wijesuriya 2018: 47).
8. This also spilt over into other UNESCO activities such as the Intangible Cultural Heritage Convention, where World Heritage often became the precedent to be avoided. See the US internal report from the 2006 Committee session of that body, for example: 'A recurring theme in Asian-Pacific States' interventions was a sense of grievance or disappointment about some aspects of the World Heritage Committee process and the need to avoid repeating that experience on this Committee … India energetically led the charge in urging the Committee to avoid the experience of the World Heritage Committee that was limited to just two advisory bodies that were 'overworked and understaffed.' India took the floor to note that, in the World Heritage Committee context, '"when Western NGOs seek to collaborate with NGOs from the South, it often does not work and tends to lead to sharp North/South divisions and the impression that the North is telling the South what to do" … India took the floor to note that, in the World Heritage Committee context, "when Western NGOs seek to collaborate with NGOs from the South, it often does not work and tends to lead to sharp North/South divisions and the impression that the North is telling the South what to do". This statement resonated poorly with European and some other delegates and, at the end of the day, gave the meeting a sour overtone … some Committee members seem intent on ensuring that the Intangible Cultural Heritage Convention will give to the "South" the world-renowned cultural heritage listings that the World Heritage Convention has already given to the "North"' (https://wikileaks.org/plusd/cables/06PARIS7553_a.html (retrieved 5 September 2020)).
9. Cf. https://wikileaks.org/plusd/cables/06PARIS5726_a.html.

10. Curiously, roughly the same ratios hold for surface areas.
11. In contrast to Bertacchini et al. (2017), I could not observe a dominance of, or a tacit pact between, the four BRICS states serving on the Committee in the Brasilia session. Brazil and Russia were indeed vocal, but China was so mostly for its own national concerns and while non-BRICS Mexico and Egypt were very active, BRICS member South Africa was not. The mutual exchange of verbal support in favour of softer decisions included *all* Committee members except the three European dissenters. Bertacchini and colleagues claim to find a close alliance between BRICS states for the entire period of 2003–13, but what the employed statistical method – cluster analysis – yields are similarities between the behaviour of these states in relation to *all* other states rather than mutual support within their own circle (which would come closer to what is commonly understood as an alliance). The claim is also at odds with Claudi – strangely cited in support – who concludes that BRIC cooperation in the Committee sessions of 2002–10 was 'inconsistent and weak' (2011: 67; her omission of South Africa alone does not account for the discrepancy). The other statistical analyses published by Bertacchini, Meskell and others (Bertacchini et al. 2016; Bertacchini, Liuzza and Meskell 2017; Meskell et al. 2014), while observing power shifts to emerging countries and a rise of Asia, likewise lump together the period before and after Brasilia, making it difficult to engage with their findings in detail – my central problem of explanation, the transformation starting with the 2010 session, simply does not feature in these studies, despite the salience it had for anyone who was present at the sessions. The approach of these scholars – looking for a country's special ties and affinities and relating its weight in the global economy and international relations with its success with inscriptions – might therefore be more relevant for the period up until 2009, preceding my fieldwork, but it does not do much for 2010 and after, when almost everyone was united in a single grand coalition and, by and large, got what they wanted. Explanatory factors such as delegation size or a country's dependency on tourism (Bertacchini et al. 2016) appear far-fetched to me for this latter period; nothing in my ethnographic observation of the sessions suggested that they played a role (see also note 11 in Chapter 5).
12. Cf. WHC-15/39.COM/INF.8B, pp. 152–59.
13. Cf. ibid., pp. 37, 43, 44, 45, 47.
14. Cf. ibid., pp. 154–55, 159.
15. http://whc.unesco.org/en/list/stat#s6 (retrieved 5 September 2020).
16. Cf. http://www.imf.org/external/pubs/ft/weo/2017/01/weodata/weoselgr.aspx (retrieved 5 September 2020), with options set to 'All countries' and 'Gross domestic product based on purchasing-power-parity (PPP) per capita GDP' for the year 2015.
17. Cf. WHC-15/39.COM/7; WHC/16/40.COM/7; WHC/17/41.COM/7; WHC/18/42.COM/7; WHC/19/43.COM/7.
18. Even under this condition, Europeans and North Americans were overrepresented in four of the five sessions, whereas Asia and the Pacific was underrepresented throughout. Cf. http://www.icomos.org/images/DOCUMENTS/World_Heritage/ICOMOS_World_Heritage_Panel_2014.pdf, https://www.icomos.org/images/DOCUMENTS/World_Heritage/20160613_ICOMOS_2015_Panel.pdf, https://www.icomos.org/en/home-wh/9074-icomos-world-heritage-panel-2016, https://www.icomos.org/en/home-wh/42724-icomos-world-heritage-panel-2017-2018 and https://www.icomos.org/en/home-wh/59209-icomos-world-heritage-panel-2018-2019 (retrieved 5 September 2020).
19. https://www.iucn.org/sites/dev/files/content/documents/iucn_panel_members_2016-2017_cycle.pdf and https://www.iucn.org/sites/dev/files/iucn_world_heritage_panel_2018-2019.pdf (retrieved 5 September 2020).

20. A total of 45.5 per cent (35 out of 77 nominations recommended for inscription), whereas it was 66.7 per cent for both Asia and the Pacific (24 out of 36) and Latin America and the Caribbean (8 out of 12), 57.1 per cent for Africa (4 out of 7) and, for reasons I do not know, 0 per cent for the Arab countries (0 out of 12).
21. WHC-13/37.COM/INF.8B, p. 119.
22. Ibid., p. 163.
23. WHC-11/35.COM/INF.8B, p. 191.
24. Ibid., p. 192.
25. Ibid., p. 196.
26. This has also occurred to arena insiders: already by 1994, the report of the Global Strategy expert meeting mused that 'it would be necessary to proceed not by subcontracting the work exclusively to a single NGO which could not guarantee the diversity of approaches and disciplines required' (WHC-94/CONF.001/INF.4, p. 5).
27. Cf. the synoptic tables on the first pages of documents 8B for the Committee sessions 2005–15 (retrieved 5 September 2020 from http://whc.unesco.org/en/sessions). Since transboundary nominations are exempt from the national quotas of all except one of the participating countries, I exclude them from the count, as well as the nominations for mere extensions of listed properties, since these are almost always approved. It should be noted that four out of the six Israeli 'double nominations' concerned the same site, the Triple Arch-Gate at Dan, that – standing on contested territory – was withdrawn or postponed each time (see Chapter 5, note 57).
28. That coefficient (calculated at http://shlegeris.com/gini) would be 1 if all World Heritage properties were in a single treaty state and 0 if all states had an equal number of properties. If only the 2010–19 period is considered, the Gini coefficient is even higher (0.730).
29. Numbers retrieved from http://www.imf.org/external/pubs/ft/weo/2017/01/weodata/weoselgr.aspx (retrieved 5 September 2020, with options set to 'All countries' and 'Gross domestic product based on purchasing-power-parity (PPP) per capita GDP' for the years 2009 and 2019; Gini coefficients for these numbers calculated at http://shlegeris.com/gini.
30. As already mentioned, Meskell does not accord these dates special relevance, and her work with colleagues lumps together the pre- and post-Brasilia periods (cf. e.g. Meskell et al. 2014: 6–7).

Conclusion

Utopian Remnants and the Logic of Growth

The Bonn Committee session of 2015, one last time. Under the State of Conservation agenda item, the German chairwoman opens discussion of 'Great Barrier Reef' (Australia), which had received a threat of Danger Listing the previous year. WWF volunteers have been present during the preceding days outside the building, displaying a mini reef with cute-looking corals and sea fauna, and distributing hand-knitted toy fish to arriving participants, and more than 200,000 emails are said to have reached the World Heritage Centre regarding the site. However, IUCN's draft decision foresees no Danger Listing and its representative acknowledges that the '2050 Long-Term Sustainability Plan' adopted by the Australian government to stop the deterioration of the reef is ambitious. This plan includes a ban on the dumping of dredge spoil in reef waters, a halt on port development and a reduction of the pollution of the rivers feeding into the reef by 80 per cent, supported by A$200 million over a decade. In a rare display of support, all Committee member delegations take the floor and praise these plans, even when some of them emphasize that implementation is what really counts. The Australian Minister of the Environment bends over backwards to praise the Committee and specific countries for their insistence and support; it made them achieve in eighteen months what would usually have taken a decade for what, he says, is 'the world's Great Barrier Reef'. NGO observers unusually take the floor too, with Greenpeace urging the Committee to keep a close eye on Australia and WWF approving the Australian plan it also wishes to see realized. A representative of Queensland – part of a new state government much more inclined towards conservation – appreciates the NGOs' and everyone else's engagement, and welcomes the draft decision, which is then adopted. Applause rises and the chair highlights what she sees as a great moment for the Committee: Australia has drawn its lessons from last year's controversial debate, she says, and has acted upon them in a constructive way. She declares herself delighted with the power of the Convention. Later, I manage to talk to the Greenpeace speaker outside the hall. She tells me that Australia's expanding coal production is the real problem, as global warming accelerates the coral reef's decline, yet still, they too welcome the Committee decision.

27 September 2016: after three years of proceedings, the International Criminal Court in The Hague convicts Ahmad Al Faqi Al Mahdi, a Malinese national with Tuareg ethnic background, of the war crime of mutilating religious and historical buildings in Timbuktu, of which he has pleaded guilty, in 2012. For the first time, the destruction of cultural heritage is the principal charge in an international criminal case, not just auxiliary to other, more serious charges. It is precisely the challenge to which the World Heritage Committee had trouble reacting on the Saint Petersburg session day: the defendant led a morality brigade installed by the Ansar Dine fundamentalist insurgents, overseeing the destruction of nine Sufi mausoleums and a mosque entrance, which, with one exception, were part of the Timbuktu World Heritage property (Casaly 2016). Intentionally attacking religious buildings and historic monuments that are not military objectives constitutes a war crime in ICC statutes.[1] However, in order to be admissible, such attacks must be 'of sufficient gravity'. Such gravity might have been constructed on the grounds of the impact on the local community alone, yet both the Prosecutor and the Trial Chamber cite World Heritage status and the impact on humankind as well. In their eyes, the fact that the destruction hit World Heritage aggravates the crime, and it prolongs the prison sentence handed down (Casaly 2016).

The Naked Emperor

These two incidents demonstrate that World Heritage does live up to its promise of owning humanity's most important sites in common at times. We see a rich country on the southern margin of the North defer to the authority of the World Heritage Committee, asserting that global interference has sped things up many times over. We also see an international law court basing a war crime case on damage to cultural heritage, largely because a property on the World Heritage List was the target. For similar nods to the power of World Heritage, I already noted the withdrawal of the Serengeti highway plan being announced at the World Heritage Committee session, not elsewhere. The corporate world sometimes follows suit, not just with the no-go pledge of the major mining firms but also when British oil and gas company SOCO International (now Pharos Energy) made its announcement of not using its prospecting rights in Virunga National Park in the Democratic Republic of the Congo four days before the World Heritage Committee session of 2014 opened.[2] As already mentioned, an IUCN official I talked to considers the World Heritage title to be the most powerful designation for nature conservation, and I am convinced that no single other factor has contributed more to the global boom in cultural heritage than the World Heritage Convention, even when some critics accuse it of diverting attention and resources away from lesser sites. Certainly, World Heritage is not to be slighted.

But certainly, the World Heritage Committee is like the naked emperor in Andersen's fairytale. The Australian minister's performance as a committed multilateralist was for the benefit of Australians, not just of the Committee or UNESCO. Countries choose to bow the Committee when it suits them or when the domestic constellation creates an interest in World Heritage being posited as a major force. However, as soon as someone proclaims that the emperor wears no clothes and that the Committee is weak, it loses its powers and multilateral agency, and becomes a free-for-all for countries wishing to shine on the global stage while not being overly bothered with protective demands. I have shown how this arises from individuals in the arena being national individuals: most of them pursue national interests and expect others to do so too, with all of them getting out of the way when strong national interests collide, rather than daring to step in (see Chapter 5). I have shown how the Advisory Bodies' and the Committee's procedures for evaluating World Heritage nominations, monitoring inscribed properties and taking decisions are vulnerable, offering multiple entry points for these vested interests to play out (see Chapter 6). I have shown how this is fostered by a lack of clarity in conceiving and of consistency in applying fundamental ideas about the authenticity, integrity and value of World Heritage properties, particularly of cultural heritage (see Chapter 7). And, most importantly, I have shown how reforms meant to more actively showcase the cultural heritage of the Global South failed to have a decisive effect, creating a strong sense of exclusion among Southern participants. In a favourable geopolitical situation, this led to major Southern countries turning the tables on Northern experts and their Northern supporters. However, with the transformed mode of operations sealed in the peaceful session of Bonn, these leading Southern countries more or less get what they want now. Therefore, they do not bother pushing for the eradication of the remnants of Northern hegemony and Northern bias, with World Heritage inequality persisting as a result (see Chapter 8).

I do not think that there is a way back to the Committee that in 2009 united to punish a leading Northern power by deleting one of its properties from the World Heritage List. On the contrary, national pushing has become even more blatant after Bonn, as Meskell vividly describes for the inscription of 'Archaeological Site of Ani' – the old Armenian site on Turkish territory – in 2016 (2018: 245–49). In 2018, the Committee approved its most egregious 'upgrades' so far: 'Al-Ahsa' (Saudi-Arabia) and then 'Naumburg Cathedral' (Germany) were listed, despite a noninscription recommendation for Al-Ahsa (cf. also Hølleland and Wood 2019) and no recommendation at all for Naumburg: in a unique move, ICOMOS had refused to give one, as two earlier noninscription recommendations that had declared this Gothic cathedral to be inferior to others had been ignored and the property's OUV already confirmed by the Committee in a prior session. The inscription of 'Historic Centre of Sheki with the Khan's Palace' nominated by the 2019 host nation Azerbaijan continued this new trend of entirely overriding the Advisory Bodies' judgements.

This means that the Committee is exactly what the States Parties of the respective properties allow it to be, and there is a tacit pact of noninterference in place now that effectively rules out another Dresden. Humankind is only being empowered if this suits the concerned country. I cannot see any indication that, and how, this state of affairs might change anytime soon.

Inflating Distinctions

Could this all have gone down a different route? I think that earlier and more effective countermeasures against both perceptions and realities of bias and exclusion – such as restrictions on Northern nominations and more financial and other support for Southern ones – would have allowed for a more gradual transition, with the Advisory Bodies and the World Heritage Centre retaining greater weight. Yet, I am less sure whether the development could have been stopped in the long run.

Here, a comparison with the allocation of other titles, prizes and awards is instructive, even when these most often honour people rather than places. As acts of 'cultural consecration' (Allen and Parsons 2006; Bourdieu 1984: 6), they are an increasingly important part of contemporary social life (Best 2008; English 2002). Successful consecration projects, Allen and Parsons maintain, are characterized by four conditions. These are the cultural authority of the organization awarding the honour, which often employs a group of experts; rigorous selection procedures; the selectivity of the award such that only a small number of potential recipients actually receive it; and – most importantly in their assessment – the existence of objective differences between those receiving the honour and those who do not (Allen and Parsons 2006: 810–12). Seen in this light, World Heritage is certainly selective: nomination quotas prevent too rapid an expansion and even with upgrades having become rampant, the average annual number of inscriptions in 2010–19 has not been more than twenty-three, up somewhat from the preceding years, but below the all-time average of twenty-six. A burdensome procedure of selection with voluminous nomination files and an intricate process of evaluation continue to play a restraining role. However, the cultural authority of the awarding body is more questionable, given that in the last decade, the Committee made almost 40 per cent of inscriptions against expert advice (see Table 5.4). The Committee is supposed to be an expert body itself, as the Convention asks for delegates specializing in heritage conservation, but member delegations freely ignore this requirement. Yet, even when looking at the authorized experts, I see a difference between the members of the ICOMOS and IUCN World Heritage Panels and, say, the professors of Sweden's leading medical school who award the Nobel Prize in Medicine, the seasoned baseball journalists who cast their votes on the induction of retired players into the US National Baseball Hall of Fame (Allen and Parsons 2006), the team coaches and captains of the world's national football teams voting for the FIFA World Player of the Year, or the members of the Academy of Motion Picture Arts and Sciences who cast their ballots for the Academy Awards. All these voting individuals are empowered because of defined qualities or achievements confirmed by external recognition and/or a process of competitive screening by employers (the professors and journalists). With the exception of the medical school, they also do not usually function as a social circle outside elections, or at least not in their entirety. In addition, votes are cast independently and secretly. All this is distinct from the World Heritage Panels where ICOMOS and IUCN are free to invite whom they see fit, where there seem to be no set standards for the expertise of the panellists, which is left unexplained, where several of these individuals cooperate intensely on a different body, the ICOMOS Board, and where decisions are usually taken openly and by consensus, just like in the World Heritage Committee, which creates pressure to go with the majority. Finally and to Allen and Parsons' fourth point, doubts are regularly voiced as to whether the selection of properties for the World Heritage List reflects objective differences, less so perhaps those between approved and rejected sites than the differences *between* inscriptions, with worries about unworthy additions diluting the value and reputation built by the undisputed ones.

However, the crucial difference is that other awards make nominees compete: each year, there is only one Nobel Prize in Medicine shared by a maximum of three individuals, one Oscar for Best Picture and one winner of the FIFA World Player award for each gender. This means that voting individuals have to choose. Yet, in the case of World Heritage, all candidate sites could theoretically go on the List if they are found to have OUV. Inscribing the Danish hunting park does not diminish the chances of Kenyan and Ugandan cultural landscapes. Therefore, the temptation to talk or lobby the World Heritage standard into existence is huge, and Southern candidates not making it this time round – everyone can console themselves – can come back anytime in the future, given that the List stays open: the stone money sites and Ugandan rock art have only been postponed, not rejected for good. Meanwhile, however, the reality of European hegemony continues.

In the world of distinctions, the closest comparative case I can find is the canonization process of the Roman Catholic Church, starting with the fact that it likewise builds on a presumably inherent quality: saintliness is construed as just as objective as OUV and even if humans fail to recognize particular saints, God knows them all (Woodward 1996: 17). Just like World Heritage but different, say, from literary or artistic awards, which tend to multiply (Best 2008), it also has no serious competitors. Recent developments of the canonization process show a number of striking parallels. For a long time, the recognition of saints remained a more or less local process with not more than episcopal involvement, and it took until the seventeenth century for the papacy to take exclusive control, with formal procedures adopted in 1735. The process bears some resemblance to that of World Heritage: saints are also nominated in a bottom-up process by a group of supporters to a local bishop. Should the latter convert their request into a 'cause' (similar to what a State Party does by nominating a site for the World Heritage List), the ensuing process – to be funded by the supporters – involves examination by the Congregation for the Causes of Saints (CCS), consisting of twenty-five cardinals and bishops (comparable, even in size, to the ICOMOS and IUCN World Heritage Panels). This then makes a recommendation to the pope, who is the only authority to grant beatification and confer the title 'Blessed' (just like World Heritage inscription is the prerogative of the Committee) and, after proof of a further miracle, that of saint, which is irrevocable (as the World Heritage title usually also is). However, in a crucial difference, the examination of prospective saints famously involved an adversarial process in the past: it was the task of the Promoter of the Faith or 'Devil's Advocate' to raise all conceivable objections to the specific 'cause' (Barro and McCleary 2011: 3–4; Bennett 2011: 442–43). Thus, until 1978, the popes did not canonize more than around 350 saints, a fraction only of the more than 10,000 saints the Catholic church recognizes (Bennett 2011: 442).

Yet, the Brasilia of canonization did come with the election of John Paul II in 1978. Under his rule, the definition of martyrdom was widened to now also embrace 'martyrs of charity' sacrificing their lives for the benefit of others rather than being killed for not betraying their Christian faith,[3] and the posthumous miracles required for canonization – proof of the saint's nearness to God, which allows for supernatural interference – were reduced in number or, in the case of martyrs, completely waived. However, most importantly for what followed, the office of the Promoter of the Faith was abolished. Instead, a College of Relators now oversees the (largely local) collection of documentary evidence and preparation of a biography. The case is then assessed by the CCS (Barro and McCleary 2011: 4–5) or first by a committee of theologians and historians and then the CCS (Bennett 2011: 443–45). Still, this has taken the critical edge from a process in which everyone's interests tend to converge now (Woodward 1996: 3–4). Therefore, during the twenty-seven years of his papacy, John Paul II performed no fewer than 482 canonizations – more than all his predecessors put together – and 1,300 beatifications, acquiring himself such epithets as the 'sainting machine' or 'saint factory' (Bennett 2011: 441). His successors have continued at the same brisk pace. As Bennett shows for John Paul II's beatifications and canonizations, these reflected his religious agenda, such as strict sexual morality and the glorification of celibacy, the wish to fight the image of the Catholic Church as the willing collaborator of fascism and National Socialism (Bennett 2011: 447–51), and a conscious globalization of both martyrs (Woodward 1996: 7) and confessor saints (the other major type), with the Eastern European and non-European share growing as a consequence (Barro and McCleary 2011: 9).

All this has clearly made selection more permissive: Bennett is sure that a hugely popular but nevertheless controversial figure such as Padre Pio, considered a fraud by two prior popes, would not have made it past a Protector of the Faith, or at least not as quickly. Also, pressure by John Paul II, who was an admirer, would not have accelerated the process so much (2011: 445-446). Yet still and in another parallel to World Heritage, Europeans continue to dominate, with three-quarters of all new confessor saints coming from the continent and more than half of these from Italy (Barro and McCleary 2011: 29), blessed with saints as much as with World Heritage. Again, the lack of a numerical limit appears to have encouraged the door to be opened more widely at some point; again, once leniency is firmly established, there appears to be no way back to the earlier selectivity; and again, accelerated growth does not prevent a European-based and European-dominated institution from continuing to employ a European optic. (More than half of the cardinals electing Pope Francis in 2013 were from this one continent and almost half of these were Italians,[4] despite only one-fifth of Catholics living in Europe.)[5]

This means that the heritage canonization process of the UNESCO World Heritage Committee is not so special after all.

Hooked on Growth

However, at the heart of all this is the expectation of growth, and I consider this the root cause of the World Heritage arena's transformation. Let us assume for a moment that a not too generous numerical limit had been imposed on annual World Heritage listings in the beginning, such as, say, five a year, a number well in line with initial thoughts of inscribing around 100–200 properties on the List (Batisse and Bolla 2005: 74, 78, 94; Cameron and Rössler 2013: 48–49). Lobbying and deal-making might still occur, but Committee members would have to choose between competing requests for support, making secret ballots likely. Complaints about bias would perhaps lead to regional slots, just like in the allocation of Committee seats. But this would still make it improbable for the Global North and the handful of usual suspects from the Global South to get as large a piece of the pie as with a presumably absolute definition of OUV.

However, one only rarely hears such proposals in the World Heritage arena, and I have yet to meet anyone who argues for closing the List anytime soon. Also, while many a State Party has not submitted any new candidate property for years, none has officially declared itself done with World Heritage nominations. This is astonishing – assuming that treaty states have been working their way down from what they saw as their worthiest candidates, rather than holding back the best, a presumably constant OUV should make them run out of sites at some point. Yet in what is a contrary logic, the fact that ever more treaty states submit Tentative Lists is praised as a positive development in the sessions. The 1,722 entries on these 178 national lists (as of 2019) create pressure for expansion, given that an entry is taken as a promise of future nomination in most cases. Yet, too much depends on growth not to have it: new inscriptions provide for scenes of jubilation in the hall, with delegates cheering, hugging, waving national flags and receiving the congratulations of their peers. New listings also make for happy global news, unlike developments at already inscribed sites, which are often enough sobering and, even for what are perceived as the success stories, rather complex to tell. Without the promise of further World Heritage titles, quite a few states would lose interest, threatening in turn the voluntary contributions on which the poorly funded system depends so much, and the World Heritage Centre, IUCN, ICCROM and especially ICOMOS would lose public stature.

List growth is further accelerated by the fact that nothing else in the arena – funds, personnel or organizational capacities – increases at even remotely the same rate, making the title the only significant reward to be collected. While for the 2010–11 biennium, the World Heritage Fund, the UNESCO regular budget (mostly for Centre personnel) and 'extrabudgetary funding' (additional voluntary State Party contributions) together would still make for World Heritage-related expenses of US$41.1 million, this had fallen to US$25.0 million in the 2016–17 biennium, mainly because the voluntary contributions dropped from US$19.1 million to a mere US$5.4 million. As a consequence, the expenses for conservation (monitoring, international assistance and the like) halved, from US$19.0 million to US$9.5 million, and in the 2015 session, it was announced that 19 per cent of the approved international assistance requests could not be funded, despite their modest sizes of around US$20,000 each. However, what hardly contracted was the sum paid to the Advisory Bodies for their services, also given that demand for these has expanded with the new feedback options (it went from US$5.5 million to US$5.4 million in the same period, of which the largest part – US$3.7 million, down from US$4.0 million – was spent on evaluations).[6] It is not just that the World Heritage Fund is increasingly used for paying the working costs of the Convention (Sheppard and Wijesuriya 2018: 68); these costs are also ever more focused on nominations – that is, getting further properties on the List – not on other activities.

This state of affairs should suggest restraint, but as already outlined and despite affecting only a minority of countries, even a reduction to one nomination per state and year was brought down in Bonn, to be adopted only a year later where it became the more lenient '35/1' instead of '25/1'. The history of World Heritage is littered with attempts to introduce limitations, which were then fought down by Committee majorities or, as consensus is usually sought, vocal minorities, starting at the very first sessions (cf. Cameron and Rössler 2013: 50–54). States also do not practise what they preach: in 1999, an informal working group summoned by the Bureau recommended rigorous selection and both voluntary restraint and a focus on underrepresented heritage categories to the well-endowed countries.[7] Of the group members, France and Japan did indeed present no inscriptions for three years in a row; however, Italy simply continued with its annual inscription rhythm and the United Kingdom had no fewer than six properties listed in 2000 and 2001. When asked to comment on limitation proposals, States Parties issue statements that are almost comical in their effort to dress up the thirst for more titles, such as by making the earliest possible listing of sites with OUV an almost sacred duty. Similar arguments keep being deployed today, even when the countries making them change. The Turkish ambassador in the 2015 session, for example, stated that many sites on the Tentative List are vulnerable and require protection as soon as possible so that they should be listed, rather than trying to save a mere US$100,000 or so by slowing nominations. A Vietnamese delegate backed him up, saying that after seeing so many happy scenes of celebration in the sessions, one should not really consider putting so many people on hold. An Egyptian representative probably made the point most succinctly when telling me that the List is merely instrumental, not valuable in itself, and that every inscription yields added control of heritage sites. Yet, actual efforts to receive the title belie the idea of a valueless List, and none of these individuals reflected on how liberal inscription practices bolster Eurocentric List contents and dilute the limited financial resources.

As, in addition, the World Heritage title can be freely multiplied, there is little to control State Party desires or even to distract them from this goal even for a moment. On the Saint Petersburg session day, we already saw that the challenge of the Timbuktu destructions was gently but firmly postponed – 'we've got work to do' – until the day's inscription business was done. The very prominent global news coverage highlighted the provocative nature of the acts – 'UNESCO is what?', an Ansar Dine spokesperson had asked the journalists (Joy 2016), yet the global agency in charge or, rather, the World Heritage Committee was slow to take a stance. Very clearly, this was not a strategy of deliberately withholding the attention the insurgents were craving; rather, nominations were on, and this meant that many Committee members and other participants were absorbed by their self-interests. As already described, the Committee did opt for a public appeal and after discussion of a draft on the next morning – still squeezed in between further nominations – the Russian chairwoman, lining up with the Malian Minister of Culture, representatives of all Committee member states and other UNESCO ambassadors, did read a solemn condemnation in several languages to the assembled press the following day. The chosen location in front of the Bronze Horseman, the statue of Peter the Great on the Senate Square, was certainly dignified and part of a World Heritage property itself. Yet still, the event was, in a way, outsourced: it took place in the lunch break, several kilometres away from the meeting venue and with only a small group of people in attendance. Nothing could mark it more clearly as a mere sideshow to the main business, the routine Committee affairs of the plenary session, which, by using the break, could continue uninterrupted. Taking plenary time for reading the appeal, perhaps with *all* participants standing together for what might make for more impressive photos, did not occur to anyone.

All this means that the growth of the World Heritage List is the one thing that must never be questioned. Almost everyone knows unlisted sites that deserve inscription in their view; almost everyone can name properties that should not be on the List, not just from among the recent overrulings – properties such as 'Castles of Augustusburg and Falkenlust at Brühl', 'Town Hall and Roland on the Marketplace of Bremen' (both Germany), 'Provins, Town of Medieval Fairs' (France) and 'Pirin National Park' (Bulgaria) were given by my interlocutors. Yet, no initiative to clear the List of unworthy cases has ever been launched and the likeliness of States Parties letting this happen is nil. This means that the only way to 'get the List right' is through further additions and a fuller List – much fuller, in fact, if current imbalances are to be corrected to any significant degree. But once growth is thus turned into a necessity, attempts to harness it to one's own purposes will continue as well.

I wish to stress here that this is a structural factor and not anyone's 'fault', such as that of the diplomats, Northern countries or Southern countries letting down their Southern peers; rather, it is an unstoppable dynamic with a 'tragedy of the commons' (Hardin 1968) logic to it. The ongoing dilution of the World Heritage List through what even in the nominating countries' eyes are not their first tier of sites is a very gradual process. If considered undesirable (which, as we saw, not all participants would agree to), the reputational damage of an undeserving inscription is shared by all treaty states and more than a thousand properties on the List. It therefore becomes small for everyone in comparison to the perceived benefits for the nominating country and local supporters of the listed site. Restraining access numerically might help, but while it could have been an option initially, it is difficult to see a majority forming for its introduction now. All other efforts of slowing down the title hunt, such as by rigid expert screening, have long been superseded. However, benefits still appear to outweigh costs for the participants in the venture – and despite occasional threats, no State Party has denounced the Convention so far, not even those that have felt compelled to leave UNESCO. So whatever else will happen to the World Heritage List, it will continue to grow and while the modalities of growth will continue to be debated, it will be interest-driven growth.

Reproducing Global Inequality

For those hoping for more fairness and equity in global governance – a flatter world – the story I have told will be sobering. In the World Heritage Committee, a body that is less central to Northern powers than other global fora and should therefore offer favourable conditions, reforms meant to strengthen the Global South failed to undo Northern hegemony. Northern countries were not stopped from serving themselves, new World Heritage listings – the major perks on offer – continued to go to European countries, and Southern frustration with presumably biased advisory organizations and procedures mounted. In 2010, Southern countries banded together to overrule the experts and realize desired outcomes, leading to the normalization of the pursuit of national interests. Yet the resulting free-for-all for everyone with some basic knowledge of the diplomatic lobbying game continues to disproportionately reward the North. Stronger Southern countries team up with the leading Northerners to keep their pieces of the World Heritage pie large, while the poorer and weaker Southern countries are left to fend for themselves. As an unintended side-effect, fundamental reform expunging Northern bias is postponed, and heritage 'theory from the South' (Comaroff and Comaroff 2012) – new perspectives coming from or inspired by heritage in the non-Western world – does not arise.

This closely parallels recent developments in more consequential global bodies where, despite all talk of increasing multipolarity, leading Southern countries contribute their share to sustaining Northern hegemony. However, these are relatively open struggles over power and resources, whereas the World Heritage Committee – with its UNESCO tradition of consensus decisions and sympathies for the South – has returned to ostensible peace, never clearly addressing that the fault lines of the early 2010s were dividing North and South. Yet while fragmented decision-making and the absorption of state delegates by their own national concerns distract everyone's attention from the outcomes, these do not stray any less from the proclaimed targets than the IMF and World Bank reforms. This speaks strongly against the autonomy of heritage diplomacy and calls for examining the role of Southern countries, which the dominant 'Eurocentric conception of world politics' (Hobson 2012) tends to neglect.

Time lags exacerbate this tendency. They are constitutive for heritage: things are so labelled when they have presumably ceased to evolve at some point, thus standing for the past rather than the present and moving from history time (subjected to change) to heritage time (subjected to arrested change). No wonder then that in the World Heritage Committee, multiple time lags – in terms of rewards, influence on concepts and decisions, and production of the things singled out as heritage later on – increased Southern participants' sense of not receiving their due. Heritage research has amply reflected about the underlying temporalities, with such concepts as monumental versus social time (Herzfeld 1991: 6–14) or reflective versus affective 'past presencing' (Macdonald 2013: 233), but the temporalities of global heritage appropriation need to be taken into account as well.

The larger lesson I draw from this study is that the reproduction of global inequalities and Northern rule has a symbolic dimension that is not reducible to material factors alone. It needs to be understood in its own right if advancing the intellectual decolonization of the contemporary world is a serious objective. It is too simple to just blame ongoing Northern imperialism and the unwillingness to share benefits and resources – Southern complicity is important too, and so are more subtle dimensions such as the time lags, disciplinary biases in global organizations such as ICOMOS and 'culture chaos' in contemporary intellectual life. But without growth as the unquestioned premise, inhibited neither by a numerical bottleneck nor by clearly operationalized standards, different constellations might still have developed in the World Heritage arena. I think that anthropology is crucially placed to unravel such complex dynamics.

As to explaining global governance, anthropology should not aim too low. Anthropologists looking at the UN and international organizations have often focused on what the dominant disciplines such as international relations neglect – seemingly marginal aspects such as the generative power of textual form (Billaud 2014: 64–65; Riles 1998), the language of negotiations (Groth 2012), the ritualistic nature of public sessions (Cowan 2014), the role of boredom (Billaud 2014: 68–69; Fresia 2013: 53–54; Kelly 2011: 733, 740) or the appropriation of anthropological key terms (Eriksen 2001; Nielsen 2011). Yet ethnographic observation of the seemingly marginal, such as participants' informal views, discourses and strategies, combined with a sensitivity to (post)colonial global power relations and the view from the South that I see as constitutive of a critical and reflective anthropology, also qualify us to contribute to the 'great game' of explaining the power shifts within these bodies, as I hope I have demonstrated.

The survival of Northern hegemony in the contemporary world has occupied anthropologists for a long time. We have been particularly attentive to its intellectual and symbolic dimensions, such as by reflecting on the micropolitics of the ethnographic encounter (Abu-Lughod 1991; Clifford and Marcus 1986; Dwyer 1982), Anglo-American dominance of the discipline (Lins Ribeiro 2014) and constructions of 'the West' by those outside it (Carrier 1995), and by delivering countless ethnographic studies that show local communities across the world naturalizing, or being made to naturalize, Northern supremacy. Yet, the world-making of global organizations plays its own part in maintaining a skewed planet, however universalist their aspirations and anthropological their claimed perspectives. It should therefore concern us accordingly.

World Heritage will continue its growth course, with ever more being brought under its umbrella, but with its official guardians and governing bodies increasingly less in control of what is being done in its name. It is surprising how unaffected World Heritage seems to be in the public consciousness – yet again, news coverage of the devastating floods that hit Venice in the autumn of 2019 never failed to mention the World Heritage title, hard-pressed as one would be for a city whose fame depends less on this distinction. World Heritage has become, and will continue to be, a global affair, but making it One World Heritage will remain a challenge.

Notes

1. US President Donald Trump was alerted of this fact when he threatened to destroy Iranian cultural heritage sites in January 2020 (https://www.bbc.com/news/world-middle-east-51014237 (retrieved 7 September 2020)).
2. https://www.theguardian.com/environment/2014/jun/11/soco-oil-virunga-national-park-congo-wwf (retrieved 7 September 2020).
3. Such as Maximilian Kolbe (1894–1941), the Polish Franciscan friar who volunteered to take the place of a father of a family in an Auschwitz death cell.
4. https://en.wikipedia.org/wiki/Cardinal_electors_for_the_2013_papal_conclave (retrieved 7 September 2020). This is a much-viewed and properly referenced list, so I consider it reliable.
5. https://press.vatican.va/content/salastampa/en/bollettino/pubblico/2019/03/06/190306b.html (retrieved 7 September 2020).
6. WHC/16/40.COM/15, pp. 29–31 of the PDF file; WHC-12/36.COM/15.Rev, pp. 32–34 of the PDF file.
7. WHC-99/CONF.204/15, pp. 92–93.

References

Abdulla, Ameer Awad, David Obura, Bastian Bertzky and Yichuan Shi. 2013. *Marine Natural Heritage and the World Heritage List: Interpretation of World Heritage Criteria in Marine Systems, Analysis of Biogeographic Representation of Sites, and a Roadmap for Addressing Gaps.* Gland, Switzerland: IUCN. Retrieved 10 September 2020 from https://portals.iucn.org/library/sites/library/files/documents/2013-033.pdf.

Abélès, Marc. 1992. *La vie quotidienne au Parlement Européen.* Paris: Hachette.

_____ (ed.). 2011. *Des anthropologues à l'OMC: Scènes de la gouvernance mondiale.* Paris: CNRS Éditions.

Abu-Lughod, Lila. 1991. 'Writing against Culture', in Richard G. Fox (ed.), *Recapturing Anthropology: Working in the Present.* Santa Fe: School of American Research Press, pp. 137–62.

Allais, Lucia. 2013. 'Integrities: The Salvage of Abu Simbel', *Grey Room* 50(Winter): 6–45.

Allen, Michael Patrick, and Nicholas L. Parsons. 2006. 'The Institutionalization of Fame: Achievement, Recognition, and Cultural Consecration in Baseball', *American Sociological Review* 71(5): 808–25.

Andersen, Casper. 2011. 'The Philae Controversy: Muscular Modernization and Paternalistic Preservation in Aswan and London', *History and Anthropology* 22(2): 203–20.

Anderson, Benedict. 1983. *Imagined Communities: Reflections on the Origin and Spread of Nationalism.* London: Verso Editions & NLB.

Appadurai, Arjun. 1996. *Modernity at Large: Cultural Dimensions of Globalization.* Minneapolis: University of Minnesota Press.

Arizpe, Lourdes (ed.). 2011. *Compartir el patrimonio cultural inmaterial: Narrativas y representaciones.* México: CONACULTA, UNAM, CRIM.

Arizpe, Lourdes, and Cristina Amescua (eds). 2013. *Anthropological Perspectives on Intangible Cultural Heritage.* Cham: Springer.

Asahi Shinbun Digital. 2015. 'Shasetsu: Sekai isan tōroku: "Gōi" fumae mae e susume', 8 July. Retrieved 10 September 2020 from http://digital.asahi.com/articles/DA3S11846692.html.

Aykan, Bahar. 2015. '"Patenting" Karagöz: UNESCO, Nationalism and Multinational Intangible Heritage', *International Journal of Heritage Studies* 21(10): 949–61.

Badman, Tim, Bastian Bomhard, Annelie Fincke, Josephine Langley, Pedro Rosabal and David Sheppard. 2008. *Outstanding Universal Value: Standards for Natural World Heritage.* Retrieved 10 September 2020 from https://portals.iucn.org/library/efiles/documents/2008-036.pdf.

Barro, Robert J., and Rachel M. McCleary. 2011. 'Saints Marching in, 1590–2009', *National Bureau of Economic Research Working Paper* 16769. Retrieved 10 September 2020 from http://www.nber.org/papers/w16769.

Batisse, Michel, and Gérard Bolla. 2005. *The Invention of 'World Heritage'.* Paris: Association of Former UNESCO Staff Members.

Bellier, Irène. 2013. '"We Indigenous Peoples …": Global Activism and the Emergence of a New Collective Subject at the United Nations', in Birgit Müller (ed.), *The Gloss of Harmony: The Politics of Policy-Making in Multilateral Organisations.* London: Pluto Press, pp. 177–201.

Bendix, Regina F., Aditya Eggert and Arnika Peselmann (eds). 2012. *Heritage Regimes and the State.* Göttingen: Universitätsverlag Göttingen.

Benesch, Oleg. 2014. *Inventing the Way of the Samurai: Nationalism, Internationalism, and Bushido in Modern Japan.* Oxford: Oxford University Press.

Bennett, Oliver. 2011. 'Strategic Canonisation: Sanctity, Popular Culture and the Catholic Church', *International Journal of Cultural Policy* 17(4): 438–55.

Berliner, David. 2018. *Perdre sa culture.* Brussels: Zones Sensible Editions.

_____. 2020. *Losing Culture: Nostalgia, Heritage, and Our Accelerated Times.* New Brunswick, NJ: Rutgers University Press.

Bertacchini, Enrico, Claudia Liuzza and Lynn Meskell. 2017. 'Shifting the Balance of Power in the UNESCO World Heritage Committee: An Empirical Assessment', *International Journal of Cultural Policy* 23(3): 331–51.

Bertacchini, Enrico, Claudia Liuzza, Lynn Meskell and Donatella Saccone. 2016. 'The Politicization of UNESCO World Heritage Decision Making', *Public Choice* 167(1): 95–129.

Bertacchini, Enrico, and Donatella Saccone. 2012. 'Toward a Political Economy of World Heritage', *Journal of Cultural Economics* 36(4): 327–52.

Berthod, Olivier. 2011. 'On Institutions, Paths, and Routes Set in Stone: The Construction of a Bridge as Path Instantiation', Ph.D. thesis. Berlin: Free University of Berlin.

Bertzky, Bastian, Yichuan Shi, Alice Hughes, Barbara Engels, Mariam Kenza Ali and Tim Badman. 2013. *Terrestrial Biodiversity and the World Heritage List: Identifying Broad Gaps and Potential Candidate Sites for Inclusion in the Natural World Heritage Network.* Gland, Switzerland: IUCN. Retrieved 10 September 2020 from https://portals.iucn.org/library/sites/library/files/documents/2013-016.pdf.

Best, Joel. 2008. 'Prize Proliferation', *Sociological Forum* 23(1): 1–27.

Betts, Paul. 2015. 'The Warden of World Heritage: UNESCO and the Rescue of the Nubian Monuments', *Past & Present* 226(Supplement 10): 100–25.

Billaud, Julie. 2014. 'Keepers of the Truth: Producing "Transparent" Documents for the UN Universal Periodic Review', in Hilary Charlesworth and Emma Larking (eds), *Assessing the Universal Periodic Review: Rights, Ritual, Ritualism.* Cambridge: Cambridge University Press, pp. 63–83.

Boggs, James P. 2004. 'The Culture Concept as Theory, in Context', *Current Anthropology* 45(2): 187–209.

Borofsky, Robert, Fredrik Barth, Richard A. Shweder, Lars Rodseth and Nomi Maya Stolzenberg. 2001. 'WHEN: A Conversation about Culture', *American Anthropologist* 103(2): 432–46.

Bortolotto, Chiara. 2007. 'From Objects to Processes: UNESCO's Intangible Cultural Heritage', *Journal of Museum Ethnography* 19(March): 21–33.

_____. 2011a. 'Il patrimonio immateriale e l'autenticità: Una relazione indissolubile', *La Ricerca Folklorica* 64: 7–17.

_____ (ed.). 2011b. *Le patrimoine culturel immatériel: Enjeux d'une nouvelle catégorie.* Paris: Maison de Sciences de l'Homme.

_____. 2016. 'Placing ICH, Owning a Tradition, Affirming Sovereignty: The Role of Spatiality in the Practice of the ICH Convention', in Peter Davis and Michelle L. Stefano (eds), *The Routledge Companion to Intangible Cultural Heritage*. Abingdon: Routledge, pp. 46–58.

_____. 2020. 'Le patrimoine immatériel et le tabou de l'authenticité: De la pérennisation à la durabilité', in Julia Csergo, Christian Hottin and Pierre Schmit (eds), *Le patrimoine culturel immatériel au seuil des sciences sociales: Actes du colloque de Cerisyla-Salle, septembre 2012*. Paris: Éditions de la Maison des sciences de l'homme, pp. 1–18.

Bortolotto, Chiara, Philipp Demgenski, Panas Karampampas and Simone Toji. 2020. 'Proving Participation: Vocational Bureaucrats and Bureaucratic Creativity in the Implementation of the UNESCO Convention for the Safeguarding of the Intangible Cultural Heritage', *Social Anthropology* 28(1): 66–82.

Bourdieu, Pierre. 1984. *Distinction: A Social Critique of the Judgement of Taste*. Cambridge, MA: Harvard University Press.

Braun, Nikola. 2007. *Globales Erbe und regionales Ungleichgewicht: Die Repräsentativitätsprobleme der UNESCO-Welterbeliste*. Hamburg: Verlag Dr. Kovač.

Brightman, Robert. 1995. 'Forget Culture: Replacement, Transcendence, Relexification', *Cultural Anthropology* 10(4): 509–46.

Brisson, Thomas. 2015. 'Western and Non-Western Views of the Social and Behavioral Sciences: De-Westernizing the Social and Behavioral Sciences', in James D. Wright (ed.), *International Encyclopedia of the Social and Behavioral Sciences*, 2nd edn, vol. 25. Oxford: Elsevier, pp. 541–50.

Brumann, Christoph. 1998a. 'The Anthropological Study of Globalization: Towards an Agenda for the Second Phase', *Anthropos* 93: 495–506.

_____. 1998b. *Die Kunst des Teilens: Eine vergleichende Studie zu den Überlebensbedingungen kommunitärer Gruppen*. Hamburg: Lit.

_____. 1999. 'Writing for Culture: Why a Successful Concept Should Not Be Discarded', *Current Anthropology* 40(S1): S1–S27.

_____. 2000. 'The Dominance of One and Its Perils: Charismatic Leadership and Branch Structures in Utopian Communes', *Journal of Anthropological Research* 56: 425–51.

_____. 2001. 'Religious Consensus and Secular Dissent: Two Alternative Paths to Survival for Utopian Communes', *Anthropos* 96: 87–103.

_____. 2002. 'On Culture and Symbols', *Current Anthropology* 43: 509–10.

_____. 2003. '"All the Flesh Kindred That Ever I See": A Reconsideration of Family and Kinship in Utopian Communes', *Comparative Studies in Society and History* 45: 395–421.

_____. 2005. 'Kyotos Dilemma: Das Stadtbild als "commons"', in Werner Pascha and Cornelia Storz (eds), *Wirkung und Wandel von Institutionen: Das Beispiel Ostasien*. Stuttgart: Lucius & Lucius, pp. 133–68.

_____. 2009. 'Outside the Glass Case: The Social Life of Urban Heritage in Kyoto', *American Ethnologist* 36(2): 276–99.

_____. 2011. 'Unser aller Kulturgut: Eine ethnologische Annäherung an das UNESCO-Welterbe', *Sociologus* 61(1): 19–43.

_____. 2012a. 'Multilateral Ethnography: Entering the World Heritage Arena', *Max Planck Institute for Social Anthropology Working Papers* 136. Retrieved 10 September 2020 from http://www.eth.mpg.de/cms/en/publications/working_papers/pdf/mpi-eth-working-paper-0136.pdf.

_____. 2012b. *Tradition, Democracy and the Townscape of Kyoto: Claiming a Right to the Past*. London: Routledge.

_____. 2013. 'Comment le patrimoine mondial de l'Unesco devient immatériel', *Gradhiva* 18: 5–29.

_____. 2014a. 'Heritage Agnosticism: A Third Path for the Study of Cultural Heritage', *Social Anthropology* 22(2): 73–88.

_____. 2014b. 'Shifting Tides of World-Making in the UNESCO World Heritage Convention: Cosmopolitanisms Colliding', *Ethnic and Racial Studies* 37(12): 2176–92.

_____. 2015. 'Community as Myth and Reality in the UNESCO World Heritage Convention', in Nicolas Adell, Regina F. Bendix, Chiara Bortolotto and Markus Tauschek (eds), *Between Imagined Communities and Communities of Practice: Participation, Territory and the Making of Heritage*. Göttingen: Universitätsverlag Göttingen, pp. 273–86.

_____. 2016a. 'Conclusion. Imagining the Ground from Afar: Why the Sites Are So Remote in World Heritage Committee Sessions', in Christoph Brumann and David Berliner (eds), *World Heritage on the Ground: Ethnographic Perspectives*. Oxford: Berghahn Books, pp. 294–317.

_____. 2016b. 'UNESCO-Welterbe, ostasiatische Nachbarn und japanische Altlasten', in David Chiavacci and Iris Wieczorek (eds), *Japan 2016: Politik, Wirtschaft und Gesellschaft*. Munich: Iudicium, pp. 93–115.

_____. 2017a. 'The Best of the Best: Positing, Measuring and Sensing Value in the UNESCO World Heritage Arena', in Ron Niezen and Maria Sapignoli (eds), *Palaces of Hope: The Anthropology of Global Organizations*. Cambridge: Cambridge University Press, pp. 245–65.

_____. 2017b. 'How to Be Authentic in the UNESCO World Heritage System: Copies, Replicas, Reconstructions, and Renovations in a Global Conservation Arena', in Corinna Forberg and Philipp Stockhammer (eds), *The Transformative Power of the Copy: A Transcultural and Interdisciplinary Approach*. Heidelberg: Heidelberg Publishing, pp. 269–87.

_____. 2018. 'Anthropological Utopia, Closet Eurocentrism and Culture Chaos in the UNESCO World Heritage Arena', *Anthropological Quarterly* 91(4): 1203–34.

_____. 2019a. 'Global Linkages, Connectivity and the Indian Ocean in the UNESCO World Heritage Arena', in Burkhard Schnepel and Tansen Sen (eds), *Travelling Pasts: The Politics of Cultural Heritage in the Indian Ocean World*. Leiden: Brill, pp. 21–38.

_____. 2019b. 'Slag Heaps and Time Lags: Undermining Southern Solidarity in the UNESCO World Heritage Committee', *Ethnos* 84(4): 719–38.

Brumann, Christoph, and David Berliner. 2016a. 'Introduction. UNESCO World Heritage – Grounded?', in Christoph Brumann and David Berliner (eds), *World Heritage on the Ground: Ethnographic Perspectives*. Oxford: Berghahn Books, pp. 1–34.

_____ (eds). 2016b. *World Heritage on the Ground: Ethnographic Perspectives*. Oxford: Berghahn Books.

Brumann, Christoph, and Rupert A. Cox (eds). 2010. *Making Japanese Heritage*. London: Routledge.

Brumann, Christoph, and Lynn Meskell. 2015. 'UNESCO and New World Orders', in Lynn Meskell (ed.), *Global Heritage: A Reader*. Hoboken, NJ: Wiley-Blackwell, pp. 22–42.

Bruner, Edward M. 1994. 'Abraham Lincoln as Authentic Reproduction: A Critique of Postmodernism', *American Anthropologist* 96(2): 397–415.

Buzzini, Gionata P., and Luigi Condorelli. 2008. 'Article 11 List of World Heritage in Danger and Deletion of a Property from the World Heritage List', in Francesco Francioni (ed.), *The 1972 World Heritage Convention: A Commentary*. Oxford: Oxford University Press, pp. 51–62.

Cameron, Christina. 2008. 'From Warsaw to Mostar: The World Heritage Committee and Authenticity', *Bulletin of the Association for Preservation Technology* 39(2–3): 19–24.

Cameron, Christina, and Judith Herrmann. 2018. *Guidance and Capacity Building for the Recognition of Associative Values Using World Heritage Criterion (vi): Final Report*. Retrieved 10 September 2020 from https://whc.unesco.org/document/167760.

Cameron, Christina, and Mechtild Rössler. 2011. 'Voices of the Pioneers: UNESCO's World Heritage Convention 1972–2000', *Journal of Cultural Heritage Management and Sustainable Development* 1(1): 42–54.

_____. 2013. *Many Voices, One Vision: The Early Years of the World Heritage Convention.* Farnham: Ashgate.

Carducci, Guido. 2008. 'Articles 4–7 National and International Protection of the Cultural and Natural Heritage', in Francesco Francioni (ed.), *The 1972 World Heritage Convention: A Commentary.* Oxford: Oxford University Press, pp. 103–45.

Carrier, James G. (ed.). 1995. *Occidentalism: Images of the West.* Oxford: Oxford University Press.

Casaly, Paige. 2016. 'Al Mahdi before the ICC: Cultural Property and World Heritage in International Criminal Law', *Journal of International Criminal Justice* 14(5): 1199–220.

Chang, Ha-Joon. 2002. *Kicking away the Ladder: Development Strategy in Historical Perspective.* London: Anthem.

Cheung, Ah Li Leah. 2019. *Where the Past Meets the Future: The Politics of Heritage in Xi'an.* Berlin: Lit.

Claudi, Ida Breckan. 2011. 'The New Kids on the Block: BRICs in the World Heritage Committee', MA thesis. Oslo: University of Oslo.

Clifford, James, and George Marcus (eds). 1986. *Writing Culture: The Poetics and Politics of Ethnography.* Berkeley: University of California Press.

Comaroff, Jean, and John L. Comaroff. 2012. *Theory from the South: Or, How Euro-America Is Evolving toward Africa.* Boulder, CO: Paradigm Publishers.

Courrier, Yves. 2005. *L'UNESCO sans peine.* Paris: L'Harmattan.

Cowan, Jane K. 2014. 'The Universal Periodic Review as a Public Audit Ritual: An Anthropological Perspective on Emerging Practices in the Global Governance of Human Rights', in Hilary Charlesworth and Emma Larking (eds). *Assessing the Universal Periodic Review: Rights, Ritual, Ritualism.* Cambridge: Cambridge University Press, pp. 42–62.

De Cesari, Chiara. 2014. 'World Heritage and the Nation-State: A View from Palestine', in Chiara de Cesari and Ann Rigney (eds). *Transnational Memory: Circulation, Articulation, Scales.* Berlin: De Gruyter, pp. 247–70.

De Giosa, Pierpaolo. 2021 (forthcoming). *World Heritage and Urban Politics in Malaysia: A Cityscape below the Winds.* Amsterdam: Amsterdam University Press.

Denham, T.P., S.G. Haberle, C. Lentfer, R. Fullagar, J. Field, M. Therin, N. Porch and B. Winsborough. 2003. 'Origins of Agriculture at Kuk Swamp in the Highlands of New Guinea', *Science* 301(5630): 189–93.

Denyer, Susan. 1978. *African Traditional Architecture: An Historical and Geographical Perspective.* London: Heinemann.

Di Giovine, Michael A. 2009. *The Heritage-Scape: UNESCO, World Heritage, and Tourism.* Lanham, MD: Lexington Books.

Disko, Stefan, and Helen Tugendhat (eds). 2014. *World Heritage Sites and Indigenous Peoples' Rights.* Copenhagen: IWGIA, Forest Peoples Programme, Gundjeihmi Aboriginal Corporation.

Dwyer, Kevin. 1982. *Moroccan Dialogues: Anthropology in Question.* Baltimore: Johns Hopkins University Press.

Elliott, Michael A., and Vaughn Schmutz. 2016. 'Diffusion and Decoupling in the World Heritage Movement: Exploring Global/Local Tensions in Africa', *European Journal of Cultural and Political Sociology* 3(2–3): 152–76.

English, James F. 2002. 'Winning the Culture Game: Prizes, Awards, and the Rules of Art', *New Literary History* 33(1): 109–35.

Eriksen, Thomas Hylland. 2001. 'Between Universalism and Relativism: A Critique of the UNESCO Concept of Culture', in Jane K. Cowan, Marie-Bénédicte Dembour and Richard Wilson (eds), *Culture and Rights: Anthropological Perspectives*. Cambridge: Cambridge University Press, pp. 127–48.

Falenczyk, Tanya. 2017. 'All für alle – bis es alle ist', *Die Tageszeitung*, 15 August. Retrieved 10 September 2020 from https://taz.de/Bergbau-im-Weltraum/!5433961.

Fichtner, Ullrich. 2008. 'Im Palast der Unsichtbaren', *Der Spiegel*, 10 May. Retrieved 10 September 2020 from http://www.spiegel.de/spiegel/print/d–56898815.html.

Fitzpatrick, Scott N. 2010. *Transboundary Nomination for Stone Money Sites in Palau and Yap*. No place: No publisher.

Foucault, Michel. 1991. 'Governmentality', in Graham Burchell, Colin Gordon and Peter Miller (eds), *The Foucault Effect: Studies in Governmentality*. Chicago: University of Chicago Press, pp. 87–104.

Fox, Richard G., and Barbara J. King (eds). 2002. *Anthropology Beyond Culture*. Oxford: Berg.

Foyer, Jean (ed.) 2015. *Regards croisés sur Rio+20: La modernisation écologique à l'épreuve*. Paris: CNRS Éditions. Retrieved 10 September 2020 from https://books.openedition.org/editionscnrs/26253.

Francioni, Francesco (ed.). 2008. *The 1972 World Heritage Convention: A Commentary*. Oxford: Oxford University Press.

Fresia, Marion. 2013. 'The Making of Global Consensus: Constructing Norms on Refugee Protection at UNHCR', in Birgit Müller (ed.). *The Gloss of Harmony: The Politics of Policy-Making in Multilateral Organisations*. London: Pluto Press, pp. 50–74.

Friedman, Jonathan. 1994. *Cultural Identity and Global Process*. London: Sage.

Furness, William Henry. 1910. *The Island of Stone Money: Uap of the Carolines*. Philadelphia: J.B. Lippincott.

Gable, Eric, and Richard Handler. 1996. 'After Authenticity at an American Heritage Site', *American Anthropologist* 98(3): 568–78.

Galla, Amareswar (ed.). 2012. *World Heritage: Benefits Beyond Borders*. Cambridge: Cambridge University Press.

Gfeller, Aurélie Elisa. 2013. 'Negotiating the Meaning of Global Heritage: "Cultural Landscapes" in the UNESCO World Heritage Convention, 1972–1992', *Journal of Global History* 8(3): 483–503.

______. 2015. 'Anthropologizing and Indigenizing Heritage: The Origins of the UNESCO Global Strategy for a Representative, Balanced and Credible World Heritage List', *Journal of Social Archaeology* 15(3): 366–86.

______. 2017. 'The Authenticity of Heritage: Global Norm-Making at the Crossroads of Cultures', *American Historical Review* 122(3): 758–91.

Goody, Jack. 1994. 'Culture and Its Boundaries: A European View', in Robert Borofsky (ed.), *Assessing Cultural Anthropology*. New York: McGraw-Hill, pp. 250–60.

Grabowsky, Volker, and Sok Deth. 2016. 'Heritage and Nationalism in the Preah Vihear Dispute', *SEATIDE Online Paper* 3. Retrieved 10 September 2020 from https://hal.archives-ouvertes.fr/hal-01963845/document.

Graham, Brian J., and Peter Howard (eds). 2008. *The Ashgate Research Companion to Heritage and Identity*. Burlington, VT: Ashgate.

Greenough, Paul R., and Anna Lowenhaupt Tsing (eds). 2003. *Nature in the Global South: Environmental Projects in South and Southeast Asia*. Durham, NC: Duke University Press.

Groth, Stefan. 2012. *Negotiating Tradition: The Pragmatics of International Deliberations on Cultural Property*. Göttingen: Göttingen University Press.

The Guardian. 2015. 'Japan Sites Get World Heritage Status after Forced Labour Acknowledgement', 6 July. Retrieved 10 September 2020 from http://www.theguardian.com/world/2015/jul/06/japan-sites-get-world-heritage-status-after-forced-labour-acknowledgement.

Hafstein, Valdimar T. 2004. 'The Making of Intangible Cultural Heritage: Tradition and Authenticity, Community and Humanity', Ph.D. thesis. Berkeley: University of California, Berkeley.

_____. 2007. 'Claiming Culture: Intangible Heritage Inc., Folklore©, Traditional Knowledge™', in Dorothee Hemme, Markus Tauschek and Regina Bendix (eds), *Prädikat 'Heritage': Wertschöpfungen aus kulturellen Ressourcen*. Berlin: Lit, pp. 75–100.

_____. 2009. 'Intangible Heritage as a List: From Masterpieces to Representation', in Laurajane Smith and Natsuko Akagawa (eds), *Intangible Heritage*. London: Routledge, pp. 93–111.

_____. 2018. *Making Intangible Heritage: El Condor Pasa and Other Stories from UNESCO*. Bloomington: Indiana University Press.

Hale, Thomas N., and David Held. 2011. 'Editors' Introduction: Mapping Changes in Transnational Governance', in Thomas N. Hale and David Held (eds), *The Handbook of Transnational Governance: Institutions and Innovations*. London: Polity, pp. 1–36.

Hall, Melanie (ed.). 2011. *Towards World Heritage: International Origins of the Preservation Movement 1870–1930*. Farnham: Ashgate.

Handler, Richard, and Jocelyn Linnekin. 1984. 'Tradition, Genuine or Spurious', *Journal of American Folklore* 97(385): 273–90.

Hannerz, Ulf. 1989. 'Culture between Center and Periphery: Towards a Macroanthropology', *Ethnos* 54: 200–16.

_____. 1996. *Transnational Connections: Culture, People, Places*. London: Routledge.

_____. 1999. 'Reflections on Varieties of Culturespeak', *European Journal of Cultural Studies* 2(3): 393–407.

Hansen, Sven. 2015. 'Streit um Industriedenkmäler: Tokio verschweigt Zwangsarbeit', *taz*, 23 May. Retrieved 10 September 2020 from http://www.taz.de/Streit-um-Industriedenkmaeler/!5203613.

Hanson, Allan. 1989. 'The Making of the Maori: Culture Invention and Its Logic', *American Anthropologist* 91(4): 890–902.

Hardin, Garrett. 1968. 'The Tragedy of the Commons', *Science* 162: 1243–48.

Harrison, Rodney. 2012. *Heritage: Critical Approaches*. London: Routledge.

Hart, Keith, and John Sharp (eds). 2015. *People, Money, and Power in the Economic Crisis: Perspectives from the Global South*. New York: Berghahn Books.

Hassan, Fekri. 2007. 'The Aswan High Dam and the International Rescue Nubia Campaign', *African Archaeological Review* 24(3): 73–94.

Hauser-Schäublin, Brigitta. 2011. 'Preah Vihear: From Object of Colonial Desire to a Contested World Heritage Site', in Brigitta Hauser-Schäublin (ed.), *World Heritage Angkor and Beyond: Circumstances and Implications of UNESCO Listings in Cambodia*. Göttingen: Göttinger Universitäts-Verlag, pp. 33–56.

Hauser-Schäublin, Brigitta, and Sven Missling. 2014. 'The Enduring Agency of Borderland Regimes: The Aftermath of Serial Regulations with Different Scopes and Temporal Scales at Preah Vihear, Cambodia', *Journal of Legal Pluralism and Unofficial Law* 46(1): 79–98.

Hedlund, Anna. 2019. *Exile Warriors: Violence and Community among Hutu Rebels in the Eastern Congo*. Philadelphia: University of Pennsylvania Press.

Heinich, Nathalie. 2009. *La fabrique du patrimoine: De la cathédrale à la petite cuillère*. Paris: Maison des sciences de l'homme.

Herzfeld, Michael. 1991. *A Place in History: Social and Monumental Time in a Cretan Town.* Princeton: Princeton University Press.

Hobsbawm, Eric, and Terence Ranger (eds). 1983. *The Invention of Tradition.* Cambridge: Cambridge University Press.

Hobson, John M. 2012. *The Eurocentric Conception of World Politics: Western International Theory, 1760–2010.* New York: Cambridge University Press.

Höhler, Sabine 2014. 'Exterritoriale Ressourcen: Die Diskussion um die Tiefsee, die Pole und das Weltall um 1970', in Isabella Löhr and Andrea Rehling (eds), *Global Commons im 20. Jahrhundert: Entwürfe für eine globale Welt.* Berlin: De Gruyter, pp. 53–82.

Hølleland, Herdis, Evan Hamman and Jessica Phelps. 2019. 'Naming, Shaming and Fire Alarms: The Compilation, Development and Use of the List of World Heritage in Danger', *Transnational Environmental Law* 8(1): 35–57.

Hølleland, Herdis, and Marit Johansson. 2019. '"... To Exercise in All Loyalty, Discretion and Conscience": On Insider Research and the World Heritage Convention', *International Journal of Cultural Policy* 25(3): 309–21.

Hølleland, Herdis, and Elisabeth Niklasson. 2020. 'How (Not) to "Study up": Points and Pitfalls When Studying International Heritage Regimes', *Journal of Field Archaeology* 45(3): 140–52.

Hølleland, Herdis, and Morgaine Wood. 2019. 'An Emotional Plea for Al-Ahsa: A Case Study on How Discourses of Representativeness, Climate and Discord Are Strategized in the World Heritage Regime', *International Journal of Cultural Policy*: 1–15. Retrieved 10 September 2020 from https://doi.org/10.1080/10286632.2019.1646734.

Holston, James. 1989. *The Modernist City: An Anthropological Critique of Brasília.* Chicago: University of Chicago Press.

Human, Helen. 2015. 'Democratising World Heritage: The Policies and Practices of Community Involvement in Turkey', *Journal of Social Archaeology* 15(2): 160–83.

ICOMOS. 2004. *The World Heritage List: Filling the Gaps – An Action Plan for the Future.* Paris: ICOMOS. Retrieved 10 September 2020 from whc.unesco.org/document/102409.

_____. 2018. *Evaluations of World Heritage Nominations Related to Sites Associated with Memories of Recent Conflicts.* Retrieved 10 September 2020 from https://whc.unesco.org/document/167810.

International Coalition of Sites of Conscience. 2018. *Interpretation of Sites of Memory.* Retrieved 10 September 2020 from https://whc.unesco.org/document/165700.

Istasse, Manon. 2016. 'Affects and Senses in a World Heritage Site: People–House Relations in the Medina of Fez', in Christoph Brumann and David Berliner (eds), *World Heritage on the Ground: Ethnographic Perspectives.* Oxford: Berghahn Books, pp. 37–59.

James, Luke. 2016. 'The Symbolic Value of Expertise in International Heritage Diplomacy', *Future Anterior* 13(1): 82–96.

James, Luke, and Tim Winter. 2015. 'Expertise and the Making of World Heritage Policy', *International Journal of Cultural Policy* 23(1): 36–51.

Japan Times News. 2015. 'Japan, South Korea Mark 50 Years of Postwar Ties', 25 March. Retrieved 10 September 2020 from https://www.japantimes.co.jp/news/2019/03/25/national/south-korea-court-approves-seizure-mitsubishi-heavy-assets-wartime-labor.

_____. 2019. 'South Korea Court Approves Seizure of Mitsubishi Heavy Trademarks and Patents over Wartime Forced Labor', 25 March. Retrieved 10 September 2020 from https://www.japantimes.co.jp/news/2019/03/25/national/south-korea-court-approves-seizure-mitsubishi-heavy-assets-wartime-labor.

Jokilehto, Jukka. 1999. *A History of Architectural Conservation.* Amsterdam: Elsevier Butterworth-Heinemann.

_____. 2006. *What Is OUV? Defining the Outstanding Universal Value of Cultural World Heritage Properties.* Berlin: Bäßler Verlag.
Jones, Siân, and Thomas Yarrow. 2013. 'Crafting Authenticity: An Ethnography of Conservation Practice', *Journal of Material Culture* 18(1): 3–26.
Joy, Charlotte. 2016. '"UNESCO Is What?" World Heritage, Militant Islam and the Search for a Common Humanity in Mali', in Christoph Brumann and David Berliner (eds), *World Heritage on the Ground: Ethnographic Perspectives.* Oxford: Berghahn Books, pp. 60–77.
June, Raymond, and Janine Wedel. 2010. 'The Avalanche of Indicators in Global Governance: Implications for the Anthropology of Policy', *Anthropology News* 51(7): 25–26.
Kalb, Don, and Luisa Steur. 2015. 'Global South', in James D. Wright (ed.), *International Encyclopedia of the Social and Behavioral Sciences*, 2nd edn, vol. 10. Oxford: Elsevier, pp. 186–91.
Kearney, Michael. 1995. 'The Local and the Global: The Anthropology of Globalization and Transnationalism', *Annual Review of Anthropology* 24: 547–65.
Keesing, Roger M. 1994. 'Theories of Culture Revisited', in Robert Borofsky (ed.), *Assessing Cultural Anthropology.* New York: McGraw-Hill, pp. 301–10.
Kelly, Tobias. 2011. 'The Cause of Human Rights: Doubts about Torture, Law, and Ethics at the United Nations', *Journal of the Royal Anthropological Institute* 17(4): 728–44.
Keyuan, Zou. 1991. 'The Common Heritage of Mankind and the Antarctic Treaty System', *Netherlands International Law Review* 38(2): 173–98.
Khalaf, Roha W. 2017. 'A Viewpoint on the Reconstruction of Destroyed UNESCO Cultural World Heritage Sites', *International Journal of Heritage Studies* 23(3): 261–74.
_____. 2019. 'Continuity: A Fundamental Yet Overlooked Concept in World Heritage Policy and Practice', *International Journal of Cultural Policy*: 1–15. Retrieved 10 September 2020 from https://doi.org/10.1080/10286632.2019.1696782.
_____. 2020. 'Authenticity or Continuity in the Implementation of the UNESCO World Heritage Convention? Scrutinizing Statements of Outstanding Universal Value, 1978–2019', *Heritage* 3: 243–74. Retrieved 10 September 2020 from https://www.mdpi.com/2571-9408/3/2/15.
Koester, David. 2005. 'Global Movements and Local Historical Events: Itelmens of Kamchatka Appeal to the United Nations', *American Ethnologist* 32(4): 642–59.
Kormos, Cyril F., Tim Badman, Tilman Jaeger, Bastian Bertzky, Remco van Merm, Elena Osipova, Yichuan Shi and Peter Bille Larsen. 2017. *World Heritage, Wilderness and Large Landscapes and Seascapes.* Gland, Switzerland: IUCN. Retrieved 10 September 2020 from https://portals.iucn.org/library/sites/library/files/documents/2017-028.pdf.
Kuper, Adam. 1999. *Culture: The Anthropologists' Account.* Cambridge, MA: Harvard University Press.
Kuutma, Kristin. 2007. 'The Politics of Contested Representation: UNESCO and the Masterpieces of Intangible Cultural Heritage', in Dorothee Hemme, Markus Tauschek and Regina Bendix (eds), *Prädikat 'Heritage': Wertschöpfungen aus kulturellen Ressourcen.* Berlin: Lit, pp. 177–96.
Labadi, Sophia. 2010. 'World Heritage, Authenticity and Post-authenticity: National and International Perspectives', in Sophia Labadi and Colin Long (eds), *Heritage and Globalisation.* Oxford: Routledge, pp. 66–84.
_____. 2012. *UNESCO, Cultural Heritage, and Outstanding Universal Value: Value-Based Analyses of the World Heritage and Intangible Cultural Heritage Conventions.* Lanham, MD: Rowman & Littlefield.
Laïdi, Zaki. 2014. 'Towards a Post-Hegemonic World: The Multipolar Threat to the Multilateral Order', *International Politics* 51(3): 350–65.

Larkham, Peter J. 1996. *Conservation and the City.* London: Routledge.

Larkin, Craig, and Michael Dumper. 2009. 'UNESCO and Jerusalem: Constraints, Challenges and Opportunities', *Jerusalem Quarterly* 39: 16–28. Retrieved 10 September 2020 from https://www.palestine-studies.org/sites/default/files/jq-articles/39_Larkin_UNESCO.pdf.

Larsen, Peter Bille (ed.). 2018. *World Heritage and Human Rights: Lessons from the Asia-Pacific and Global Arena.* London: Routledge.

Larsen, Peter Bille, and Kristal Buckley. 2018. 'The World Heritage Committee and Human Rights: Learning from Event Ethnography', in Peter Bille Larsen (ed.), *World Heritage and Human Rights: Lessons from the Asia-Pacific and Global Arena.* London: Routledge, pp. 29–48.

Le Saout, Soizic, Michael Hoffmann, Yichuan Shi, Adrian Hughes, Cyril Bernard, Thomas M. Brooks, Bastian Bertzky, Stuart H.M. Butchart, Simon N. Stuart, Tim Badman and Ana S.L. Rodrigues. 2013. 'Protected Areas and Effective Biodiversity Conservation', *Science* 342(6160): 803–5.

Lenclud, Gérard. 1987. 'La tradition n'est plus ce qu'elle était: Sur la notion de "tradition" et de "société traditionnelle" en ethnologie', *Terrain* 9: 110–23.

Lentz, Carola. 2016. 'Culture: The Making, Unmaking and Remaking of an Anthropological Concept', *Arbeitspapiere des Instituts für Ethnologie und Afrikastudien der Johannes Gutenberg-Universität Mainz* 166. Retrieved 10 September 2020 from http://www.ifeas.uni-mainz.de/Dateien/AP166.pdf.

Lins Ribeiro, Gustavo. 2014. 'World Anthropologies: Anthropological Cosmopolitanisms and Cosmopolitics', *Annual Review of Anthropology* 43: 483–98.

Logan, William. 2013. 'Australia, Indigenous Peoples and World Heritage from Kakadu to Cape York: State Party Behaviour under the World Heritage Convention', *Journal of Social Archaeology* 13(2): 153–76.

Logan, William, Máiréad Nic Craith and Ullrich Kockel (eds). 2016. *A Companion to Heritage Studies.* Chichester: Wiley-Blackwell.

Logan, William, and Keir Reeves (eds). 2008. *Places of Pain and Shame: Dealing with 'Difficult Heritage'.* London: Routledge.

Macdonald, Sharon. 2013. *Memorylands: Heritage and Identity in Europe Today.* London: Routledge.

Mainichi Shinbun. 2015. 'Shasetsu: Sekai isan to Nikkan: Kongo ni shikori o nokosu na', 7 July. Retrieved 20 December 2015 from http://mainichi.jp/articles/20150707/ddm/005/070/042000c (no longer available).

Malinowski, Bronislaw. 1953 [1922]. *Argonauts of the Western Pacific: An Account of Native Enterprise and Adventure in the Archipelagoes of Melanesian New Guinea.* New York: E.P. Dutton.

Marcus, George E. 1995. 'Ethnography in/of the World System: The Emergence of Multi-sited Ethnography', *Annual Review of Anthropology* 24: 95–117.

Marquart, Vivienne. 2015. 'Monuments and Malls: Heritage Politics and Urban Struggles in Istanbul', Ph.D. thesis. Halle: Martin Luther University Halle-Wittenberg.

Maswood, S. Javed. 2000. 'Kakadu and the Politics of World Heritage Listing', *Australian Journal of International Affairs* 54(3): 357–72.

Maurel, Chloé. 2007. 'La question des races: Le programme de l'Unesco', *Gradhiva* 5: 114–31. Retrieved 10 September 2020 from https://journals.openedition.org/gradhiva/815.

Merriman, Nick C. 1989. 'Heritage from the Other Side of the Glass Case', *Anthropology Today* 5(2): 14–15.

Merry, Sally Engle. 2006. *Human Rights and Gender Violence: Translating International Law into Local Justice.* Chicago: University of Chicago Press.

Meskell, Lynn. 2012a. *The Nature of Heritage: The New South Africa*. Malden, MA: John Wiley & Sons.

_____. 2012b. 'The Rush to Inscribe: Reflections on the 35th Session of the World Heritage Committee, UNESCO Paris, 2011', *Journal of Field Archaeology* 37(2): 145–51.

_____. 2013a. 'UNESCO's World Heritage Convention at 40: Challenging the Economic and Political Order of International Heritage Conservation', *Current Anthropology* 54(4): 483–94.

_____. 2013b. 'UNESCO and the Fate of the World Heritage Indigenous Peoples Council of Experts (WHIPCOE)', *International Journal of Cultural Property* 20(2): 155–74.

_____. 2014. 'States of Conservation: Protection, Politics, and Pacting within UNESCO's World Heritage Committee', *Anthropological Quarterly* 87(1): 217–44.

_____ (ed.). 2015a. *Global Heritage: A Reader*. Hoboken, NJ: Wiley-Blackwell.

_____. 2015b. 'Gridlock: UNESCO, Global Conflict and Failed Ambitions', *World Archaeology* 47(2): 225–38.

_____. 2015c. 'Transacting UNESCO World Heritage: Gifts and Exchanges on a Global Stage', *Social Anthropology* 23(1): 3–21.

_____. 2016. 'World Heritage and WikiLeaks', *Current Anthropology* 57(1): 72–95.

_____. 2018. *A Future in Ruins: UNESCO, World Heritage, and the Dream of Peace*. Oxford: Oxford University Press.

Meskell, Lynn, Claudia Liuzza, Enrico Bertacchini and Donatella Saccone. 2014. 'Multilateralism and UNESCO World Heritage: Decision-Making, States Parties and Political Processes', *International Journal of Heritage Studies* 21(5): 423–40.

Meskell, Lynn, Claudia Liuzza and Nicholas Brown. 2015. 'World Heritage Regionalism: UNESCO from Europe to Asia', *International Journal of Cultural Property* 22(4): 437–70.

Mitchell, Nora, Mechtild Rössler and Pierre-Marie Tricaud (eds). 2009. *World Heritage Cultural Landscapes: A Handbook for Conservation and Management*. Paris: UNESCO.

Muehlebach, Andrea. 2001. '"Making Place" at the United Nations: Indigenous Cultural Politics at the UN Working Group on Indigenous Populations', *Cultural Anthropology* 16(3): 415–48.

Müller, Birgit (ed.). 2013. *The Gloss of Harmony: The Politics of Policy-Making in Multilateral Organisations*. London: Pluto Press.

Nader, Laura. 1972. 'Up the Anthropologist: Perspectives Gained from Studying up', in Dell H. Hymes (ed.), *Reinventing Anthropology*. New York: Pantheon, pp. 284–311.

Nas, Peter J.M. 2002. 'Masterpieces of Oral and Intangible Culture: Reflections on the UNESCO World Heritage List', *Current Anthropology* 43(1): 139–48.

Nielsen, Bjarke. 2011. 'UNESCO and the "Right" Kind of Culture: Bureaucratic Production and Articulation', *Critique of Anthropology* 31(4): 273–92.

Niezen, Ron, and Maria Sapignoli (eds). 2017. *Palaces of Hope: The Anthropology of Global Organizations*. Cambridge: Cambridge University Press.

Oldham, Paul, and Miriam Anne Frank. 2008. '"We the Peoples": The United Nations Declaration on the Rights of Indigenous Peoples', *Anthropology Today* 24(2): 5–9.

Peterson, M.J. 2010. 'How the Indigenous Got Seats at the UN Table', *Review of International Organizations* 5(2): 197–225.

Ranger, Terence. 1993. 'The Invention of Tradition Revisited: The Case of Colonial Africa', in Terence Ranger and Olufemi Vaughan (eds), *Legitimacy and the State in Twentieth-Century Africa: Essays in Honour of A.H. Kirk-Greene*. Oxford: Macmillan, pp. 62–111.

Rao, Kishore. 2010. 'A New Paradigm for the Identification, Nomination and Inscription of Properties on the World Heritage List', *International Journal of Heritage Studies* 16(3): 161–72.

Reed, Steven R. 2002. 'Impersonal Mechanisms and Personal Networks in the Distribution of Grants', in Michio Muramatsu, Farrukh Iqbal and Ikuo Kume (eds), *Local Government Development in Postwar Japan*. Oxford: Oxford University Press, pp. 112–31.

Rehling, Andrea. 2011. 'Universalismen und Partikularismen im Widerstreit: Zur Genese des UNESCO-Welterbes', *Zeithistorische Forschungen/Studies in Contemporary History* 8: 414–36.

_____. 2014. '"Kulturen unter Artenschutz"? Vom Schutz der Kulturschätze als Gemeinsames Erbe der Menschheit zur Erhaltung kultureller Vielfalt', in Isabella Löhr and Andrea Rehling (eds), *Global Commons im 20. Jahrhundert: Entwürfe für eine globale Welt*. Berlin: De Gruyter, pp. 109–38.

Rehling, Andrea, and Isabella Löhr. 2014. '"Governing the Commons": Die *global commons* und das Erbe der Menschheit im 20. Jahrhundert', in Isabella Löhr and Andrea Rehling (eds), *Global Commons im 20. Jahrhundert: Entwürfe für eine globale Welt*. Berlin: De Gruyter, pp. 3–32.

Reyes, Victoria. 2014. 'The Production of Cultural and Natural Wealth: An Examination of World Heritage Sites', *Poetics* 44: 42–63.

Ricca, Simone. 2007. *Reinventing Jerusalem: Israel's Reconstruction of the Jewish Quarter after 1967*. London: I.B. Tauris.

Riles, Annelise. 1998. 'Infinity within the Brackets', *American Ethnologist* 25(3): 378–98.

Rodseth, Lars. 1998. 'Distributive Models of Culture: A Sapirian Alternative to Essentialism', *American Anthropologist* 100(1): 55–69.

Rößler, Maren. 2008. *Zwischen Amazonas und East River: Indigene Bewegungen und ihre Repräsentation in Peru und bei der UNO*. Bielefeld: Transcript.

Rössler, Mechtild. 2005. 'Weltkulturerbe und Globalisierung: Vom Weltwunder zum Erbe der Menschheit', in Iris Schröder and Sabine Höhler (eds), *Welt-Räume: Geschichte, Geographie und Globalisierung seit 1900*. Frankfurt: Campus, pp. 235–57.

_____. 2006a. 'World Heritage Cultural Landscapes: A UNESCO Flagship Programme 1992–2006', *Landscape Research* 31(4): 333–53.

_____. 2006b. 'World Heritage: Linking Biological and Cultural Diversity', in Barbara T. Hoffman (ed.), *Art and Cultural Heritage: Law, Policy, and Practice*. Cambridge: Cambridge University Press, pp. 201–5.

Rudolff, Britta. 2010. *'Intangible' and 'Tangible' Heritage: A Topology of Culture in Contexts of Faith*. Bonn: Scientia Bonnensis.

Ruggles, Clive (ed.). 2017. *Heritage Sites of Astronomy and Archaeoastronomy in the Context of the UNESCO World Heritage Convention: Thematic Study*, vol. 2. Bognor Regis and Paris: Ocarina Books and ICOMOS. Retrieved 10 September 2020 from http://openarchive.icomos.org/1856/1/Astronomy2%20Final%20low%20res.pdf.

Ruskin, John. 1920 [1849]. *The Seven Lamps of Architecture*. London: Waverley.

Sahlins, Marshall. 1999. 'Two or Three Things That I Know about Culture', *Journal of the Royal Anthropological Institute* 5(3): 399–421.

Sankei Nyūsu. 2015a. '"Daigyo o son shita" "Gaimushō no tochiri" hihan zokushutsu: Jimin sōmukai, Kankoku no gōi hogo de gaimushō kara keii chōshu', 10 July. Retrieved 10 September 2020 from http://www.sankei.com/politics/news/150710/plt1507100026-n1.html.

_____. 2015b. 'Gaishō kaidan gōi mushi shita Kankoku, misshitsu de tobikau dogō: Kōshō sekininsha "Kankoku no akui ni michita kōi kakushin shita"', 11 July. Retrieved 10 September 2020 from http://www.sankei.com/politics/news/150711/plt1507110003-n1.html, http://www.sankei.com/politics/news/150711/plt1507110003-n2.html, http://www.sankei.com/politics/news/150711/plt1507110003-n3.html, http://www.sankei.com/politics/news/

150711/plt1507110003-n4.html, http://www.sankei.com/politics/news/150711/plt1507110003-n5.html, http://www.sankei.com/politics/news/150711/plt1507110003-n6.html.

_____. 2015c. 'Kankoku gaimushō, saito de jisetsu o tenkai: Kyōsei sareta "rekishiteki jijitsu" ni Nihon ga hajimete genkyū to setsumei', 7 July. Retrieved 10 September 2020 from http://www.sankei.com/world/news/150707/wor1507070039-n1.html, http://www.sankei.com/world/news/150707/wor1507070039-n2.html.

_____. 2015d. 'Minshu, Kōno danwa to kondō saseru izu? Abe shushō "Kyōsei rōdō o imi shinai" to hitei', 10 July. Retrieved 10 September 2020 from http://www.sankei.com/politics/news/150710/plt1507100022-n1.html.

_____. 2015e. 'Suishin-yaku no Katō Kōko-shi, Kankoku no hantai kōsaku ni "Suna kamu yō na kuyashisa"', 13 July. Retrieved 10 September 2020 from http://www.sankei.com/life/news/150713/lif1507130013-n1.html.

_____. 2015f. 'Yakusoku ga chigau! Jimintō ga "Kyōsei rōdō nashi" ketsugi e: "Kōno danwa ya ianpu mondai no yō ni gokai shōrai ni nokosu"', 10 July. Retrieved 10 September 2020 from http://www.sankei.com/politics/news/150710/plt1507100014-n1.html.

Sapignoli, Maria. 2017. 'A Kaleidoscopic Institutional Form: Expertise and Transformation in the UN Permanent Forum on Indigenous Issues', in Ron Niezen and Maria Sapignoli (eds), *Palaces of Hope: The Anthropology of Global Organizations.* Cambridge: Cambridge University Press, pp. 78–105.

_____. 2020. *Hunting Justice: Displacement, Law, and Activism in the Kalahari.* Cambridge: Cambridge University Press.

Savova, Nadezhda Dimitrova. 2009. 'Heritage Kinaesthetics: Local Constructivism and UNESCO's Intangible-Tangible Politics at a Favela Museum', *Anthropological Quarterly* 82(2): 547–85.

Schäfer, Hilmar. 2016. '"Outstanding Universal Value": Die Arbeit an der Universalisierung des Wertvollen im UNESCO-Welterbe', *Berliner Journal für Soziologie* 26: 353–75.

Schmitt, Thomas M. 2009. 'Global Cultural Governance: Decision Making Concerning World Heritage between Politics and Science', *Erdkunde* 63(2): 103–21.

_____. 2011. *Cultural Governance: Zur Kulturgeographie des UNESCO-Welterberegimes.* Stuttgart: Franz Steiner Verlag.

Schmitt, Thomas, and Andreas Schweitzer. 2007. 'Welterbe oder Stadtentwicklung in Gefahr? Zu deutschen Debatten und zur global-lokalen Governanz von UNESCO-Welterbestätten am Beispiel des Kölner Doms', *Berichte zur deutschen Landeskunde* 81(4): 329–52.

Schmutz, Vaughn, and Michael A. Elliott. 2017. 'World Heritage and the Scientific Consecration of "Outstanding Universal Value"', *International Journal of Comparative Sociology* 58(2): 140–59.

Sheppard, David, and Gamini Wijesuriya. 2018. *Strenghtening the Effectiveness of the World Heritage Reactive Monitoring Process: Final Report.* Retrieved 10 September 2020 from https://whc.unesco.org/document/174948.

Shore, Cris. 2000. *Building Europe: The Cultural Politics of European Integration.* London: Routledge.

Siebert, Ute. 1997. 'Die Bedeutung des Indigenenbegriffs im UNO-System: Stationen in der Diskussion um eine formale Definition', *Tsantsa* 2: 76–91.

Siemons, Mark. 2015a. 'Insel des Grauens', *Frankfurter Allgemeine Sonntagszeitung*, 17 May, p. 38.

Siemons, Mark, with an introduction by William Underwood. 2015b. 'Island of Horror: Gunkanjima and Japan's Quest for UNESCO World Heritage Status', *Asia-Pacific Journal* 13 (26, 3). Retrieved 10 September 2020 from https://apjjf.org/Mark-Siemons/4333.html.

Silverman, Helaine. 2011. 'Border Wars: The Ongoing Temple Dispute between Thailand and Cambodia and UNESCO's World Heritage List', *International Journal of Heritage Studies* 17(1): 1–21.

Sinding-Larsen, Amund. 2012. 'Our Common Dignity: Rights-Based Approaches to Heritage Management', *World Heritage Review* 62: 58–60.

Smith, Laurajane. 2006. *Uses of Heritage.* New York: Routledge.

Smith, Nathaniel. 2011. 'Right-Wing Activism in Japan and the Politics of Futility', Ph.D. thesis. New Haven: Yale University.

Stolcke, Verena. 1995. 'Talking Culture: New Boundaries, New Rhetorics of Exclusion in Europe', *Current Anthropology* 36(1): 1–24.

Stott, Peter H. 2011. 'The World Heritage Convention and the National Park Service, 1962–1972', *The George White Forum* 28(3): 279–90.

_____. 2012. 'The World Heritage Convention and the National Park Service: The First Two Decades, 1972–1992', *The George White Forum* 29(1): 148–75.

_____. 2013. 'The World Heritage Convention and the National Park Service, 1993–2009', *The George White Forum* 30(1): 18–44.

Swenson, Astrid. 2016. 'The First Heritage International(s): Conceptualizing Global Networks before UNESCO', *Future Anterior: Journal of Historic Preservation, History, Theory, and Criticism* 13(1): 1–15.

Swenson, Astrid, and Peter Mandler (eds), 2013. *From Plunder to Preservation: Britain and the Heritage of Empire, c.1800–1940.* Oxford: Oxford University Press.

Tabet, Jade. 2010. *Review of ICOMOS' Working Methods and Procedures for the Evaluation of Cultural and Mixed Properties Nominated for Inscription on the UNESCO World Heritage List.* Paris: ICOMOS. Retrieved 10 September 2020 from https://www.icomos.org/world_heritage/WH_Committee_34th_session_Brasilia/JT_Final_report_en.pdf.

Tauschek, Markus. 2013. *Kulturerbe: Eine Einführung.* Berlin: Reimer.

Terrill, Greg. 2012. '40th Anniversary of the World Heritage Convention: 2002–2012 Highlights', *World Heritage Review* 65: 62–69.

_____. 2014a. 'Politics before Lunch? A Morning's Work by the World Heritage Committee', *Historic Environment* 26(2): 24–36.

_____. 2014b. '"Surprise!" Is Not Good System Design: The Upstream Process for Nominations to the World Heritage List', *Heritage & Society* 7(1): 59–71.

Titchen, Sarah M. 1995. 'On the Construction of Outstanding Universal Value: UNESCO's World Heritage Convention (Convention Concerning the Protection of the World Cultural and Natural Heritage, 1972) and the Identification and Assessment of Cultural Places for Inclusion in the World Heritage List', Ph.D. thesis. Canberra: Australian National University.

Trouillot, Michel-Rolph. 2002. 'Adieu, Culture: A New Duty Arises', in Richard G. Fox and Barbara J. King (eds), *Anthropology Beyond Culture.* Oxford: Berg, pp. 37–60.

Tsing, Anna. 2000. 'The Global Situation', *Cultural Anthropology* 15(3): 327–60.

Tunbridge, J.E., and G.J. Ashworth. 1996. *Dissonant Heritage: The Management of the Past as a Resource in Conflict.* Chichester: J. Wiley.

Turtinen, Jan. 2000. *Globalising Heritage: On UNESCO and the Transnational Construction of a World Heritage.* Retrieved 10 September 2020 from http://www.score.su.se/polopoly_fs/1.26651.1320939806!/200012.pdf.

_____. 2006. 'Väldsarvets villkor: Intressen, förhandlinger och bruk i internationell politik', Ph.D. thesis. Stockholm: University of Stockholm.

Underwood, William. 2015. 'History in a Box: UNESCO and the Framing of Japan's Meiji Era', *Asia-Pacific Journal* 13 (26, 2). Retrieved 10 September 2020 from https://apjjf.org/William-Underwood/4332.html.

UNESCO. 2011. *Preparing World Heritage Nominations*, 2nd edn. Paris: UNESCO. Retrieved 10 September from http://whc.unesco.org/uploads/activities/documents/activity–643–1.pdf.

_____. 2016. *World Heritage in the High Seas: An Idea Whose Time Has Come.* Paris: UNESCO. Retrieved 10 September from whc.unesco.org/document/143493.

Van der Aa, Bart. 2005. 'Preserving the Heritage of Humanity? Obtaining World Heritage Status and the Impacts of Listing', Ph.D. thesis. Groningen: University of Groningen.

Vestergaard, Jakob, and Robert H. Wade. 2013. 'Protecting Power: How Western States Retain the Dominant Voice in the World Bank's Governance', *World Development* 46: 153–64.

Viollet-le-Duc, E. 1875. *Dictionnaire raisonné de l'architecture française du XIe au XVIe siècle*, vol. 8. Paris: A. Morel.

Wade, Robert H. 2011. 'Emerging World Order? From Multipolarity to Multilateralism in the G20, the World Bank, and the IMF', *Politics & Society* 39(3): 347–78.

_____. 2013. 'The Art of Power Maintenance', *Challenge* 56(1): 5–39.

Walsh, Kevin. 1992. *The Representation of the Past: Museums and Heritage in the Post-modern World.* London: Routledge.

Waterton, Emma, and Steve Watson (eds). 2015. *The Palgrave Handbook of Contemporary Heritage Research.* Basingstoke: Palgrave Macmillan.

Wells, Clare. 1987. *The UN, UNESCO, and the Politics of Knowledge.* New York: St Martin's Press.

Whitby-Last, Kathryn. 2008. 'Article 1 Cultural Landscapes', in Francesco Francioni (ed.), *The 1972 World Heritage Convention: A Commentary.* Oxford: Oxford University Press, pp. 51–62.

Wikan, Unni. 1999. 'Culture: A New Concept of Race', *Social Anthropology* 7: 57–64.

Williams, Tim. 2011. 'The Curious Tale of Preah Vihear: The Process and Value of World Heritage Nomination', *Conservation and Management of Archaeological Sites* 13(1): 1–7.

Williams, Timothy. 2014. *The Silk Roads: An ICOMOS Thematic Study.* Charenton-le-Pont: ICOMOS. Retrieved 10 September 2020 from http://discovery.ucl.ac.uk/1356660/1/ICOMOS_WHThematicStudy_SilkRoads_final_lv_201406.pdf.

Wimmer, Andreas. 2005. *Kultur als Prozess. Zur Dynamik des Aushandelns von Bedeutungen.* Wiesbaden: Verlag für Sozialwissenschaften.

Winter, Tim. 2015. 'Heritage Diplomacy', *International Journal of Heritage Studies* 21(10): 997–1015.

Winter, Tim, and Emma Waterton (eds). 2013. 'Special Issue: Critical Heritage Studies', *International Journal of Heritage Studies* 19(6): 529–609.

Wolfrum, Rüdiger. 2009. 'Common Heritage of Mankind', in Rüdiger Wolfrum (ed.), *Max Planck Encyclopedia of Public International Law.* Retrieved 10 September 2020 from https://opil.ouplaw.com/view/10.1093/law:epil/9780199231690/law-9780199231690-e1149.

Wong, Edward. 2015. 'China Fears Loss of Great Wall, Brick by Brick', *Sinosphere*, 29 June. Retrieved 10 September 2020 from https://sinosphere.blogs.nytimes.com/2015/06/29/disappearance-of-chinas-great-wall-prompts-concerns.

Woodward, Kenneth L. 1996. *Making Saints: How the Catholic Church Determines Who Becomes a Saint, Who Doesn't, and Why*, 2nd edn. New York: Simon & Schuster.

Yan, Haiming. 2018. *World Heritage Craze in China: Universal Discourse, National Culture, and Local Memory.* New York: Berghahn Books.

Index

Aapravasi Ghat 64, 107
Abe Shinzō 132–34
Abu Mena 171, 220n83
Abu Simbel. *See* Nubian monuments
Academy Awards 6, 261
Advisory Bodies 98–99, 104, 127, 155, 188, 265; criticised 90, 91, 93–94, 103, 163, 171, 176, 232, 238, 254n8; and World Heritage Centre 66, 124, 167–68. *See also* ICCROM, ICOMOS, IUCN
advocacy. *See* national delegations
Afghanistan 35, 37n4, 215–16n40
Africa 28–29, 35, 54, 84, 95–96, 102, 105, 106, 112, 147n34, 224, 232, 233, 239, 240, 243, 244, 248, 249, 254n4, 254n7; as UNESCO region/electoral group 34, 90, 146n19, 146n22, 153n71, 226–27, 228, 230, 231, 236–38, 239, 256n20
African World Heritage Fund 84, 230
Al-Ahsa oasis 260
Albania 178n12, 187, 190, 210
Aleppo 193, 194
Algeria 25, 34, 74, 93, 233–34, 237
Al Mahdi, Ahmad Al Faqi 258
ambassadors 6, 24, 25, 27, 29, 30, 31, 33, 35, 70, 76, 85, 87, 89, 90, 93, 99, 101, 102, 104, 105, 106–15, 116, 127, 131, 137, 140–1, 144n4, 144n8, 146n18, 149n53, 150n59, 151–52n66, 153n70, 167, 178n20, 225, 226, 265, 266. *See also* diplomats
amendments. *See* World Heritage Committee sessions
Angkor 5, 7, 78, 80n56, 151n64
Angola 127, 209
Ani archaeological site 152n68, 260
Antarctica 39–41
Antarctic Treaty 40
anthropology 217n54, 233, 244; and global governance 9–10; and World Heritage 10–12, 19–20, 31–32, 53, 58–59, 63–64, 179–80, 210, 227, 228, 242, 243, 268–69
Antonine Wall 186
Arabian Oryx Sanctuary 1, 2, 74
Arab countries 2, 109, 114, 232; as UNESCO region/electoral group 25–26, 137, 141, 149–50nn55–57, 157, 228, 230, 231, 238, 240, 256n20
archaeological sites 31, 43–45, 46n13, 55, 57, 60–61, 63, 65, 74, 92, 135–37, 140, 149–50n57, 152n68, 170, 181, 199, 208, 214n34, 216n42, 219n78, 220n83, 228, 237, 260
archaeology 8, 18, 39, 44, 53, 58–59, 114, 239, 244–45
Argentina 29, 108, 163, 187–88, 189–90, 214nn23–25, 214n27, 215–16n40, 216–17n43, 247. *See also* World Heritage Committee session 1984
Armenia 62, 65, 152n68, 260
Asia 104, 113, 133, 147n34, 148n50, 177n8, 186, 196, 215n40, 224, 240, 255n11
Asia and the Pacific as UNESCO region/electoral group 119, 146n19, 231, 238, 239, 254n8, 255n18, 256n20
astronomical heritage 205
Aswan dam 43–45
Atsinanana, rainforests of the 85, 154n73

audits 14, 16, 54, 119, 146n23, 156, 169, 176, 194, 242
Augustusburg castle at Brühl 51, 266
Auschwitz concentration camp 50, 129, 132, 140, 149n56, 153n70, 209, 270n3
Australia 5, 7, 15, 51, 53, 58–59, 60, 71–73, 81n60, 82n79, 87, 88, 89–90, 92–93, 97n2, 105, 108, 117, 124–25, 126, 127, 137, 144n4, 145n14, 154n73, 162, 183, 196, 213n11, 213n18, 219n65, 229, 234, 235, 236, 240, 241, 247, 257–58, 259. *See also* World Heritage Committee session 1981, 2000
Austria 7, 61, 125, 147n30, 165, 187, 190, 193–94, 210, 218n61, 218n63
authenticity 30, 46n14, 50, 65, 79n18, 174, 195–201, 211; and continuity 195–96, 200; and integrity 180–81, 201, 215–16n40, 217–18n54, 218n56, 221n101, 236; Nara Document on 54–60, 181, 196–201, 228. *See also* reconstructions
Ayers Rock. *See* Uluṟu-Kata Tjuṯa National Park
Azerbaijan 260. *See also* World Heritage Committee session 2019

Bagrati cathedral 154n73, 199, 201
Bahá'i holy places 63
Bahrain 82n79, 87, 90, 96–97, 235. *See also* World Heritage Committee session 2018
Balkan war 71, 138, 200
Bamiyan 35, 37n4
Ban Ki-moon 146n56
Bangladesh 145–46n14, 216–16n43
Barbados 76, 82n79, 90, 235
Baroque palaces and parks 24, 36, 51, 61, 59, 183, 201, 219n77, 236–37, 243–44, 266
Baseball Hall of Fame 261
Battir 137, 149n55, 218n62
beech forests, European ancient and primeval 187, 190, 210
Belarus 182, 183, 184–85, 188, 192, 216n43
belfries of Belgium and France 186–87, 190
Belgium 27, 29, 58, 67, 86–87, 88, 108, 151n64, 185, 186–87, 187–88, 190, 206, 209, 210. 214n20, 214nn23–25, 214n27, 248
Bergpark Wilhelmshöhe 243
Berlin, 132, 195; Museumsinsel in 181; modern housing estates in 61, 64. *See also* World Heritage Committee session 1995
Berliner, David x, 8, 9
Bethlehem, Church of the Nativity in 109–10, 137
Białowieża forest 183, 216n43
Bikini nuclear test site 140
biodiversity 114, 180, 204, 206, 213n18, 226
Bokova, Irina 149n56
Bolgar 208, 222–23n123
Bolivia 163, 189–90, 215–16n40
Bonn. *See* World Heritage Committee session 2015
Bortolotto, Chiara x, 17, 217–18n54
Bosnia and Herzegovina 200–1, 220n86, 221n99
boundaries. *See* World Heritage properties
Brasilia 60. *See also* World Heritage Committee session 1988, 2010
Brazil 13, 18, 28, 30, 36, 60, 82n79, 83–90, 106, 115, 117, 121, 146n21, 147n30, 191, 206, 216–17n43, 217n51, 221n101, 229, 231–32, 233, 234, 235, 240, 248, 248, 255n11. *See also* World Heritage Committee session 1998, 2010
Bremen 266
BRICS alliance 18, 146n21, 233, 234, 255n11
bridges 136, 189, 215–16n40, 220n86. *See also* Dresden, Istanbul, Mostar
Buddhism 35, 55, 57, 190, 193, 215n37, 221n100, 243
buffer zones. *See* World Heritage properties
Bulgaria 51, 187, 190, 198, 210, 266–67
Bureau. *See* World Heritage Committee
Burkina Faso 144n6, 233

Cablegate 108, 117, 167
Cambodia 5, 7, 35, 65, 78, 80n56, 138–42, 145–46n14, 150–52nn60–67, 178n19, 235. *See also* World Heritage Committee session 2013

Cameroon 32–33, 127, 206, 216–17n43
Camino Real de Tierra Adentro 86, 87, 88, 186, 214n34
Canada 51, 58–59, 60, 67, 82n79, 90, 106, 111, 113, 117, 124, 126, 137, 190, 215–16n40, 218n62, 219n68, 229, 234, 240. *See also* World Heritage Committee session 1990, 2008
Canal du Midi 61, 215–16n40
canals 61, 188, 208, 215–16n40
canonization in Catholicism 262–63
Cape Verde 108
Carcassonne 199, 201, 219n77, 221n93
Caribbean 87, 93, 102, 103. *See also* Latin America and the Caribbean as UNESCO region/electoral group
Çatalhöyük 31, 206, 245
Category 2 Centres of UNESCO 84, 230
cathedrals 4, 5, 7, 31, 57, 61, 73–74, 81n72, 92, 154n73, 183–84, 189, 192–93, 194, 198, 199, 201, 208, 219–20n79, 228, 244, 260
Central African Republic 32–33, 127, 193–94, 206, 216–17n43, 244
Chad 32, 206
chairpersons. *See* World Heritage Committee/General Assembly sessions
Chengjiang fossil site 33
Chief Roi Mata's domain 10, 64–65, 80n44, 192, 219n68
Chile 23, 163, 189–90, 201, 215–16n40
China 18, 33, 60, 61, 62, 63, 65, 73, 86, 87, 88, 89, 92–93, 97n6, 106, 107, 108, 112, 117, 126, 127, 128, 132, 140, 144–45n11, 146n21, 147n34, 147–48n40, 148n50, 151n64, 151n66, 152–53n69, 154n73, 157, 163, 181, 185, 186, 189–90, 215–15nn37–40, 218n62, 220n82, 220–21n89, 226, 232, 233, 234, 238, 239, 235, 240, 246, 248–49, 255n11. *See also* World Heritage Committee session 2004
China Danxia 86, 157, 186
Christianity 4, 149n55, 189, 246. See also canonization, cathedrals, churches, hidden Christians, monasteries
churches 53, 61, 63, 205. *See also* cathedrals
Closing Event of the Celebration of the Fortieth Anniversary of the World Heritage Convention 2012 in Kyoto 12, 24, 80n52, 115, 127
Cologne Cathedral 4–5, 7, 73–74, 81n72, 183–84, 194
Colombia 25, 28, 30, 93, 117, 145–46n14, 163, 189–90, 215–16n40, 218n62, 247. *See also* World Heritage Committee session 1993
colonial heritage 45, 46n13, 59, 61–63, 64, 65, 140, 152n67, 196, 229
common heritage of mankind 2, 18, 38–45, 73, 74, 101, 142–43, 258–59
communities. *See* local communities
comparative analysis. *See* nominations
confusion: at World Heritage Committee sessions 1, 25–27, 91, 172; at World Heritage General Assembly sessions 175–76, 178n20
Congo 32–33, 127, 206, 216–17n43. *See also* Democratic Republic of the Congo
Convention Concerning the Protection of the World Cultural and Natural Heritage (1972). *See* World Heritage Convention
Convention for the Safeguarding of the Intangible Cultural Heritage (2003) 10, 12, 217–18n54, 254n8; links with World Heritage Convention 7
Convention on the Law of the Sea 39–40
Corbusier, Le: the architectural work of 29, 108, 187–88, 190, 214nn23–25, 214n27
Costa Rica 145–146n14
conflicts of interests: at World Heritage Committee sessions 100–1, 103, 115–16, 123, 164; in ICOMOS World Heritage Panel 100–1, 162, 242–43
corruption 107, 187
Côte d'Ivoire 35, 80n56, 81n68
Crimea 37n2, 150n59
criterion (vi). *See* Outstanding Universal Value
critical heritage studies 8, 19
Croatia 61, 71, 80n55, 187, 190, 210, 237
Cuba 23, 60–61, 65, 117, 137–38, 149n56, 197, 224
cultural consecration 260–61

cultural landscapes 1, 10, 28, 52–53, 54, 57–60, 61, 63–65, 78–79nn12–14, 79n17, 82n80, 85, 90, 91, 97nn5–6, 115, 126, 137, 149n55, 159, 160, 161, 176, 192, 196–97, 199, 200–1, 208, 212n1, 214–15n36, 217n49, 217n52, 218–19nn60–68, 220–1n89, 228, 230, 236–37, 243, 250, 254n5, 262
cultural routes 10, 61, 79n13, 86, 109, 149–50n57, 163–64, 188–90, 208, 214–16nn29–40, 219n67, 220n83
cultural World Heritage. *See* World Heritage properties
culture, anthropological concept of 10–11, 18, 77, 227, 228, 243–45
culture chaos 238–46, 269
Cuzco 214–15n36
Czech Republic 145–146n14, 187, 219n66, 222–23n123, 227, 229, 240

Dan, triple arch-gate at 149–50n57, 256n27
Danger, List of World Heritage in 3, 34, 47, 50, 69, 70–77, 80–82nn54–81, 84–85, 90, 92, 97n4, 110, 124–25, 136, 137, 138, 140, 144n3, 147n30, 153n71, 154n73, 165–67, 168, 170, 173–74, 183, 193–94, 199, 230–31, 233–34, 238; Desired State of Conservation for the removal of a property (DSOCR) from 194; threat of inscription in 3, 31, 74, 75, 81n74, 85, 90–91, 92, 108, 125, 153n71, 166–67, 170, 173–74, 183, 194, 212–13n10, 213n13, 257
Danube delta 153n71
Darjeeling Himalayan Railway 186
Darwin's house at Down 88, 174–75, 192, 208, 210
Declaration of the Rights of Indigenous People, United Nations (UNDRIP) 32, 125–26, 127
deferrals. *See* World Heritage candidate properties
delisting. *See* World Heritage properties
Democratic Republic of the Congo 7, 80–81n57, 259
Denmark 108, 190, 208, 218n62, 219n65, 236–37, 240, 243, 262
destruction of World Heritage properties, deliberate 2, 24, 26–27, 34–35, 71, 72, 73, 77, 140, 148n52, 172, 198, 200, 219n74, 258, 266, 270n1
diplomats 14–15, 16–17, 26, 30, 37, 69–70, 85, 86–87, 90, 98, 99, 102, 104–5, 106–15, 117, 119, 128, 132–35, 139, 141, 143, 144–45nn10–11, 145–46n14, 147n30, 151–52n66, 154n73, 167, 171, 232, 251–52, 253, 267–68; home ministries of 69, 103, 111–12, 252–52. *See also* ambassadors
Director-General. *See* UNESCO
documents. *See* ethnographic fieldwork, World Heritage Committee sessions
draft decisions. *See* World Heritage Committee sessions
Dresden Elbe valley 1–2, 3, 74, 75–77, 82nn76–81, 90, 101, 105, 109, 124, 183, 184, 198, 199, 201, 219n77, 220–21n89, 260
Droste, Bernd von 66
Dubrovnik, old city of 71, 80n55

earthen architecture 55–56, 60, 63, 196, 228
Eastern Europe 95, 104, 196; as UNESCO electoral group 118, 163, 231, 263. *See also* Europe and North America UNESCO region
Ecuador 47, 71, 74, 80n56, 84–85, 89, 106, 163, 189–90, 215–16n40, 205, 220n80
Edison National Historic Site 50–51, 175
Egypt 43–45, 46nn11–14, 48–49, 73, 76, 82n79, 117, 149n54, 149–50n57, 171, 181, 220n81, 220n83, 227, 229–30, 232, 234, 235, 247, 255n11, 266. *See also* World Heritage Committee session 1979
Eidsvoll building 202
electoral groups. *See* national delegations
Erzgebirge/Krušnohoří mining region 219n66, 222–23n123, 227
Estonia 28, 30, 33, 87, 90, 91, 93, 107, 113, 114, 116, 141, 146n23, 184–85, 188, 192, 220n88, 234, 235
Ethiopia 34, 47, 80–81n57, 85, 90, 91, 115, 127, 153n71, 178n19, 218n62, 234, 235

ethnographic fieldwork 12–18; access during 12–13, 16–17; and documents 14–15; funding of 13; informants' response to 15–16; and interviews 14–15; languages used in 14; multilateral/multisited 12; self-identification during 15; social panic during 15; on the United Nations 9–10; on World Heritage 10, 17–18
Eurocentrism 4, 38, 51, 53, 58, 227–46, 250–51, 259, 260, 264, 266, 267–69. *See also* ICOMOS, IUCN, World Heritage List: biased in favour of Europe/the Global North
Europe 9, 36, 44, 48, 55–57, 58, 59, 62–63, 73, 77, 87–89, 106, 111–12, 125, 147n34, 149n54, 177, 182, 187, 190, 206, 208, 224, 226; and North America UNESCO region 230, 231, 236, 238, 239, 243, 248–49, 253, 254n8, 255, 262, 263. *See also* Western Europe; Eastern Europe; World Heritage List: biased in favour of Europe/the Global North
European Union (EU) 10, 87, 111, 138, 233, 234, 247
evaluations of candidate properties 17, 30–32, 68, 88, 92, 93, 103, 104, 113, 127, 155–56, 158–64, 168, 171, 212, 213n11, 259, 261, 265; conflicts of interests in 162–63; desk reviews for 159; by ICOMOS 47–48, 52–53, 69, 96, 137, 158–64, 196, 201, 205–9, 215–16n40, 222n115, 222–23n123, 236–37, 239–41; by IUCN 31, 47–48, 52–53, 69, 158–64, 205–6, 239–40. *See also* missions, nominations, Outstanding Universal Value, World Heritage candidate properties
Everglades 124
experts. *See* Advisory Bodies, World Heritage Centre, World Heritage Committee: expert meetings

factual error letters. *See* World Heritage Committee sessions
fieldwork. *See* ethnographic fieldwork
FIFA World Player 261
financial crisis of post-2008 247, 252
Finland 184–85, 188, 192. *See also* World Heritage Committee session 2001
forced labour 2, 128–35, 148n50
forests 64, 85, 92, 97n4, 154n73, 171, 182, 187, 190, 210, 216–17n43, 219n68, 244
fossil sites 33, 60, 78n8, 153n71, 206
France 15, 24, 25, 29, 37n5, 43, 51, 52, 58, 61, 64, 87, 93, 100, 105, 108, 111, 117, 126, 129, 130, 132, 140, 141, 145–46n14, 147n34, 150n60, 151n64, 151–52n66, 158, 162, 175, 181, 183, 184, 186–87, 187–88, 189, 190, 199, 201, 209, 211, 214n20, 214nn23–25, 214nn27–28, 214n30, 215–16n40, 218–19nn63–64, 219n68, 219n77, 221n93, 225–26, 229, 232, 234, 235, 240, 241, 243, 246, 265, 266–67. *See also* UNESCO headquarters
Francke Foundation 9, 187
frontiers of the Roman Empire 61, 186, 189, 220n86
Fujian *tulou* 63, 65
Future of the Convention process 12

G7/G8 237, 247–48, 249
G20 234, 237, 247, 248, 249
Galápagos islands 47, 71, 74, 84–85, 89, 205
Gaudí, works of Antoni 199, 214n20
George Town 61–63
Georgia (country) 154n73, 199, 201, 220n81, 220n87
Germany 1–2, 3, 4–5, 7, 9, 13, 23–27, 29, 30, 31–32, 35–36, 51, 60, 61, 64, 66–67, 73–74, 75–77, 80n52, 81n72, 82nn76–81, 90, 92–93, 100, 101, 102, 103, 105, 108, 109, 111, 117, 124, 132, 153n70, 157, 162, 172, 175, 177n10, 181, 183–84, 185, 186, 187–88, 190, 194, 195, 198, 199, 201, 202, 206, 208, 210, 214nn23–25, 214n27, 216n42, 217n50, 219nn66–67, 219n77, 219–20n79, 220n81, 220n86, 220–21nn88–91, 222n110, 222–23n123, 225, 227, 229, 230, 236, 237, 240, 241, 243, 246, 257–58, 260, 266. *See also* World Heritage Committee session 1995, 2015

Gfeller, Aurélie Élisa x, 56
Gibe III Dam 127, 153n71
Gini coefficient 247, 256nn28–29
globalization 252, 263; anthropological studies of 11–12
Global North and South 3–4, 51, 53, 58, 76, 199–201, 205, 224–256, 254n5, 264, 267–69; time lags in relations of 227, 250–51, 253, 268
Global Strategy for a Representative, Balanced and Credible World Heritage List (1994) 53–55, 57–60, 61–66, 119, 126, 146n26, 156, 188, 217n49, 227–28, 230, 243–44, 250, 251, 256n26
Görlitz 157
Goslar 185
governmentality 77
Great Barrier Reef 5, 51, 105, 257–58
Great Rift Valley lakes 32–33, 127
Great Wall 61, 189
Greece 61, 67, 181, 198, 219n78, 229, 240
Greenpeace 105, 257–58
Guinea Bissau 244

Hadrian's Wall 61, 186, 189, 220n86
Hague Convention for the Protection of Cultural Property in the Event of Armed Conflict (1954) 41, 43
Hälsingland, decorated farmhouses of 27–28, 107, 185, 206, 245
Hamburg, Speicherstadt of 201
Hannerz, Ulf 11–12
Hashima 129, 148n42, 148n50. *See also* Meiji industrial sites
Hatra 7, 140
Hebron 137–38, 149nn55–56
Heidelberg 187
Heinich, Natalie 211
heritage, cultural: 'agnosticism' 19; 'belief' 19, 178; boom of 9; critical studies of 8, 19; diplomacy 268; research trends 8–9, 19
hidden Christian properties of the Nagasaki region 133, 192–93, 222–23n123
high-rises 5, 7, 73, 81n72, 125, 165, 174, 183, 193, 194
Hiroshima peace memorial 60, 140
historic towns 18, 60, 161, 181, 212n1
Holy See 35
Honduras 145–46n14
Hōryūji 55–56, 57
human evolution 60, 78n8
Hungary 35, 67, 218n61, 248. *See also* World Heritage Committee session 2002

ICCROM (International Centre for the Study of the Preservation and Restoration of Cultural Property) 158–59, 264
Iceland 108, 190, 208, 218n61
ICOMOS (International Council of Monuments and Sites) 55, 98–99, 158, 264; European and North American dominance in 232, 236–37, 238–46, 268–69; General Assemblies 57, 242–43, 244; lobbying inside 162, 207, 222–23n123; national committees 99–100, 101, 158, 242–43; office-holders 16, 24, 29, 58–59, 89, 90, 95, 100, 107, 115, 157, 158, 159, 160–62, 176, 210, 242–44, 261; President 100, 135, 147–48n40, 161, 177n10, 180–81, 242; thematic studies by 204–5, 208–10; World Heritage Advisors 24, 25, 51, 76, 101, 131, 159, 161, 162, 183, 203, 215–16n40, 225, 239–41; World Heritage Panel 94, 101, 147–48n40, 148n50, 160–63, 241–44, 261, 262. *See also* Advisory Bodies, evaluations, monitoring
Ilisu dam 105, 153n71
Incense Route 149–50n57, 214n34, 220n83
India 18, 24, 25, 26, 29, 30, 31, 32, 33–34, 36, 51, 60, 61–62, 73, 80n56, 93, 101, 106–7, 108, 112, 115, 117, 126, 127, 144n4, 146n21, 147, 151n46, 178n20, 186, 187–88, 190, 206, 207–8, 214nn23–25, 214n27, 215–16n40, 216–17n43, 220n83, 229, 231–33, 234, 237, 239, 240, 245, 246, 248–49, 254n8, 255n11
Indigenous Peoples' Forum on World Heritage (IIPFWH) 127
indigenous rights and World Heritage 10, 31–34, 37, 53, 58–59, 60, 63–64, 71,

105, 125–27, 143, 147n34, 147n39, 197, 217n49, 244–45. *See also* Kakadu National Park
Indonesia 97n4, 129, 145–46n14, 218n62
industrial sites 2, 10, 36, 58, 60, 128–35, 161, 178n12, 185, 186, 215–16n40, 218–19n63, 228
inscriptions. *See* World Heritage candidate properties
intangible cultural heritage. *See* Convention for the Safeguarding of the Intangible Cultural Heritage (2003)
integrity 180–94, 213n18; and authenticity 58, 180–81; visual 183–84
international assistance. *See* World Heritage Fund
International Centre for the Study of the Preservation and Restoration of Cultural Property. *See* ICCROM
International Committee for the Conservation of the Industrial Heritage, The (TICCIH) 161
International Coordinating Committee for the Safeguarding and Development of the Historic Site of Angkor (ICC-Angkor) 78
International Council for Monuments and Sites. *See* ICOMOS
International Council on Minings and Metal (ICMM) 73, 88, 259
International Criminal Court (ICC) 258
International Court of Justic (ICJ) 138, 139, 151n64
International Labor Organization (ILO) 132
International Monetary Fund (IMF) 248–51, 253, 268
International Union for Conservation of Nature. *See* IUCN
International Workgroup for Indigenous Affairs 32
Iran 65, 119, 141, 153n71, 157, 215–16n40, 218n60, 219n64, 220n80, 238, 240, 246, 249, 270n1
Iraq 2, 7, 105, 140, 141, 145–46n14, 153n71, 178n19, 220n83, 235
Ireland 145–46n14, 190
Ise shrine 56–57
Islam 24, 26–27, 36, 135, 136, 137, 138, 149n55, 215n37. *See also* Timbuktu
'Islamic State' 7, 140
Israel 2, 29, 34, 70, 78n4, 82n79, 90, 95, 102, 110, 114, 137–38, 141, 146n24, 149n53, 149–50nn55–57, 153n70, 214n34, 220n83, 229, 246, 256n27; leaving UNESCO 138, 253. *See also* Jerusalem, Palestine
Istanbul, historic areas of 92, 125, 183, 185, 220n80. *See also* World Heritage Committee session 2016
Italy 27, 36, 44, 46n13, 51, 60, 61, 65, 206, 218–19n63, 44, 51, 79n13, 79n25, 96, 108, 117, 124, 143n1, 144n7, 149n53, 157, 162, 170, 187, 190, 206, 210, 214n20, 217n52, 218–19nn61–63, 219n67, 228, 229, 230, 231, 238, 240, 246, 247, 263, 265, 269. *See also* Holy See, San Marino, World Heritage Committee session 1983, 1997
IUCN (International Union for Conservation of Nature) 42, 73, 89, 98–99, 158–59, 264; European and North American dominance in 240–1; thematic studies 203–5; World Heritage Panel 160–61, 203–4, 261, 262; World Heritage Unit and officials 31, 88, 106, 159, 210, 211, 223n128, 240–1, 254n5, 259. *See also* Advisory Bodies, evaluations, monitoring

Jamaica 145–46n14, 238
Japan 2, 9, 12, 13, 16, 18–19, 23, 24, 29, 44, 55–59, 60, 66–67, 80n52, 93, 100, 108, 114–15, 117, 127, 129, 133, 140, 148n50, 151n64, 152–53n69, 183–84, 185, 186, 187–88, 190, 192–93, 194, 201, 214nn23–25, 214n27, 214n34, 215–16n40, 218n57, 218n61, 219n67, 222–23n123, 229, 233, 236, 237, 238, 240, 243, 246, 265; and South Korea 95, 105, 106, 128–35, 147–48n40, 148n42, 148nn47–50, 148n50, 185–86. *See also* World Heritage Committee session 1998
Jerusalem, old city of 2–3, 29, 34, 36, 70, 84, 95, 135–38, 141, 144n10,

148–50nn52–57, 177n3, 218n62. *See also* Battir
John Paul II. 262–63
Jordan 2, 29, 34, 70, 82n79, 95, 141, 149–50n57, 235, 238, 246. *See also* Jerusalem

Kakadu National Park 7, 71–73, 81n60, 125, 126
Kamakura 148n50, 243
Kassel 243
Kasubi tombs 197
Kathmandu valley 55–56, 72–73, 193–94
Kazakhstan 145–46n14, 163–64, 189–90, 212n5, 215n37
Kenya 1, 32–33, 36, 64–65, 68n219, 82n79, 85, 90, 105, 111, 127, 153n71, 171, 218n59, 219n68, 236–37, 238, 262
Kiev, Saint Sophia Cathedral in 31
Kii mountain range routes 214n34, 219n67
Kiribati 89–90
Komi forests 85, 92, 171
Konso cultural landscape 85, 90, 91, 115, 218n62
Kosovo, medieval monuments in 84, 125, 138, 141
kremlins, Russian 30–31, 33, 107, 186, 206
Krušnohoří/Erzgebirge mining region 219n66, 222–23n123, 227
Kuk early agricultural site 10, 63
Kurdistan 105, 153n71
Kyoto 13, 16, 18–19, 243; fortieth-anniversary event in 12, 24, 80n52, 114–15, 127. *See also* World Heritage Committee session 1998
Kyrgyzstan 163–64, 189–90, 215n37, 219n68

Lake District 52, 208, 218n62, 239
Lakes of Ounianga 32, 206
Lake Turkana 153n71
Laos 79n29, 97n5, 218n59
Lascaux cave 214n28, 232
Latin America 108, 240; and the Caribbean UNESCO region/electoral group 28, 118, 120–21, 146n19, 231, 238, 239, 241, 256n20
Latvia 184–85, 188, 190, 192, 108, 190, 208
League of Nations 45
Lebanon 67, 73, 102, 114, 117, 149–50n57
legal advisor. *See* World Heritage Committee sessions
Lemaire, Raymond 58, 153
Le Morne 64–65, 192, 219n68
Libya 2, 140
Limes, German-Raetian 186, 220n86
Lithuania 184–85, 188, 192. *See also* World Heritage Committee session 2006
Liverpool 125, 147n30
lobbying. *See* national delegations
local communities and World Heritage 33, 124, 127, 258
London, Tower and Westminster 92, 193
Lumbini 220n81, 221n100
Luther, Martin, and memorials 175, 217n50, 219n67, 221n90, 222n110

MacBride report 252
Machu Picchu 51, 81n74, 205
Madagascar 82n79, 85, 154n73, 202
Majuli island 107, 144n4
Malaysia 61–63, 66
Malbork 199–200, 221n91
Mali 7, 24, 25, 26–27, 34–35, 36, 37n5, 79n13, 172, 175, 218n60, 225, 235, 258, 266
Manas wildlife reserve 80n56
Mandela, Nelson 175, 217n49
Manovo-Gounda St. Floris National Park 193–94
Mapungubwe cultural landscape 176–77
marine heritage 41, 84, 89–90, 181, 203, 204, 212n3
maroons. *See* slavery
Marshall Islands 140
mass media and World Heritage 11, 14, 128, 134, 148n49, 160. *See also* World Heritage Committee sessions
Matsuura Kōichirō 66
Mauritania 218n60
Mauritius 64, 79n29, 107, 64–65, 192, 219n68
Mayor, Federico 66

M'Bow, Amadou-Mahtar 60
Meiji industrial sites 128–35, 148n50, 185
Melaka 61–63
Mercury and Silver Binomial 88
Merkel, Angela 132
Meskell, Lynn, and colleagues 18, 19–20, 44, 116, 144n9, 144–45n11, 146n21, 146n23, 147n26, 147n31, 147n34, 151–52n66, 152n68, 222–23n123, 251, 253, 255n11, 256n30, 260
methodology. *See* ethnographic fieldwork
Mexico 26, 28, 63, 86, 87, 88, 93, 96, 100, 117, 144n3, 145–46n14, 186, 214n34, 214–15n36, 218–19n36, 229, 234, 235, 238, 239–40, 241–42, 246, 248–49, 255n11. *See also* World Heritage Committee session 1996
Micronesia, Federated States of 244–45, 262
Mijikenda *kaya* 64–65, 68n219
mining 40; as World Heritage 27, 86–87, 128–35, 185, 191, 196–97, 206, 219n66, 222–23n123, 225–27, 228; endangering World Heritage 7, 71–72, 73, 81n58, 81n68, 85, 88, 92, 125, 171, 174, 176, 224–26, 259
minor boundary modifications 88, 108, 182, 191, 225
missions: consultative/advisory 30–31, 92, 113, 149, 164, 222–23n123, 239; emergency 35; evaluation 34, 75, 92, 137, 159–62, 164, 174n8, 215–16n40; 222–23n123; monitoring 32, 34, 49, 73, 74, 99, 113, 137, 155, 162, 167, 168, 171, 172
modern architecture 29, 60, 61, 108, 187–88, 190, 205, 208, 214nn23–25, 214n27, 221n101, 228. *See also* industrial heritage
Moldova 184–85, 188, 192
monasteries, Christian 65
Mongolia 219n64, 238
monitoring. *See* World Heritage properties
Montenegro 47
Moon Treaty 39–40
Morocco 82n79, 218n60. *See also* World Heritage Committee session 1999
Morris, William 197–98
Mostar, old bridge of 200–1, 221n99
mountain railways of India 186
Mount Zion 136. *See also* Jerusalem
Mughrabi Gate and ascent 34, 136–37, 141. *See also* Jerusalem
multilateralism 4, 12, 20, 77–78, 89, 98, 101, 107, 111, 135, 141, 153n71, 249, 259
multisited/multilateral ethnography 12

Nagasaki *see* Hashima; hidden Christian sites
Nara 243
Nara Document on Authenticity (1994) 54–60, 181, 196–201, 228
national delegations to World Heritage Committee/General Assembly sessions 12, 24–25; advocacy by 30, 68, 86, 91, 103, 115, 117; composition of 13, 14–15, 69–70, 73–74, 76, 87, 99–103, 106, 111–15, 141, 145–46n14, 162, 261; and electoral groups 28, 85, 93, 117, 118–19, 123, 143, 146n18, 146n21, 163, 178n20, 231; interests of 2, 3, 58–59, 76–77, 95, 101–5, 110–11, 115–16, 119, 128, 143, 144n9, 151–52n66, 177, 178, 211–12, 228, 234, 249, 250, 259, 266, 267; lobbying and trading favours among 2, 16–17, 27, 31, 71, 76, 82n81, 85, 86–89, 92–93, 97n2, 102, 104–110, 116–17, 119–21, 125, 141, 143, 144nn5–6, 144n9, 144–45n11, 174, 188, 190, 211–12, 237, 245, 250, 262, 264, 268; sizes of 99, 144–45n11, 255n11. *See also* diplomats, side tables
nation states 98–154; conflicts between 128–54; individuals' association with 98–101, 110–15; internal divisions of 123–27 (*see also* Dresden; Kakadu National Park; Liverpool; Vienna); sovereignty of 37, 39–42, 49, 72–77, 99, 114, 124, 123–27, 136, 142, 225, 228, 249. *See also individual country names*
natural heritage. *See* World Heritage properties
Naumburg cathedral 208, 260
Nepal 55–56, 72–73, 193–94, 220n81, 215–16n40, 220n81, 221n100

Netherlands, the 27, 29, 129, 145–46n14, 181, 190, 233, 248
New Zealand 51, 53, 119, 124, 126, 196, 213n11, 217n52. *See also* World Heritage Committee session 2007
NGOs 31–32, 38, 100, 105–6, 124, 126, 129–30, 132, 143, 158, 169, 254n8, 256n26, 257–58
Ngorongoro 49, 52, 78n8
Nigeria 60, 82n79, 218n62, 219n67, 234, 235, 247, 248–49
Nobel Prizes 6, 261
nominations to the World Heritage List 24, 68, 99, 100, 102, 156–64, 168, 169, 174; comparative analysis in 30, 61, 205–7, 222n119, 236; cost of 157, 162, 211; documents for 18, 31, 47, 102, 113, 114, 128–29, 155, 159, 164, 171, 176, 189–90, 205, 214–15nn36–37, 215–16n40, 221n102, 261; emergency 109–10, 137–38; instructions and manual for 69, 156, 157, 197, 204, 205–6, 207, 209; media and PR campaigns for 9, 128–29, 157; numerical quotas for 102, 144n7, 157, 190, 228, 246, 250, 256n27, 261, 265; preparatory assistance for 157–58; Upstream Process for 163. *See also* evaluations, Outstanding Universal Value, World Heritage candidate properties
Nord-Pas-de-Calais mining heritage 225–26
North America 9, 52, 59, 111, 176, 241–42, 243, 247, 255n18. *See also* Europe and North America UNESCO region, Western Europe and North America electoral group
North Korea 152–53n69
North Zealand hunting landscape 236–37, 243, 262
Norway 58–59, 67, 108, 184–85, 188, 190, 192, 202, 208, 219n65
Nubian monuments and UNESCO safeguarding campaign for 43–45, 48–49, 181, 227
Nyero rock art sites 236–37, 262

Obama, Barrack 108, 249
OECD 111
oil prospecting and drilling 1, 73, 74, 81n70, 151–52n66, 259
Okinoshima 185, 186
Oman 1, 2, 74–75, 149–50n57, 220n82
Operational Guidelines of the World Heritage Convention 47, 49, 51, 52, 54–58, 60, 61, 68, 69, 75, 78n6, 124, 155, 157, 169, 170, 174, 180, 181, 182, 184, 187, 192, 195, 198, 202
Outstanding Universal Value (OUV) 2, 24, 41, 47, 56, 96, 180, 184, 201–11, 213n16, 241; criteria for 52, 59, 69, 196, 202–3, 212n1, 213n18, 236–37, 254n4; criterion (vi) for 51, 52–54, 55, 191–92, 198, 200, 203, 208–9, 212, 217nn49–50, 217n52, 222n110; defined as absolute 262, 264; intuited 209–11, 242; and physical traces 65, 174–75, 192–93, 194, 197, 217n50, 217n52; Statements of 30, 61–63, 85–86, 92, 182, 196–97, 207, 217n49, 221n101. *See also* evaluations, nominations, World Heritage candidate properties

Pacific countries 54, 65, 116, 119. *See also* Asia and the Pacific UNESCO region/electoral group
Palau 244–45, 262
Palestine 2, 29, 34, 95, 137–38, 149nn55–56, 218n62, 137–38, 149nn55–56, 246, 252; becoming full UNESCO member 109–10. *See also* Israel, Jerusalem
Palmyra 7
Pampulha (Belo Horizonte) 221n101
Panama, Panamá Viejo 92, 106
Papua New Guinea 10, 63, 66
Paris 183. *See also* UNESCO, World Heritage Committee sessions 1977, 1980, 1982, 1985, 1986, 1987, 1989, 2003, 2011, 2016; World Heritage Extraordinary Committee sessions; World Heritage General Assembly sessions
Periodic Reporting 49, 99, 164. 226. *See also* World Heritage properties
Permanent Delegates. *See* ambassadors

Permanent Forum on Indigenous Issues (PFII or UNPFII) 23, 125, 126
Peru 51, 81n74, 117, 163, 189–90, 205, 214–15n36, 215–16n40, 229, 240
Philae 43–45
Philippines 60, 63, 218n62, 220n81
Piedmont, vineyard landscape of 27, 206, 218–19n63
pile dwellings in the Alpine region, prehistoric 190
Pirin National Park 51, 266–67
poaching 74, 193
Poland 50, 129, 132, 140, 149n56, 153n70, 182–83, 198, 199–200, 200–1, 209, 212–13n10, 216n43, 218n58, 219n74, 221n91, 229, 240, 248, 270n3. *See also* World Heritage Committee session 2017
Pompei and Herculaneum 170
popes, Roman Catholic 262–63
Portugal 49, 88, 192, 218–19n63, 229, 237, 240
Preah Vihear 7, 35, 65, 138–42, 146n24, 150–52nn60–67
preparatory assistance. *See* nominations
procedures. *See* evaluations, monitoring, nominations, World Heritage Committee sessions
Provins 266–67

Qatar 25, 93, 178n19, 220n83, 237. *See also* World Heritage Committee session 2014
Qhapaq Ñan 163, 189–90, 214–15n36, 215–16n40
Queensland 124–25, 257–58

railway lines 10, 61, 186, 188, 208
rainforests 85, 97n4, 154n73
Rajasthan, hill forts of 207–8
Rammelsberg 185
rapporteur. *See* World Heritage Committee sessions
receptions 12, 13, 15, 29, 35, 36, 67, 92, 112, 116, 130, 167, 169, 224–26
Recommendation on the Historic Urban Landscape (2011) 28
reconstructions 30, 50, 55–57, 141, 154n73, 175, 181, 186, 197–201, 217n50, 219n74, 220n86, 220–21n89. *See also* authenticity
referrals. *See* World Heritage candidate properties
Regensburg 177n10
religion. *See* Buddhism, Christianity, Islam, sacred forests
restoration 55–56, 72, 137, 145n14, 158–59, 174, 193, 194, 195, 197–201, 219n77, 220n86
Réunion Island 87, 175
Rhine valley, upper middle 199, 220–21n89
Rhodes 198, 219n78
Rila monastery 198
Rio de Janeiro; cultural landscape of 28, 30, 36, 191, 206, 217n51
Robben Island 157, 217n49
rock art 60, 161, 178n12, 184, 204–5, 236–37, 262
Roi Mata. *See* Chief Roi Mata's domain
Romania 58, 145–46n14, 153n71, 187, 190, 210, 218n58
routes of Santiago: in Spain 61, 79n13, 189, 208, 214n30; in France 189
Rules of Procedure. *See* World Heritage Committee sessions
Ruskin, John 197–98
Russian Federation 7, 13, 18, 23–24, 29, 30–31, 33, 34, 36, 37n2, 85, 90, 92, 99, 106, 107, 140, 141, 146n21, 150n59, 171, 183, 184–85, 186, 188, 190, 192, 194, 206, 208, 209, 214n24, 219n65, 229, 233, 234, 235, 240, 245, 246, 247, 248, 255n11, 266. *See also* World Heritage Committee session 2012

sacred forests, landscapes and mountains 10, 53, 64, 114, 185, 214n34, 219nn67–68, 227, 228, 244
safeguarding campaigns, UNESCO 43, 44–45. *See also* Nubian monuments
Sagrada Familia, La 199
Saint Lucia 67, 112
Saint Petersburg 7, 23, 183, 194. *See also* World Heritage Committee session 2012
saints *see* canonization
San Luís Potosi 88, 214–15n36

San Marino 65–66, 221n91, 221n93
Sangha Trinational 32–33, 127, 206, 216–17n43
Santiago de Compostela. *See* routes of Santiago
Saudi Arabia 23, 65–66, 85, 140, 144n6, 148n12, 220n83, 238, 240, 247, 248, 249, 260
Schwetzingen palace and park 23–27, 29, 30, 35–36, 76, 92–93, 102, 172, 175, 206, 208
Second World War 128, 131, 209, 220n86, 220–21n89
secretariat of the World Heritage Convention. *See* World Heritage Centre
secret ballots. *See* voting
Security Council 145n14, 151–52n66, 237, 252
Selous game reserve 224–26, 230
Semmering Railway 61
Senegal 25, 27, 34, 35, 81n74, 93, 117, 219nn64–65, 226, 237
Serbia 31, 237. *See also* Kosovo
Serengeti National Park 7, 38, 51, 225, 258–59
serial properties. *See* World Heritage properties
Seville, Cathedral, Alcázar and Archivo de Indias 13, 92, 183–84, 185. *See also* World Heritage Committee session 2009
Sheki, historic centre of 260
Shōka sonjuku 129–30
side events. *See* World Heritage Committee sessions
side tables. *See* World Heritage Committee sessions
Silk Roads 163, 189–90, 215n37, 215–16n40
Simien National Park 47, 80–81n57
Singapore 238, 248
sites of conscience 60–61, 129, 208–10. *See also* Auschwitz, Bikini, Hiroshima, Robben Island, Warsaw
slavery, heritage of 10, 60–61, 64, 65, 134, 140, 192, 227
Slovakia 23, 61, 88, 187, 190, 210, 218n58
Slovenia 88, 187, 190, 210
SOCO International 259
SOC reports. *See* State of Conservation Reports
South Africa 18, 27, 28, 34, 67, 84, 146n21, 157, 175, 176–77, 217n49, 219n64, 219n68, 225, 233, 234, 235, 255n11. *See also* World Heritage Committee session 2005
South Korea 23, 82, 106, 117, 151n64, 152–53nn68–69, 181, 215–16n40, 218n57, 237–38, 240, 247, 248, 249; and Japan 2, 95, 105, 106, 128–35, 147–48n40, 148n50, 185, 246
Spain 2, 13, 36, 46n13, 61, 62–63, 66, 76, 79n13, 81n68, 82n79, 88, 92, 143n1, 145–46n14, 183–84, 185, 187, 189, 190, 192, 199, 208, 210, 214n20, 214n30, 214n32, 218n62, 219n64, 229, 231, 238, 240. *See also* World Heritage Committee session 2009
Speyer cathedral 198, 199
Sri Lanka 63, 85
Statement of Outstanding Universal Value (SOUV). *See* Outstanding Universal Value
State of Conservation Reports: by Advisory Bodies 127, 148n50, 165–66, 230–32, 238, 254n7; by States Parties 68, 99, 104, 114, 165
States Parties. *See* nation states
statutory meetings. *See* World Heritage Committee/General Assembly sessions
Stockholm Conference on the Human Environment (1972) 42
Stonehenge 195–96
Stovel, Herb 58–59
Strathern, Marilyn 10, 63–64
Struve Geodetic Arc 184–85, 188, 192
Sudan 43–45, 46n13, 48–49, 181, 227
Sumatra rainforests 97n4
summary record. *See* World Heritage Committee sessions
Suriname 218n59
Sweden 27–28, 42, 60, 85, 87, 90, 91, 93, 107, 113, 184–85, 188, 190, 192, 206, 218n61, 229, 234, 235, 240, 245, 261

Switzerland 25, 26, 28, 29, 30, 33, 61, 85, 87, 90, 91, 93, 107, 108, 109, 113, 114, 115–16, 125, 141, 144n8, 158, 187–88, 190, 206, 214nn23–25, 214n27, 218–19n63, 220n88, 225, 230, 234, 235
Sydney: Opera House 60. *See also* World Heritage Committee session 1981
Syria 2, 7, 140, 149–50n57, 193, 194, 220n81

Tajikistan 88, 215–16n40
Tanzania 7, 38, 49, 51, 52, 78n8, 224–26, 230, 258–59
Tasmanian wilderness 124–25, 127
Tauric Chersonese 150n59
Tentative Lists 136, 140, 148n50, 156–57, 163, 169, 177n2, 189, 209, 215n37, 215–16n40, 234, 246, 264, 265. *See also* World Heritage candidate properties; World Heritage List
territorial disputes. *See* nation states
Thailand 7, 35, 65, 117, 138–42, 145–46n14, 146n24, 150–52nn60–67, 178n19, 224, 234, 235. *See also* World Heritage Committee session 1994
thematic studies 53, 156, 204–5, 206
Thimlich Ohinga 236–37, 262
Three Parallel Rivers of Yunnan protected areas 88, 89, 92–93, 108, 185, 226
Timbuktu 7, 24, 25, 26–27, 34–35, 36, 37n5, 172, 175, 258, 266
Tipasa 74, 233–234
Tongariro 53, 196, 213n11, 217n52
tourism 7, 72, 84–85, 133, 187, 245, 255n11
tragedy of the commons 267
transboundary properties. *See* World Heritage properties
treaty states. *See* nation states
Tunisia 80–81n57, 82n79, 117, 214n24. *See also* World Heritage Committee session 1991
Turkey 12–13, 31, 92, 105, 125, 138, 152n68, 153n71, 157, 181, 183, 185, 206, 215–16n40, 220n80, 229, 233, 237, 238, 240, 245, 246, 248, 249, 260, 265. *See also* World Heritage Committee session 2016

Uganda 197, 236–37, 238, 250, 262
Ukraine 31, 37n2, 91, 150n59, 153n71, 184–85, 187, 188, 190, 192, 210, 218n58
Uluṟu-Kata Tjuṯa National Park 53, 213n18
UNESCO (United Nations Educational, Scientific and Cultural Organization) 38, 84, 87; Assistant Director-General for Culture 16, 34–35, 98; budget crisis 100, 119, 246; bureaucracy 13, 48, 100–1, 143n1, 252; Director-General 35, 43, 47, 60, 66, 79, 100, 107, 115, 135, 139, 143n1, 149n56, 168, 230, 251, 253; disorganisation of 252; Executive Board 69, 115, 149n56; General Conference 12, 28, 109–10, 116–17, 150n58, 172; Group of Islamic Cooperation within 27; headquarters 12, 13, 35, 38, 89, 108, 111–13, 158; heritage initiatives and conventions, other 7, 10, 12, 41, 55, 135–37, 181, 200, 217–18n54, 254n8; idealistic tone at 4, 252; member states 50, 99, 109–10, 138, 150n58, 252; National Commissions 99; national withdrawals from 138, 252–53; portfolio of 113, 252; safeguarding campaigns 43–45, 48–49, 181, 227; study of 17; World Heritage, relation to 6–7, 8, 18, 42–44, 66–67, 150n63, 163, 251–53. *See also* national delegations
United Arab Emirates 25, 141 235
United Kingdom 36, 52, 58, 60, 61, 71, 81n74, 88, 92, 102, 125, 147n30, 174–75, 186, 189, 192, 193, 195–96, 208, 210, 218n62, 220n86, 229, 239, 240, 241, 259, 265; leaving UNESCO 252
United Nations (UN) 4, 23, 32, 33, 35, 39, 41, 42, 44, 48, 50, 78, 80n56, 100, 107, 111, 117, 118, 125–26, 127, 132, 135, 138, 142, 145–46n14, 147n34, 149n53, 149–50nn56–57, 151–52n66, 200, 211, 233, 237, 247–52, 253, 268;

law courts 138, 139, 151n64, 248–51, 253, 258, 268; Development Programme (UNDP) 78; Economic and Social Council (ECOSOC) 42; ethnographic study of 9–10, 13, 17, 125, 269; General Assembly 32, 39, 125; human rights bodies 10, 125
United States of America 29, 39, 40, 42, 43–44, 46n13, 48, 50–51, 60, 71, 73, 82n79, 90, 96, 108, 113, 117, 124, 135–36, 137, 140–41, 143–43nn1–6, 145–145n16, 146n18, 149nn53–54, 151n64, 151–52n66, 152n68, 153n70, 167, 188, 210, 214n24, 214n27, 219n68, 229, 234, 236, 240, 241, 244, 247, 249, 250, 252, 254n8, 261, 270n1; leaving UNESCO 138, 252–53. *See also* World Heritage Committee session 1978, 1992
upgrading. *See* World Heritage Committee sessions
Upstream Process. *See* nominations
Uruguay 145–46n14, 178n12, 238
Uzbekistan 81n74, 212n5, 215–16n40, 220n80

Vanuatu 10, 64–65, 80n44, 192, 219n68
Vatican. *See* Holy See
Venezuela 145–46n14
Venice 44, 51, 269
Venice Charter (1964) 55–58, 154n73, 158, 198–201
Veronese, Vittorio 43
Versailles 51, 183, 243
Vienna, historical centre of 7, 125, 147n30, 165, 193–94
Vietnam 87, 218n59, 265–66
Viking age sites 108, 190, 208, 216n42
Viñales valley 197
Viollet-le-Duc, Eugène 198, 199, 219n77
Virunga National Park 7, 259
Višegrad bridge 220n86
voting 116, 118, 123, 141, 161, 170, 178n20, 241–43, 247–49, 261–62; by show of hands 26, 178n20; by secret ballot 1, 2, 74, 76, 87–89, 91, 95, 107, 109–10, 113, 115, 116, 134–35, 136–37, 144n4, 149n56, 188, 261, 264

Wadden Sea 190
Wade, Robert 248–50
Wallonia, mining sites of 27, 86–87, 185, 206
Warsaw, historic centre of 50, 198, 200–1, 219n74
Wartburg castle 175, 202, 217n50, 220n88, 222n110
Western countries. *See* Global North and South
Western Europe 25, 58, 89, 102, 103, 107, 109, 110, 111–12, 113, 137, 150n59, 210, 221n94, 229, 239; and North America UNESCO electoral group 118, 146n22, 163, 199–200, 231. *See also* Europe
Western Ghats 32, 33–34, 107, 127, 186, 206
WikiLeaks. *See* Cablegate
wine regions 27, 52, 53, 196, 197, 206, 218n61, 218–19n63, 228
wooden architecture 55–57, 58–60, 61, 72, 196
World Bank 78, 200, 233, 247–51, 253, 268
World Conservation Union. *See* IUCN
World Heritage: as 'global dream' 11; jokes 23, 35–36, 92, 95; global impact of 4–9, 24, 257–58; media presence of 11, 12, 14, 67, 80n52, 128, 134, 144n3, 148n49, 160, 174; university programmes 9, 169; relation to UNESCO 6–7, 8, 18, 42–44, 66–67, 150n63, 163, 251–53
World Heritage Advisors. *See* ICOMOS
World Heritage candidate properties; deferral of 25, 31, 32–34, 36, 91, 92, 96, 108, 122, 123, 146n23, 161, 164, 208; inscription of 23, 25, 27, 36, 47, 50, 53, 62, 68, 72, 91, 92, 96, 101–2, 120, 121, 122, 144n9, 146n25, 149–50n57, 150nn62–63, 159, 161, 165, 166, 171, 175, 237, 262, 266; non-inscription of 24, 25, 96, 102, 148, 163, 207–8, 222–23n123, 243, 260; referral of 28, 30–31, 91, 92, 94–95, 96, 108, 122, 123, 146n23, 161, 164, 222–23n123; withdrawal of 21n4, 24, 26, 96, 102,

139, 163, 168. *See also* evaluations, nominations, Outstanding Universal Value, World Heritage properties
World Heritage Centre 29, 49, 73, 75–76, 83–84, 105, 155, 156–57, 165, 166, 167–68, 169, 174, 232–33, 257; and Advisory Bodies 66, 124, 167–68, 181, 208; director of 24–25, 34, 66, 87, 101, 115, 167, 230; founding of 66–67; officials 5, 7, 15, 38, 59, 66, 85, 89, 90, 98, 100–1, 103, 113, 127, 143n1, 144n3, 149n54, 152–53n69, 170, 178n20, 209, 221, 222–23n123, 228, 241–42, 251, 252, 264
World Heritage Committee 2, 24–25, 115, 121; Bureau of the 50, 51, 61, 68, 126, 147n31, 184, 198, 254n5, 265; elections to 99, 108, 109, 110, 116–19, 139, 141, 143, 145–46n14, 146n22, 169, 175–76; expert meetings authorized by 41, 52, 53, 54, 61, 113, 115, 126, 127, 143n2, 146n22, 155, 159, 179, 192, 208, 209, 213n18, 239, 244, 256n26; tenures in 116–18. *See also* national delegations, UNESCO
World Heritage Committee session *1977 in UNESCO headquarters, Paris* 47, 50, 156, 202, 220–1n103; *1978 in Washington, D.C.* 47, 50, 51, 70, 84, 180, 227; *1979 in Luxor* 45, 47, 50, 51, 52, 70, 182, 148, 202; *1980 in UNESCO headquarters, Paris* 49, 51, 121, 175, 198, 217n49; *1981 in Sydney* 71, 198; *1982 in UNESCO headquarters, Paris* 70, 135, 226, 236; *1983 in Florence* 49, 70, 214–15n36; *1984 in Buenos Aires* 52; *1985 in UNESCO headquarters, Paris* 219–20n79; *1986 in UNESCO headquarters, Paris* 60; *1987 in UNESCO headquarters, Paris* 52, 53, 60, 61, 186, 189; *1988 in Brasilia* 83; *1989 in UNESCO headquarters, Paris*; *1990 in Banff (Canada)* 49, 52, 53, 61, 67; *1991 in Carthage (Tunisia)* 46n10, 66; *1992 in Santa Fe* 52, 55, 70, 77, 81n68, 182; *1993 in Cartagena (Colombia)* 53, 55–56, 61, 79n13, 159, 189, 217n52; *1994 in Phuket* 49, 52, 53, 227; *1995 in Berlin* 71; *1996 in Mérida (Mexico)* 49, 54, 60, 61, 71, 140; *1997 in Naples* 67, 80–81n57, 199, 217n49, 228; *1998 in Kyoto* 61, 189, 196, 228; *1999 in Marrakesh* 57, 175, 182, 186, 196, 197, 200, 217n49, 228; *2000 in Cairns (Australia)* 71, 126, 135, 228, 265; *2001 in Helsinki* 126, 129, 136, 147n31, 197, 214n30, 265; *2002 in Budapest* 69, 72, 74, 80n55, 80–81n57, 149n54; *2003 in UNESCO headquarters, Paris* 73, 81n58, 88, 169, 176, 200, 214–15n36; *2004 in Suzhou* 73, 75, 138, 149n54, 217n52; *2005 in Durban* 63, 69, 149–50n57, 186–87, 196, 199, 200, 213n17, 217n49, 246; *2006 in Vilnius* 64, 74, 75, 81n72, 89, 107, 138, 144–45n11, 177n10, 233, 234, 254n8; *2007 in Christchurch (New-Zealand)* 1, 2, 60, 63, 64, 74, 75, 84, 107, 124, 139, 140, 144–45n11, 150n63, 151–52n66, 174, 187, 196, 220n86; *2008 in Quebec City* 7, 10, 61–66, 81n74, 86, 106–7, 118, 139, 149–50n75, 150n63, 151–52n66, 186, 187, 226; *2009 in Seville* 1–3, 12, 13, 20, 24, 75–78, 87, 88, 89, 90, 101, 105, 109, 118, 154n73, 170, 172, 188, 214–15n36, 221, 228, 230–31, 260; *2010 in Brasilia* 10, 12, 13, 18, 20, 83–90, 92, 95, 100, 109, 119, 120–21, 124, 139, 140, 143, 156, 163, 172, 174, 175, 186, 199, 210, 226, 231, 233–36, 237, 238, 241–42, 247, 249, 250, 253, 255n11, 262, 267; *2011 in UNESCO headquarters, Paris* 12, 32, 35, 38, 91–93, 96–97, 97nn4–7, 103, 108, 115, 125, 126, 127, 139–40, 146n23, 147n30, 153n71, 172, 183, 187, 188, 225, 236, 241, 244–45; *2012 in Saint Petersburg* 7, 12, 23–37, 86, 90–93, 95, 101, 105, 107, 109–10, 115, 126, 127, 137, 139, 144–45n12, 147n30, 150n59, 154n73, 171, 172, 176–77, 183, 185, 186, 191, 201, 206, 207, 216–17n43, 224–26, 242, 245, 258, 266; *2013 in Phnom Penh* 35, 93, 140, 148n50, 150n59, 207–8, 222–23n123, 243; *2014 in Doha* 93, 128, 137, 163,

182, 189–90, 208, 215–16n40, 257–58, 259; *2015 in Bonn* 2–3, 12, 93–95, 99, 105–8, 128–35, 137, 140, 148n50, 153n71, 169, 174, 190, 192, 199, 201, 214n30, 222–23n123, 236–38, 241–42, 246, 253, 257–58, 259–60, 265–66; *2016 in Istanbul and UNESCO headquarters, Paris* 21n4, 140, 149n56, 150n95, 152n68, 187, 188, 212–13n10, 221n101, 222–23n123, 241, 260, 265; *2017 in Krakow* 127, 131, 137, 147n30, 149n56, 157, 159, 187, 208; *2018 in Manama (Bahrain)* 121, 127, 153n71, 159, 192–93, 208, 216n42, 222–23n123, 237, 260; *2019 in Baku* 214n26, 222–23n123, 227, 260

World Heritage Committee session, extraordinary *1981* 34, 70, 144n10; *1999* 71; *2003* 68, 72

World Heritage Committee sessions; access to 12; agenda of 21–22n6, 24, 27, 54, 68, 83, 90, 92, 149n54, 165–66, 168, 169, 170, 172, 173–74, 175; amendments in 25–26, 30, 84, 85, 87–88, 94–95, 107, 166, 169–70, 172, 176; atmosphere of 13–14, 169; attendance figures at 6, 47, 67, 127, 169; chairpersons at 24–25, 26, 31, 32, 34, 68, 91, 92, 97, 98, 101, 103, 105, 109, 115, 130–32, 134–35, 136, 150n63, 168, 171, 172–73, 231–32, 258, 266; computer screens at 26, 91, 171, 172; conflicts of interests at 100–1, 103, 115–16, 123, 164; confusion at 1, 25–27, 91, 172; cost of 29, 36; cultural performances at 13, 29, 67, 83, 169; decision-making at 169–77; decision texts at, structure of 165–67, 170; documents at 14–15, 17, 22n7, 29, 47, 69, 85–86, 104, 114, 155–56, 165, 167, 194; draft decisions at 25–26, 34, 69, 70, 131, 155–56, 165–67, 168, 169–70; duration of 12, 47, 172; electoral groups at 28, 85, 93, 117, 146n18, 146n21, 163, 231; factual error letters at 93, 155–156, 161–62; frustration with 14, 38, 88, 89, 231–33, 250, 267; as global events 6, 67, 77, 111; languages used at 25, 127; legal advisor at 25–26, 29, 35, 76, 172; mass media and press at 12, 34, 103, 130, 132, 133, 148n48, 230, 266; meals at 13, 29, 67, 129, 137, 224; rapporteurs at 26, 30, 68, 87, 98; receptions at 12–13, 29, 35–36, 67, 92, 112, 116, 130, 169, 224–25; rule-bending at 31, 37, 91, 92, 93, 156; Rules of Procedure of 21–22n6, 68, 86, 87, 104, 115, 170; seating order at 23–24, 25, 26, 34–35, 89, 149n56, 170; side events at 13, 28–29, 129, 132, 169; side tables/confidential negotiations at 29, 136, 139; speaking time at 25, 27, 105, 170; upgrading recommended decisions at 27, 30, 85, 89–90, 91, 92, 93, 94–92, 107, 108, 122–23, 146n23, 146n25, 165, 176, 236, 237, 245, 260, 261; webstreamed 103, 131, 147n30. *See also* national delegations, voting, World Heritage candidate properties

World Heritage Convention 2, 7–8, 18, 21–22n6, 38–39, 40–42, 47–48, 49, 52, 55, 72, 99, 118, 124, 135–36, 140, 146n15, 146n18, 151n64, 158, 168, 202, 251; anniversaries of 12, 24, 68, 80n52, 124; preparation of 42–45, 50; ratification figures 47, 50

World Heritage Education Programme 67

World Heritage Fund 29, 47, 48–49, 66, 67, 157–58, 162, 231, 264–65

World Heritage General Assembly of States Parties 12, 47, 123, 253; chairpersons at 96, 178n20; confusion at 175–76, 178n20; electoral groups at 118–19, 123, 143, 178n20; receptions at 15. *See* also national delegations, voting

World Heritage General Assembly session *2005* 145–46n14; *2009* 12, 141, 145–46n14, 175–76, 178n20; *2011* 12, 15, 96, 242; *2013*; *2014 (extraordinary)* 118; *2015* 146n22; *2017* 146n22; *2019* 117 (all in UNESCO headquarters, Paris)

World Heritage Indigenous Peoples Council of Experts (WHIPCOE) 126–27

World Heritage List: biased in favour of Europe/the Global North 51, 53, 199–201, 205, 224–46, 254n5, 267–69;

brand value of 5, 210, 234; growth of 51, 227–28, 264–67. *See also* Danger List, monitoring, nominations, World Heritage candidate properties, World Heritage properties
World Heritage Panels. *See* ICOMOS, IUCN
World Heritage Papers 156
World Heritage Programme. *See* IUCN
World Heritage properties; boundaries of 31, 33, 88, 108, 176, 181–83, 184, 186, 191, 199, 213n11, 217n52, 225; buffer zones of 28, 31, 32, 176, 182, 184, 191, 194, 220n86; cultural 8, 10–11, 42–43, 52–53, 70, 77, 78n8, 158–59, 167, 174, 177n2, 179–81, 195, 196, 202, 204–6, 208, 211, 221–22n103, 228, 242, 254n3; delisting of 1–2, 3, 72, 73, 74–77, 80–81n57, 82n79, 82n81, 99, 101, 166–67, 183, 199, 260; effects of listing of 7–8, 70, 210; maps of 149–50n57, 176, 181, 191, 214–15n36, 215–16n40, 225; mixed 52, 96, 159, 174; monitoring of 28, 49, 69, 70, 77, 89–90, 127, 159, 164–68, 169, 193, 194, 211, 259, 265; natural 8, 10–11, 31, 42–43, 49, 52–53, 96, 158–59, 167, 174, 177n2, 180, 203–4, 205, 208, 212n1, 213n18, 228; serial 27–28, 30, 33, 86, 88, 128, 138, 184–90, 191, 192–93, 194, 199, 208, 213n14, 213n16, 213n18, 214n34, 214–15n36, 215–16n40, 216n42, 222n119, 243; sizes of 89–90, 180, 181, 182, 187, 204, 206, 229–30; transboundary 48, 108, 116, 149–50n47, 190, 216–17n43, 256n27; TV documentaries on 23, 66–67, 80n52, 174. *See also* missions, World Heritage candidate properties, World Heritage List
World Heritage Review 156
World Trade Organization (WTO) 252
World Wildlife Fund (WWF) 105, 257–58
Wright, works of Frank Lloyd 188, 214n27

Yap stone money sites 244–45, 262
Yellowstone National Park 51, 71, 73, 124, 144n3, 210
Yemen 2, 140, 220n80
Yoshida Shōin 129–30
Yugoslavia 47, 71, 80n55

Xi'an 215n37

Zollverein coal mine 129
Zimbabwe 73, 90, 216–17n43

CPSIA information can be obtained
at www.ICGtesting.com
Printed in the USA
LVHW010008201221
706697LV00001B/42

9 781800 730441